The Practical Approach in Chemistry Series

SERIES EDITORS

L. M. Harwood
Department of Chemistry
University of Reading

C. J. Moody
Department of Chemistry
University of Exeter

The Practical Approach in Chemistry Series

Organocopper reagents
Edited by Richard J. K. Taylor

Macrocycle synthesis
Edited by David Parker

Preparation of alkenes
Edited by Jonathan M. J. Williams

Preparation of Alkenes

A Practical Approach

Edited by

JONATHAN M. J. WILLIAMS

School of Chemistry
University of Bath, UK

OXFORD NEW YORK TOKYO
OXFORD UNIVERSITY PRESS
1996

Oxford University Press, Walton Street, Oxford OX2 6DP

Oxford New York
Athens Auckland Bangkok Bombay
Calcutta Cape Town Dar es Salaam Delhi
Florence Hong Kong Istanbul Karachi
Kuala Lumpur Madras Madrid Melbourne
Mexico City Nairobi Paris Singapore
Taipei Tokyo Toronto

and associated companies in
Berlin Ibadan

Oxford is a trade mark of Oxford University Press

Published in the United States
by Oxford University Press Inc., New York

A catalogue record for this book is available from the British Library

Library of Congress Cataloging-in-Publication Data
Preparation of alkenes: a practical approach/edited by Jonathan M. J. Williams
(The practical approach in chemistry series)
Includes bibliographical references and index
1. Alkenes—Synthesis. I. Williams, Jonathan M. J. II. Series.
QD305.H7P74 1996 547'.412—dc20 96-13456

ISBN 0 19 855795 7 (hbk)
ISBN 0 19 855794 9 (pbk)

Typeset by Footnote Graphics, Warminster, Wilts
Printed in Great Britain by
Bookcraft Ltd, Midsomer Norton, Somerset

Preface

This book is one volume within the Practical Approach in Chemistry Series. Along with the other volumes in the series, this book is designed to provide hands-on help in a laboratory environment. Each chapter contains detailed Protocols, which describe the experimental procedures in considerable detail. The Protocols give a step-by-step guide to the techniques used, including how to set up apparatus, how to run the reaction, and how to isolate the product from the reaction mixture. The authors of the individual chapters have experience of the reactions which they describe, and have been able to pass on practical advice which is not always provided in the primary literature.

There are many methods for the synthesis of alkenes. This book cannot provide exhaustive coverage of all of the methods which are available. However, an effort has been made to describe the major methods for alkene synthesis, and to provide Protocols for these methods. Additionally, each chapter provides an overview of the methods used, and there are many references to research in related areas. One of the main issues with the synthesis of alkenes is the preparation of the alkene as a single isomer, and each of the chapters deals with the issue of stereoselective synthesis.

The chapters within the book describe the different methods of alkene synthesis such as Wittig reactions or catalytic coupling reactions. However, the synthesis of a particular alkene may be possible in a number of ways, and it is for this reason that a large summary chapter has been included. This final chapter considers the synthesis of particular classes of alkene (e.g. tri-substituted alkenes), and offers suggestions on which method to choose. This summary chapter provides cross-references to the rest of the book.

The synthesis of alkenes is fundamental to the preparation of many natural products and biologically active molecules, and it is hoped that this book will be of interest to anyone involved in organic synthesis in industry or in University.

I would like to thank all of the contributors to this book for parting with their 'insider' knowledge, and for providing such readable accounts of alkene synthesis.

Loughborough J. M. J. W.
June 1996

Contents

Dedicated to my mother and father.

Contributors

ALAN ARMSTRONG
Department of Chemistry, University of Nottingham, NG7 2RD, UK.

LEE BOULTON
Department of Chemistry, The University of Reading, Whiteknights, Reading RG6 2AD, UK.

DANIELE CHOUEIRY
Department of Chemistry, Purdue University, 1393 Brown Building, West Lafayette, IN 47907–1393, USA.

DAVID M. HODGSON
The Dyson Perrins Laboratory, University of Oxford, South Parks Road, Oxford, OX1 3QY, UK.

JOSHUA HOWARTH
School of Chemical Sciences, Dublin City University, Dublin 9, Ireland.

NICHOLAS J. LAWRENCE
Department of Chemistry, UMIST, PO Box 88, Manchester M60 1QD, UK.

EI-ICHI NEGISHI
Department of Chemistry, Purdue University, 1393 Brown Building, West Lafayette, IN 47907–1393, USA.

OLIVER REISER
Institut für Organische Chemie, Georg August Universität, Tammanstrasse 2, D-37077 Göttingen, Germany.

ANDREW D. WESTWELL
Department of Chemistry, Loughborough University of Technology, Loughborough, Leicestershire LE11 3TU, UK.

JONATHAN M. J. WILLIAMS
School of Chemistry, University of Bath, Claverton Down, Bath, BA2 7AY, UK.

Abbreviations

Ac	acetyl
b.p.	boiling point
Bu	butyl
c-Hx	cyclohexyl
Cp	cyclopentadienyl
dba	dibenzylideneacetone
DBN	1,5-diazabicyclo[4.3.0]non-5-ene
DBU	1,8-diazabicyclo[5.4.0]undec-7-ene
de	diastereomeric excess
DIBAL	diisobutylaluminium hydride
DMAP	4-dimethylaminopyridine
DME	1,2-dimethoxyethane
DMF	*N,N*-dimethylformamide
DMSO	dimethyl sulfoxide
dppe	1,2-bis(diphenylphosphino)ethane
dppm	bis(diphenylphosphino)methane
dppp	1,3-bis(diphenylphosphino)propane
ee	enantiomeric excess
Et	ethyl
FW	formula weight
gc	gas chromatography
gem	geminal
Hal	halide
HMPA	hexamethylphosphoric triamide
HOMO	highest occupied molecular orbital
HWE	Horner–Wadsworth–Emmons
LDA	lithium diisopropylamide
LUMO	lowest unoccupied molecular orbital
*m*CPBA	3-chloroperbenzoic acid
Me	methyl
Mes	mesityl
m.p.	melting point
PDC	pyridinium dichromate
Pr	propyl
Py	pyridine
R	alkyl or aryl group
r.t.	room temperature
TBDMS	*t*-butyldimethylsilyl
TBDPS	*t*-butyldiphenylsilyl

Tf	trifluoromethanesulfonyl
THF	tetrahydrofuran
THP	tetrahydropyranyl
TLC	thin layer chromatography
TMEDA	*N*,*N*,*N*′,*N*′-tetramethylethylenediamine
TMS	trimethylsilyl
vic	vicinal

1

Introduction and general methods

JONATHAN M. J. WILLIAMS

1. Introduction

This book is designed to provide advice on the synthesis of alkenes using some of the more common methods currently used by organic chemists. It is not possible to provide a detailed account of every method of alkene preparation that has ever been employed. However, discussion sections and literature references should enable the more adventurous chemists to devise their own practical experiments for the synthesis of alkenes.

Alkenes are compounds which contain carbon–carbon double bonds. This book considers the preparation of both simple alkenes, containing no other functional groups, and more complex structures containing a multitude of functionality. Industrially, the simple alkenes ethene and propene are prepared on a massive scale, mainly as precursors to polyethylene and polypropylene. However, the synthesis of the alkene functionality on a smaller scale remains an important objective in the synthesis of natural and unnatural products.

The basic structure of alkenes is exemplified by the simplest member of the group, ethene (Scheme 1.1). The six atoms of ethene are co-planar, and the groups attached to the alkene are disposed at angles of approximately 120°, characteristic of the sp^2 hybridisation of the carbons. The angle between two substituents attached to one end of the alkene increases as the steric bulk of the substituents increases.

The bonding in an alkene comprises two components, a σ-component and

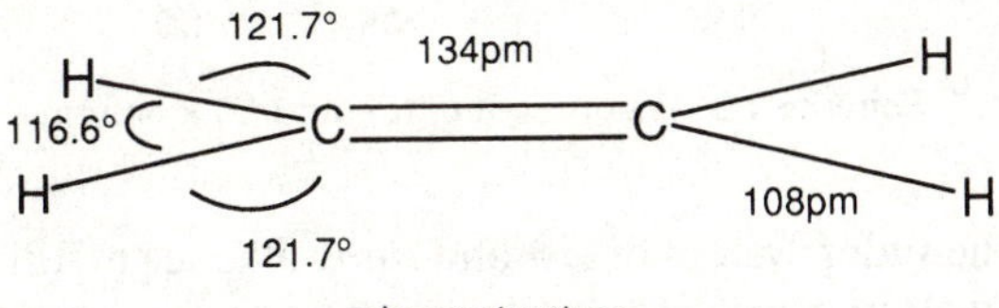

Scheme 1.1 Structure of ethene.

a π-component. The bonding σ-orbital and the bonding π-orbital are filled, and the antibonding π*-orbital and the antibonding σ*-orbital are unfilled (Scheme 1.2). It is the filled π-orbital which is largely responsible for the planar character of alkenes, and accounts for the high barrier to rotation around the carbon–carbon bond, allowing for the existence of alkene stereoisomers (*E*) and (*Z*).

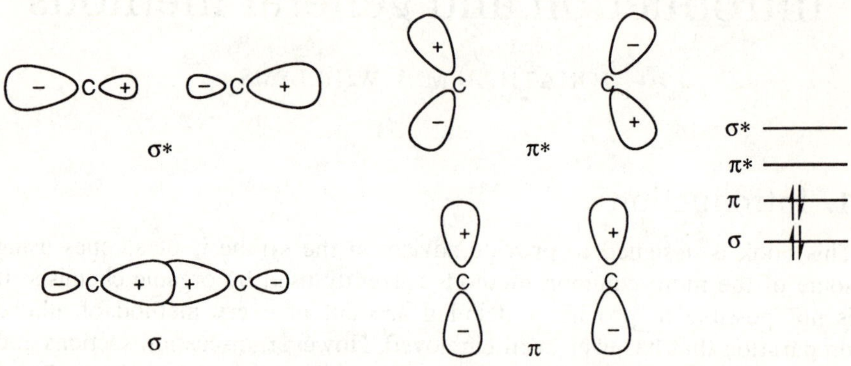

Scheme 1.2 Molecular orbitals of an alkene.

The main consideration in the nomenclature of alkenes is in describing the relative orientation of the substituents attached to the alkene unit. The use of the Cahn–Ingold–Prelog system is unambiguous, and is preferred. The priority of the substituents is based upon atomic number, and relative priorities are assigned to the groups attached to the carbons at each end of the alkene.[1] When the higher priority groups are on the same side of the alkene, this is the (*Z*)-configuration. If the higher priority groups are on opposite sides of the alkene, this is the (*E*)-configuration (Scheme 1.3). These descriptors are derived from *Z* = zusammen (German, 'together') and *E* = entgegen (German, 'opposite').[1]

lower, higher / higher, lower — (*E*)

lower, lower / higher, higher — (*Z*)

Scheme 1.3 Assignment of (*E*)- and (*Z*)-isomers.

Very often the older system of *cis-* and *trans*-alkenes is still employed. This system works well in many cases, which is why it has survived for so long. Whilst the *cis-* and *trans*-isomers of but-2-ene are easily distinguished, the introduction of a third substituent causes confusion with the *cis-* and *trans-*

nomenclature. However, these isomers are readily assigned with the appropriate (*E*)- or (*Z*)-descriptor (Scheme 1.4).

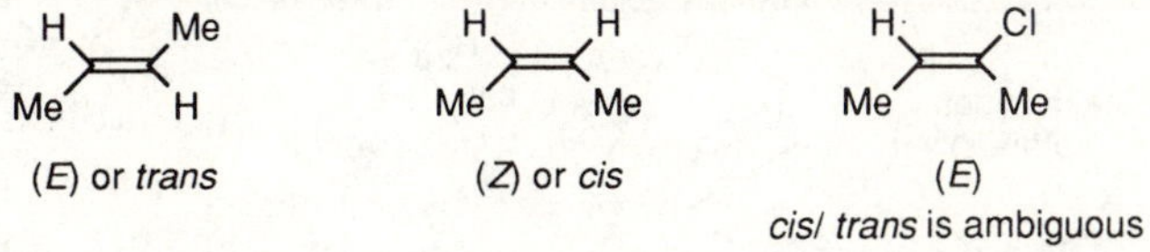

H Me Me H
(*E*) or *trans*

H H Me Me
(*Z*) or *cis*

H Cl Me Me
(*E*)
cis/ *trans* is ambiguous

Scheme 1.4 Comparison of (*E*)/(*Z*) versus *cis*/*trans*.

There are essentially four conceptually different methods for the synthesis of alkenes, and these can be arranged into the precursors required (Scheme 1.5). The alkene group can be prepared from two separate components (Type 1). This approach is described in Chapters 2, 3, and 4, and includes the Wittig reaction. Alternatively, the precursor may already contain a single carbon–carbon bond which is converted into an alkene (Type 2). Elimination reactions are described in Chapter 5. Alkenes can be prepared from other alkenes (Type 3), and it is sometimes difficult to draw the line as to what constitutes an alkene synthesis! Chapter 8 describes examples of the conversion of vinyl halides and vinyl metals into alkenes containing an additional carbon-based substituent. Finally, it is possible to convert alkynes into alkenes (Type 4). In this case the alkynes are either reduced (Chapter 6) or otherwise functionalised (Chapter 7) to afford the product.

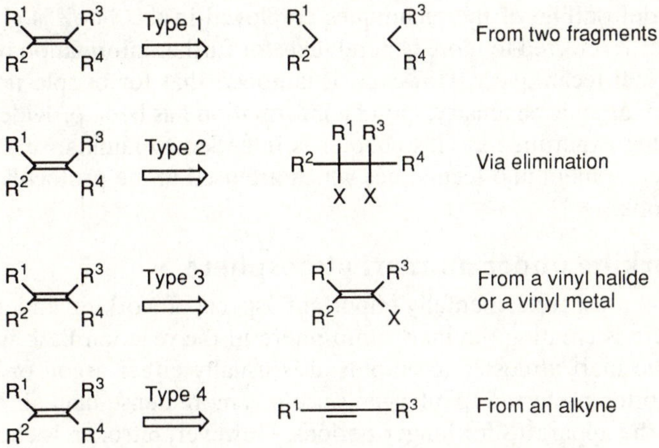

R^1 R^3 R^2 R^4 Type 1 ⟹ R^1 R^2 R^3 R^4 From two fragments

R^1 R^3 R^2 R^4 Type 2 ⟹ R^1 R^3 R^2 R^4 X X Via elimination

R^1 R^3 R^2 R^4 Type 3 ⟹ R^1 R^3 R^2 X From a vinyl halide or a vinyl metal

R^1 R^3 R^2 R^4 Type 4 ⟹ R^1—≡—R^3 From an alkyne

Scheme 1.5 Different concepts for alkene synthesis.

As well as being a synthetic end-point in their own right, alkenes may also function as precursors to a multitude of other functionalities, including those identified in Scheme 1.6.

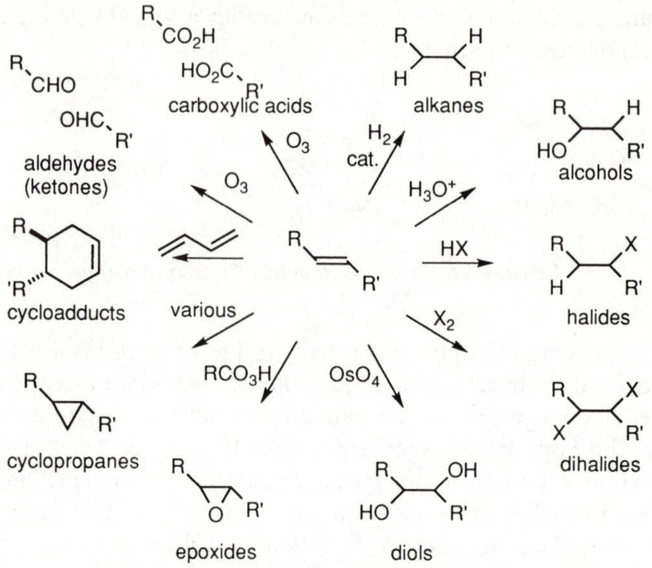

Scheme 1.6 Some synthetic uses of alkenes.

2. Experimental techniques

The techniques required in the synthesis of alkenes are the same as the standard techniques required by organic chemists for other reactions. However, a brief outline of the techniques employed in this book is given here. The reader is referred to more general texts for further information on general experimental techniques.[2] However, it is hoped that for people not experienced with organic chemistry, enough information has been provided here to perform the experiments. This chapter is intended to familiarise the reader with the equipment and techniques which are used in the protocols throughout the book.

2.1 Working under an inert atmosphere

One of the most fundamentally important aspects of working under an inert atmosphere is creating the inert atmosphere in the reaction flask in the first place.[3] The inert atmosphere employed is usually either argon or nitrogen. Argon is often preferred to nitrogen since it is more dense than air and tends to 'sit' in the apparatus for longer periods. However, nitrogen is cheaper and is often suitable for most applications. The inert gas supply may come directly from a cylinder regulator or via taps provided within the laboratory. A bubbler needs to be incorporated in order to allow the release of pressure, and this may either be attached to the nitrogen supply, the nitrogen manifold, or as an exit from the reaction vessel. This latter option may be preferred, since an indication can be gained of the rate of flow of nitrogen through the

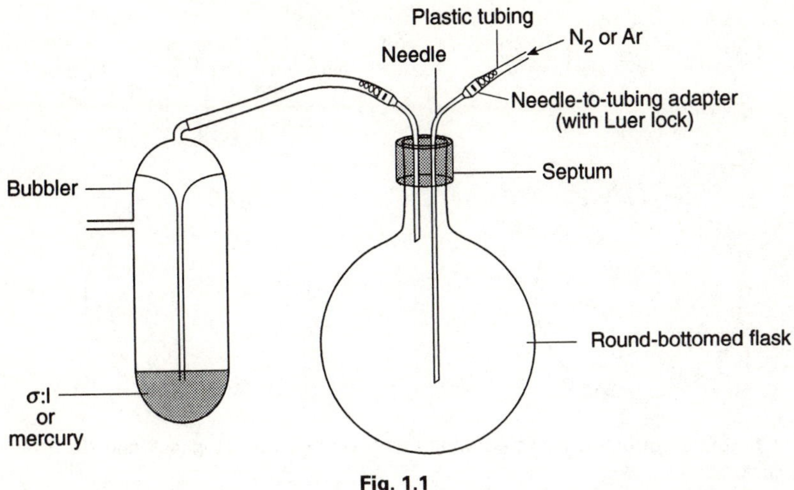

Fig. 1.1

apparatus (however, a continuous flow of nitrogen over a reaction leads to solvent loss). It is conventional to attach a bubbler to the manifold as well.

A simple way of filling a flask with a gas is to purge the flask with the gas for several minutes. Thus, introduction of nitrogen through a syringe needle into a flask equipped with a septum and an outlet needle connected to a bubbler will fill the flask with nitrogen (Fig. 1.1). However, this method probably does not entirely remove air (oxygen), and it is of course dependent upon flow rates and the size of the system employed.

A reaction vessel may also be connected to an inert gas supply via a tubing adapter, which may be either bent or straight (Fig. 1.2). When a vacuum/inert gas manifold is available, a flask equipped with a tubing adapter can be filled

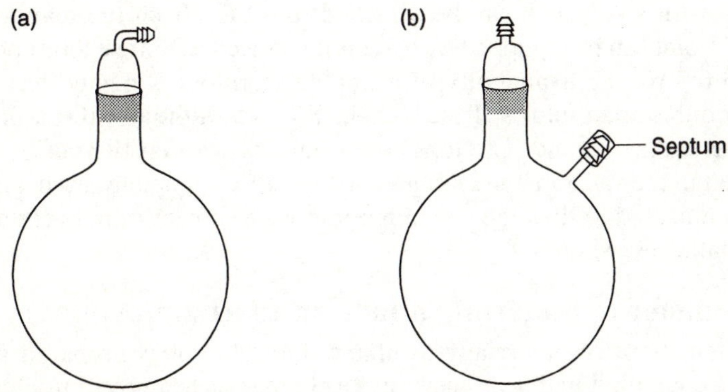

Fig. 1.2 (a) Round-bottomed flask equipped with (bent) tubing adapter. (b) Side-arm flask equipped with tubing adapter.

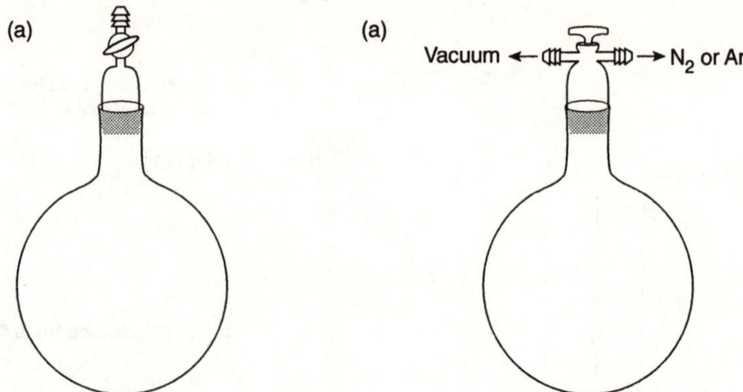

Fig. 1.3 Round-bottomed flasks equipped with (a) two-way stopcock and (b) three-way stopcock.

with an inert atmosphere as follows. First, the flask is evacuated by connecting to the vacuum. Second, the inert gas is introduced. This process is repeated three more times to ensure that there is no residual air. The same principles may be applied to reaction apparatus more complex than a simple round-bottomed flask, provided that all of the apparatus is designed to withstand high vacuum.

It is often convenient to be able to isolate the reaction vessel from the manifold, and this is most easily achieved by using a tubing adapter which contains a stopcock (Fig. 1.3). When a three-way stopcock is used, there is no need for the availability of a double manifold, since independent vacuum and inert gas supplies may be used.

If a lot of chemistry involving an inert atmosphere needs to be undertaken, it is worth investing in a range of Schlenk flasks and tubes (Fig. 1.4). The integral side-arm with a stopcock can be attached to a tube from the manifold. The Quickfit® joint can be equipped with a septum as well as more exotic apparatus.

Furthermore, the availability of a double manifold is highly desirable. A typical double manifold is illustrated in Fig. 1.5, although there are many variations on this theme. The taps are designed such that the outlets may be connected to the vacuum, the inert gas, or to neither. Generally, such a manifold will be connected to the high vacuum pump via a solvent trap. Further details are available elsewhere.[2a,b]

2.2 Transfer of materials under an inert atmosphere

Very often, both commercially available and laboratory-prepared reagents need to be handled under an inert atmosphere. For example, butyllithium in hexane and other organometallics may be transferred from the bottle into the reaction vessel either by the use of a syringe or with a metal cannula. These

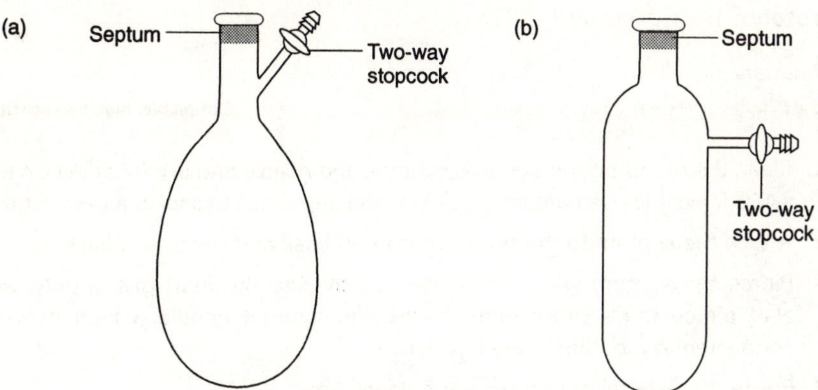

Fig. 1.4 (a) Schlenk flask; (b) Schlenk tube.

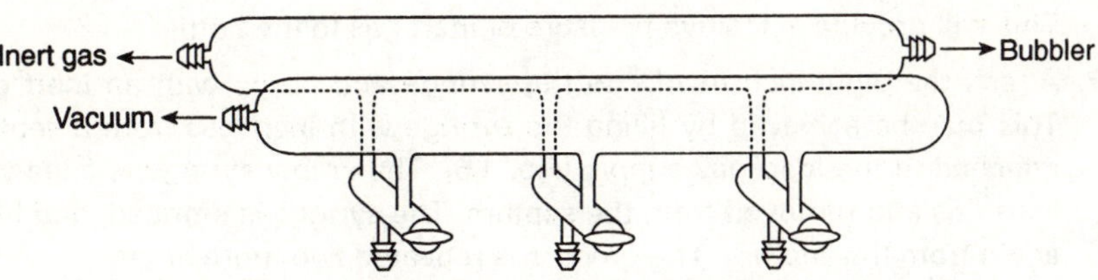

Fig. 1.5 A typical double manifold.

procedures are worthy of their own protocols, and should be referred to in detail if you are not already familiar with these techniques. There are many methods available for determining the molarity of organometallic reagents by titration.[4] This is helpful for analysis of laboratory-prepared organometallics, and for analysis of older commercial solutions.

Protocol 1.
Transfer of butyllithium in hexane from a bottle to a flask via syringe (Figs 1.7 and 1.8)

Caution! All procedures should be carried out in a well-ventilated hood, and disposable vinyl or latex gloves and chemical-resistant safety goggles should be worn.

Equipment

- One-necked, round-bottomed flask (50 mL)
- Glass syringe with needle-lock Luer (10 mL)
- Eight-inch medium-gauge needle
- Two-inch medium-gauge needle
- Source of dry nitrogen or argon attached to an outlet needle
- Septum

Protocol 1. *Continued*

Materials

- Butyllithium (FW 64.1) 1.6 M in hexane, 5 mL **flammable, moisture sensitive**

1. Clean the round-bottomed flask, syringe, and needle and dry for at least 4 h in a 120°C electric oven before use. Allow the apparatus to cool in a desiccator.
2. Attach the septum to the round-bottomed flask and clamp the flask.
3. Pierce the septum with the needle attached to the inert gas supply, and also pierce the septum with the two-inch outlet needle, which may be connected to a bubbler (see Fig. 1.1).
4. Purge the flask with inert gas for at least 5 min.
5. Clamp the bottle of butyllithium and remove the screw cap but not the Sure/Seal™ cap.
6. Pierce the Sure/Seal™ with another outlet needle from the inert gas supply. This will provide a positive pressure of inert gas to the bottle.
7. Attach the eight-inch needle to the syringe and purge with an inert gas. This may be achieved by filling the syringe with inert gas from a septum attached to the inert gas supply (Fig. 1.6). The empty syringe is filled with inert gas and removed from the septum. The syringe is emptied, and filled again from the septum. The process is repeated two more times.
8. Pierce the Sure/Seal™ with the syringe needle (Fig. 1.7). The positive pressure will be pushing the plunger out—keep it in place.
9. When the syringe needle is immersed in the liquid, bend the needle through 180° and allow the syringe to fill. Usually, there is no need to pull the plunger, since the bottle is under positive pressure.
10. Push the plunger back to remove any bubbles and excess liquid until there is 5 mL remaining in the syringe.[a]
11. Carefully holding the syringe assembly, withdraw the needle from the bottle. Quickly proceed to step 12.
12. Introduce the needle into the septum of the round-bottomed flask, (as shown in Fig. 1.8) and deliver the liquid to the flask.[b]

[a] To protect the liquid which will be at the end of the needle, a blanket of inert gas may be drawn into the syringe as follows. Raise the tip of the needle above the surface of the liquid in the bottle, and draw in 1–2 mL of inert gas prior to withdrawing the needle tip from the bottle.
[b] Once the plunger has been pushed in fully, the measured volume has been delivered. However, this means that the needle still contains air-sensitive liquid. This needs to be quenched in a suitable way. For alkyllithium reagents, the syringe may be carefully quenched with isopropanol.

Based on Protocol 1, a similar syringe transfer is possible from any other flask (i.e. through a septum) by similarly introducing a positive nitrogen pressure,

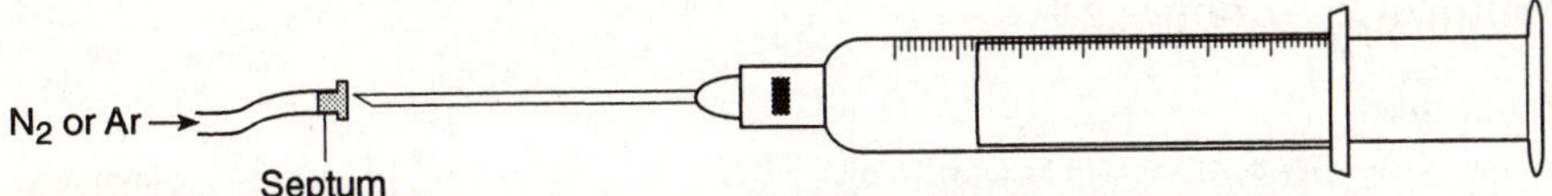

Fig. 1.6 Flusing a syringe plus needle with inert gas.

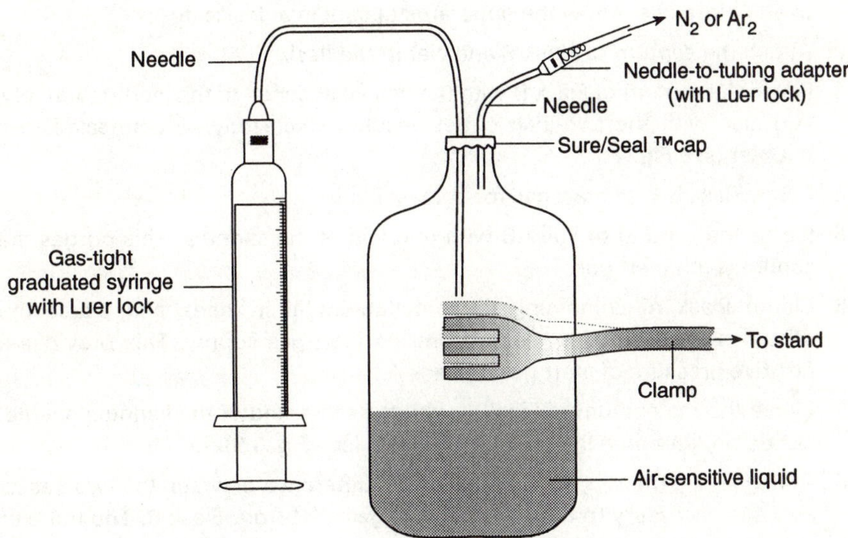

Fig. 1.7 Syringe removal of an air-sensitive liquid.

and following the steps in Protocol 1. Syringe transfer is ideal for transferring relatively small amounts and when the quantity transferred needs to be known. However, a metal cannula (tube) is often more convenient for transferring larger quantities. Cannula transfer is described in detail in Protocol 2.

Protocol 2.
Transfer of butyllithium in hexane from flask to flask via cannula (Fig. 1.9)

Caution! All procedures should be carried out in a well-ventilated hood, and disposable vinyl or latex gloves and chemical-resistant safety goggles should be worn.

Equipment

- One-necked, round-bottomed flask (50 mL). This is 'Flask B'
- Septum
- Eighteen-inch cannula with bevelled ends
- Two-inch medium-gauge needle attached to a bubbler
- Source of dry nitrogen or argon attached to an outlet needle

Protocol 2. *Continued*

Materials

- Butyllithium (FW 64.1) 1.6 M in hexane (25 mL) in a one-necked, round-bottomed flask (50 mL) sealed with a septum. This is 'Flask A' — **flammable, moisture sensitive**

1. Clean Flask B and the cannula and dry for at least 4 h in a 120°C electric oven before use. Allow the apparatus to cool in a desiccator.
2. Attach the septum to Flask B and clamp the flask.
3. Pierce the septum of Flask B with the needle attached to the inert gas supply, and also with the two-inch outlet needle, which may be connected to a bubbler (see Fig. 1.1).
4. Purge Flask B with inert gas for at least 5 min.
5. Pierce the septum of Flask B with one end of the cannula. This purges the cannula with inert gas.
6. Clamp Flask A (containing the butyllithium in hexane), and pierce the septum with the outlet needle from the inert gas supply. This provides a positive pressure of inert gas in Flask A.
7. Pierce the the septum of Flask A with the other end of the cannula needle, but do not immerse the tip into the liquid (see Fig. 1.9).
8. At this stage, there is not a big pressure difference between the two flasks, and it is necessary to remove the inert gas inlet from Flask B. The bubbler should continue to bubble, since the inert gas can pass through the cannula from Flask A.
9. By lowering the cannula tip into the liquid in Flask A, the liquid will transfer into Flask B.[a] Keep the cannula tip out of the liquid in Flask B.
10. Re-connect the inert gas supply to Flask B. Remove the cannula, and if no further chemistry is to take place remove the bubbler and then the inert gas supply.
11. Residual liquid in the apparatus needs to be carefully quenched.

[a] If the flow through the cannula needs to be regulated, this may be achieved by using a 'cannula' made up from two needles and a male/male, Luer-to-Luer stopcock (see Fig. 1.10).

The cannula transfer method may be employed to transfer liquid from one chosen system to another chosen system, using the principles described in Protocol 2.

Two of the protocols described in this book are easier with the use of a glove bag (Fig. 1.11). Glove bags are relatively inexpensive, and are much easier to use than might be imagined. A glove bag can provide a reasonable amount of protection for solids which need to be transferred. Thus, to intro-

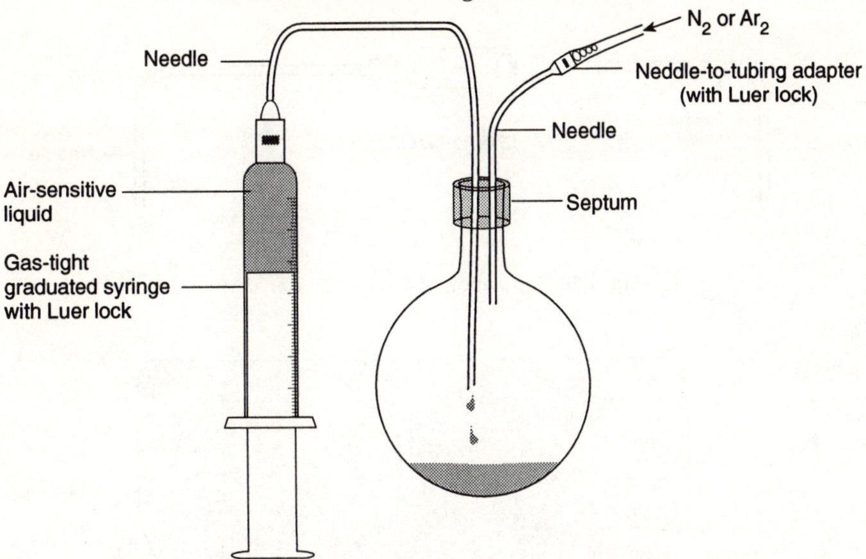

Fig. 1.8 Syringe addition of an air-sensitive liquid.

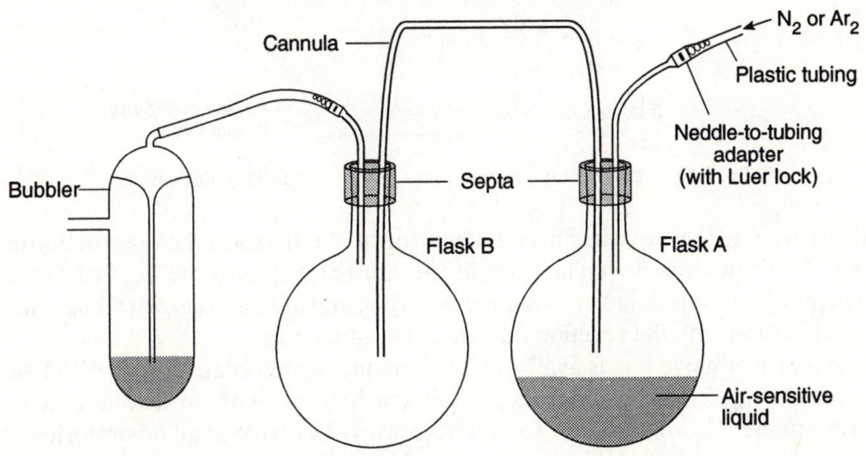

Fig. 1.9 Cannula transfer of an air-sensitive liquid.

duce an air-sensitive reagent from a bottle to a flask, the following procedure is employed. The clean, dry flask is filled with an inert atmosphere and sealed with a septum. The sealed flask, a spatula, the bottle of reagent, and a small portable balance are placed in the opened glove bag, which is then sealed, either by tape or by folding and holding in place with a clip. The manufacturer's instructions should be followed for specific details; however, in general the same principles apply. The glove bag is evacuated through the inlet, and then nitrogen (argon) introduced. This operation is repeated three more

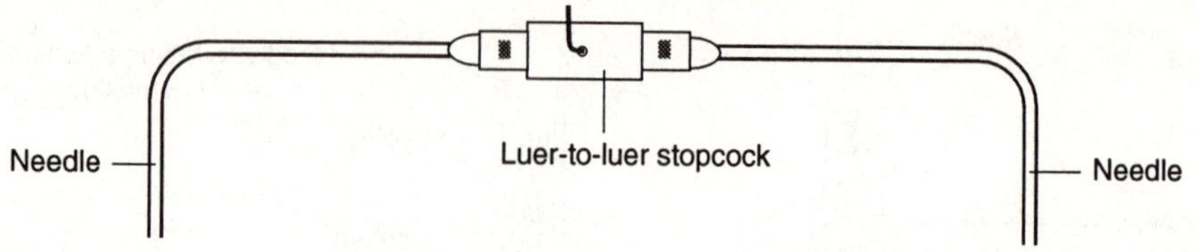

Fig. 1.10 Regulation of a 'cannula' transfer.

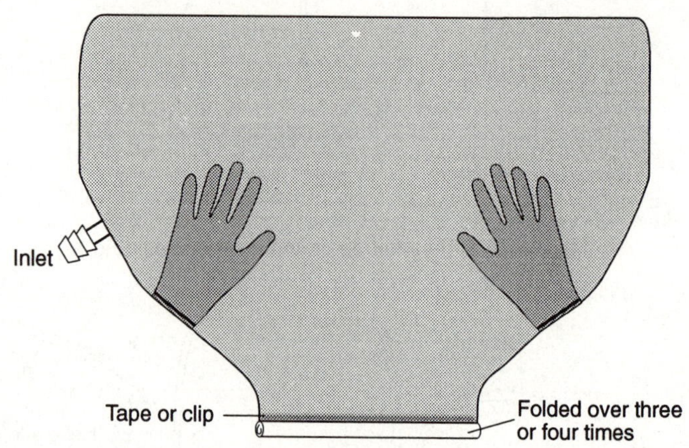

Fig. 1.11 A glove bag—a cheap and quick version of a glove box.

times, leaving the glove bag filled with nitrogen. At this stage the reagent bottle may be opened, and weighed out in the normal way into the opened flask. The bottle and flask are then sealed before removal from the glove bag, and continuation with the reaction outside of the glove bag.

If a proper glove box is available, this provides an alternative, very effective means of transferring and handling air-sensitive reagents in a totally inert atmosphere. However, these are not necessarily available in all laboratories.

2.3 Stirring, cooling, and heating

The most common method for stirring an organic reaction is with the aid of a magnetic stirrer (Fig. 1.12). The stirrer unit sometimes also has a heating facility (stirrer/hotplate). A magnetic field in the stirrer unit drives a magnetic stirrer bar (follower or flea) and thereby stirs the reaction. The stirrer bar needs to be big enough that it can efficiently stir the reaction, but small enough that it is not impeded from stirring by the reaction vessel.

For larger reactions (i.e. over 1000 mL), and for particularly viscous reaction mixtures, or those containing a lot of solid, a magnetic stirrer may not be effective. An alternative is the use of an overhead stirrer. An electric

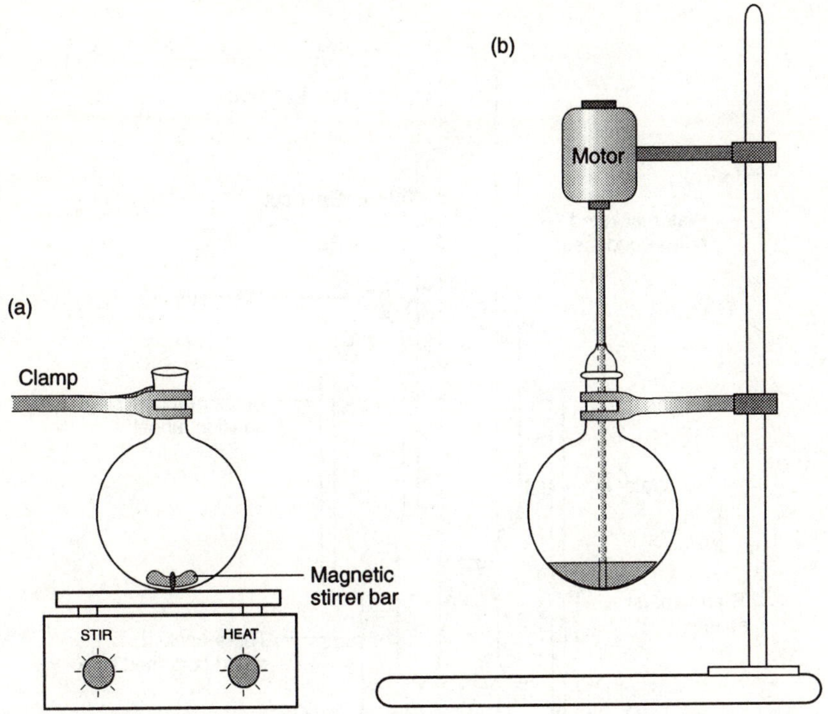

Fig. 1.12 (a) Magnetic stirrer/hotplate; (b) mechanical (overhead) stirrer.

motor turns a glass rod which is attached to a Teflon paddle (Fig. 1.12). Care needs to be taken to align the glass rod as close to the vertical as possible, in order to ensure free movement of the glass rod assembly.

Some of the protocols in this book require external cooling with either an ice bath or a dry-ice/acetone bath. In the former, the reaction vessel is immersed in a bath containing crushed ice and a small amount of water. The dry-ice/acetone bath is prepared using an insulated (Dewar) bath. Normally, the reaction flask is lowered into the bath whilst it is empty. The bath is then filled with pellets of dry ice (solid carbon dioxide). Acetone is slowly and carefully added to the bath. The acetone is cooled to −78 °C by the dry ice, and thereby cools the reaction vessel. Additional dry ice needs to be added periodically in order to maintain the cooling. Other solvents are cooled to different temperatures, and details of the different temperatures possible are available from other sources.[2a,b]

Cooling down to about −10 °C can also be achieved with commercially available stirrer coolplates, which are especially useful for maintaining the temperature of small-scale overnight reactions.

Heating a reaction can be achieved by placing the reaction vessel in an oil

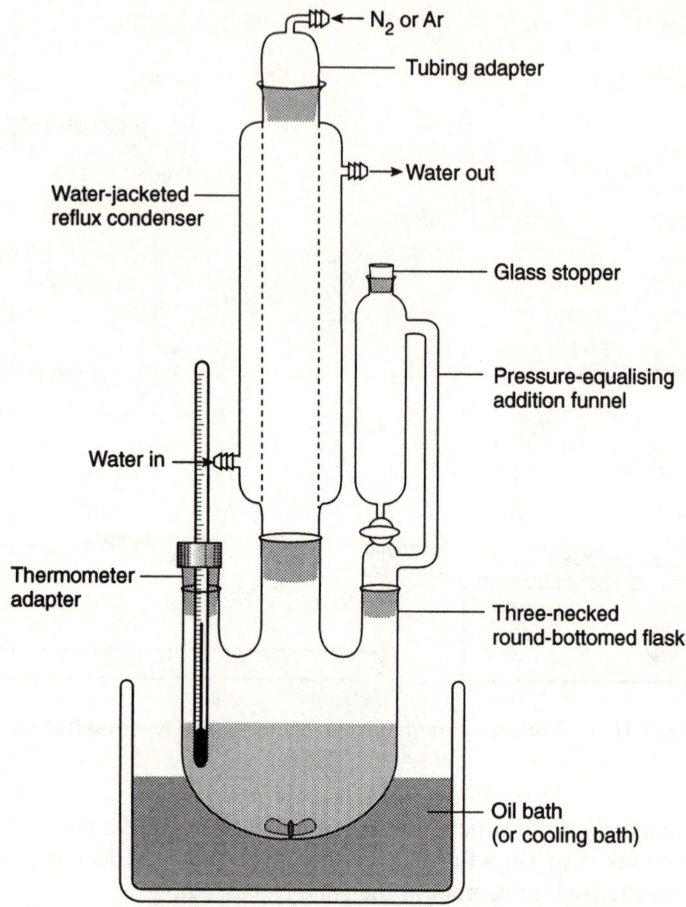

Fig. 1.13 Typical apparatus for heating a reaction under an inert atmosphere.

bath placed on top of a stirrer/hotplate. The temperature of the oil bath is controlled by the hotplate. Alternatively, an electric isomantle can be employed.

For many straightforward reactions, a simple one-necked, round-bottomed flask can be equipped with a water-jacketed reflux condenser and heated to reflux, and this arrangement will be satisfactory. However, if the internal temperature of the reaction needs to be monitored, and reagents need to be added, the arrangement shown in Fig. 1.13 should be used. A three-necked flask is equipped with a thermometer (with thermometer adapter), a water-jacketed reflux condenser with a tubing adapter, and a pressure-equalising dropping funnel. When solids need to be added, a powder-dropping funnel needs to be employed in place of the pressure-equalising dropping funnel.

When the solvent is low boiling (e.g. ammonia), a dry-ice condenser is employed.

2.4 Distillation

Starting materials, reaction products, and solvents often need to be distilled in order to enhance purity. This section gives a brief overview of the distillations encountered in the rest of the book.

For small-scale distillations, a Kugelrohr bulb-to-bulb distillation apparatus is convenient (Fig. 1.14). The sample is placed in the end bulb, and the oven temperature is raised (a vacuum may also be applied if this is appropriate). The sample is distilled into the next bulb (which is outside of the heating compartment) and collected. The operation can be repeated into the next bulb, if necessary. If the sample has a relatively low boiling point under the pressure employed, it is common practice to cool the receptor bulb with ice, dry ice, or even liquid nitrogen (absorbed onto cotton wool).

A short-path distillation procedure may be used in situations where a simple distillation is required as purification from non-volatile components—fractional distillation with this apparatus is difficult. A water-jacketed, semi-micro distillation apparatus is illustrated in Fig. 1.15. Heating can either be achieved with the aid of an oil bath, an isomantle, or a flame (**caution!** ensure that there are no flammable solvents nearby).

For larger scale distillations, the round-bottomed flask would be attached to a distillation column, a still-head, and condenser.[2d,e] The column used in the distillation is variable, and may be either a Vigreux column or a column packed with glass helices, for example. For the distillation of solvents, especially if done on a regular basis, a solvent still may be set up. This provides a convenient source of freshly dried and purified solvents. Solvent which has been standing in the still collecting-vessel should be run back into the main still, and then redistilled. The standard methods for drying solvents are identified in texts on reagent purification.[5] The most commonly encountered drying methods are as follows.

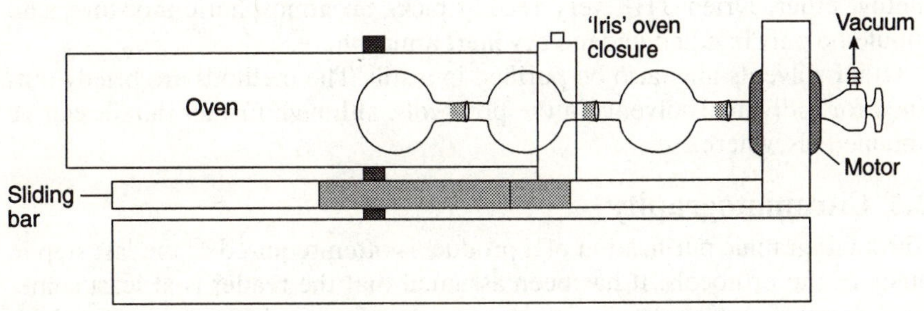

Fig. 1.14 Kugelrohr distillation apparatus.

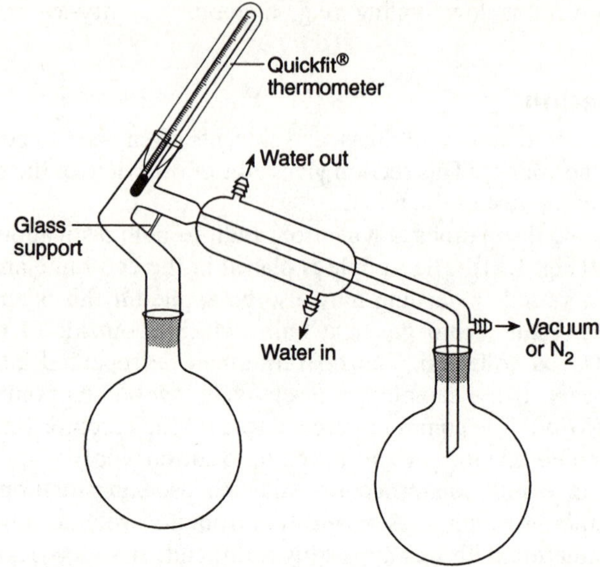

Fig. 1.15 Water-jacketed, semi-micro distillation apparatus.

Dichloromethane is distilled from calcium hydride (about 5 g of calcium hydride for every 100 mL of solvent to be distilled).

Caution! Never try to dry chlorinated solvents with sodium or with strong bases. There is a risk of explosion.

Diethyl ether undergoes preliminary drying by the addition of sodium wire. Further drying is then achieved by distilling from sodium (1 g for every 100 mL) and benzophenone (0.2 g for every 100 mL). As the ether is heated to reflux it gets drier, and when the blue colour associated with the benzophenone ketyl radical persists, the dried ether may be distilled and collected.

Caution! Diethyl ether contains peroxides which are potentially explosive. The presence of peroxides may be detected with peroxide test paper.

Tetrahydrofuran (THF) may be dried in a similar way to that described for diethyl ether. Dried THF very readily picks up atmospheric moisture, and should be only handled under a dry inert atmosphere.

Other solvents may also be purified in a still. The methods are briefly outlined for individual solvents in the protocols, although further details can be obtained elsewhere.[5]

2.5 Chromatography

Chromatographic purification of a product is often required as the last step in many of the protocols. It has been assumed that the reader is at least somewhat familiar with column chromatography. Detailed instructions and information about chromatography are available elsewhere,[2] but the following

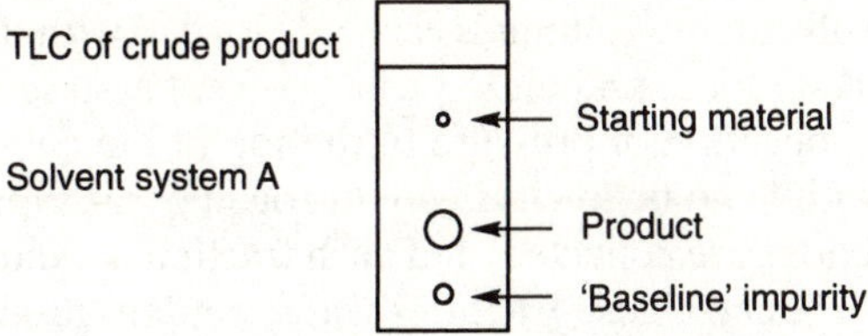

Fig. 1.16 TLC of crude product.

comments provide a brief overview of the procedure. Imagine that a reaction product contains some impurities, such as unreacted starting material and by-products. Thin layer chromatography (TLC) should be performed on the crude product to find a solvent system which provides an R_f value of approximately 0.25 (see Fig. 1.16). The R_f of the product spot can be increased by the use of a more polar solvent system, and the R_f can be decreased by the use of a less polar solvent system. It is worth spending some time optimising the TLC, as this helps with subsequent columning. The TLC may be visualised in a number of ways, but the simplest is to look at the TLC plate under a UV lamp. When this is not possible (compound UV-inactive) then various stains may be used.[2]

Silica is measured out into a beaker using approximately 30 g for every 1 g of crude product—although more may be required if spots run close to each other by TLC. (**Caution!** Silica dust is very toxic if inhaled, and should be handled in a well-ventilated fume hood.) The solvent used for the TLC (solvent system A) is then used to make up a slurry with the silica, which is added to a suitable column (Fig. 1.17). After settling, addition of a thin layer of sand,

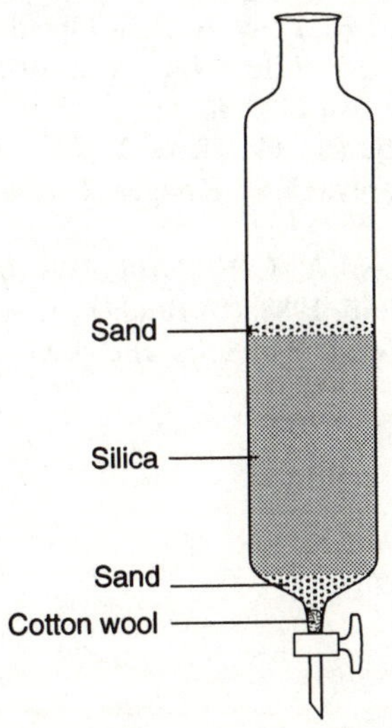

Fig. 1.17 A typical chromatography column.

and running off solvent, the column is carefully loaded with the crude product, as a concentrated solution, and eluted using solvent system A. The elution is effected by the application of pressure to the top of the column. This may be achieved either with hand bellows or with nitrogen or air supplied via a tubing adapter. The fractions are collected, and each fraction is examined by TLC for the presence of product. Those fractions which contain product are collected together and the solvent removed under reduced pressure.

References

1. Eliel, E. L.; Wilen, S. H. *Stereochemistry of Organic Compounds*; Wiley: New York, **1994**.
2. There are many guides available on general techniques, some of which are listed here: (a) Gordon, A. J.; Ford, R. A. *The Chemist's Companion*; Wiley: New York, **1972**; (b) Casey, M.; Leonard, J.; Lygo, B.; Procter, G. *Advanced Practical Organic Chemistry*; Blackie: London, **1990**; (c) Sharp, J. T.; Gosney, I.; Rowley, A. G. *Practical Organic Chemistry: A Student Handbook of Techniques*; Chapman and Hall: New York, **1989**; (d) Harwood, L. M.; Moody, C. J. *Experimental Organic Chemistry*; Blackwell: Oxford, **1989**; (e) Furniss, B. S.; Hannaford, A. J.; Smith, P. W. G.; Tatchell, A. R., eds. *Vogel's Textbook of Practical Organic Chemistry*, 5th edn; Longmans: London, **1989**; (f) Zubrik, J. W. *The Organic Chem Lab Survival Manual. A Student's Guide to Techniques*; Wiley: New York, **1992**; (g) Pavia, D. L.; Lampman, G. M.; Kriz, G. S. *Introduction to Organic Laboratory Techniques. A Contemporary Approach*, 3rd edn; Saunders College: Philadelphia, **1988**; (h) Wilcox, Jr, C. F. *Experimental Organic Chemistry. A Small Scale Approach*; Macmillan: New York, **1988**.
3. For more information on working under an inert atmosphere, see: (a) Shriver, D. S.; Drezdzon, M. A. *The Manipulation of Air-Sensitive Compounds*, 2nd edn; Wiley: New York, **1986**; (b) Kramer, G. W.; Levy, A. B.; Midland, M. M. In *Organic Syntheses via Boranes*; Brown, H. C., ed.; Wiley: New York, **1975**; Chapter 9; (c) Lane, C. F.; Kramer, G. W. *Aldrichim. Acta* **1977**, *10*, 11.
4. For example, see: (a) Kiljunen, H.; Hase, T. A. *J. Org. Chem.* **1991**, *56*, 6950; (b) Whitesides, G. M.; Casey, C. P.; Krieger, J. K. *J. Am. Chem. Soc.* **1971**, *93*, 1379.
5. (a) Perrin, D. D.; Armarego, W. L. F. *Purification of Laboratory Chemicals*, 3rd edn; Pergamon Press: Oxford, **1988**. (b) Riddick, J. A.; Bunger, W. B.; Sakano, T. K. *Organic Solvents, Physical Properties and Methods of Purification*, 4th edn; Wiley: New York, **1986**.

2

The Wittig reaction and related methods

NICHOLAS J. LAWRENCE

1. Introduction

The Wittig reaction, the reaction of an α-phosphorus-stabilised anion with a carbonyl derivative, is undoubtedly one of the most useful methods for selectively constructing carbon–carbon double bonds. This is for many reasons: the precursors are easily prepared and in many instances are available readily, the reaction is high yielding, the alkene is specifically positioned, and in many instances is stereoselective. Wittig developed the reaction of phosphonium ylides with aldehydes in the early 1950s.[1] Since this time the reaction has been further developed and modified in many ways and has now become a standard method for alkene synthesis.[2–4]

$$Ph_3P{=}CHR^1 \xrightarrow{\text{H(C=O)}R^2} R^1CH{=}CHR^2$$

There are essentially three variants of the Wittig reaction:

- The Wittig reaction of phosphonium ylides
- The Horner–Wadsworth–Emmons (HWE) reaction of phosphonate anions
- The Horner–Wittig reaction of phosphine oxide anions

In keeping with the aim of this series, a detailed mechanistic account of the Wittig reaction will not be included; this is, anyhow, more than adequately detailed elsewhere.[2–5] The aim is to help in the choice of a selective Wittig method and to provide a protocol that can be adapted for a particular task.

A detailed account of the stereoselectivity of the Wittig reaction is somewhat complicated but some generalisations can be made: stabilised phosphoryl anions will give (*E*)-alkenes whereas non-stabilised anions will give (*Z*)-alkenes. This must be borne in mind when planning a synthesis; the disconnection that aids the required selectivity should be chosen. Thus, if the

(*Z*)-alkene is required the disconnection chosen is the one where the most electron-donating group R_1 is incorporated into the ylide fragment (sense 1).

R^1–CH=CH–R^2 ⟹ [R^1CH=PPh_3 and H(C=O)R^2 — sense 1] or [R^2CH=PPh_3 and H(C=O)R^1 — sense 2]

2. The Wittig reaction of phosphonium ylides

The conventional Wittig reaction involves addition of a phosphonium ylide to an aldehyde.[6] In general, the reaction is:

- (*E*)-Selective if R_1 is anion-stabilising (CO_2Me, COMe, SO_2Ph, CN, etc.)
- (*Z*)-Selective if R_1 is electron-releasing (alkyl)
- Non-selective if R_1 is weakly anion-stabilising (phenyl, allyl)

The phosphonium salts are prepared simply by the reaction of triphenylphosphine with an equimolar quantity of an organic halide.[7] The displacement follows the normal reactivity trend of an S_N2 reaction. High yields of phosphonium salts can be obtained from 1-alkyl iodides (Protocols 1 and 2) and benzyl bromides at about 50 °C (THF or $CHCl_3$) whereas more forcing conditions are required with branched alkyl halides and 1-alkyl chlorides and bromides. In these cases the reaction should be carried out in a high-boiling-point solvent such as toluene (b.p. 111 °C) or xylene (b.p. 140 °C). Other solvents that have been used include nitromethane, formic acid, acetic acid, ethyl acetate, ethanol, acetone, butanone, cyclohexanone, acetonitrile, benzonitrile, and DMF. In some cases it may be necessary to heat the triphenylphosphine and alkyl halide in the absence of solvent.

Protocol 1.
Preparation of ethyltriphenylphosphonium iodide

Caution! All procedures should be carried out in a well-ventilated hood, and disposable vinyl or latex gloves and chemical-resistant safety goggles should be worn.

$$Ph_3P \xrightarrow[CHCl_3,\ \text{reflux}]{EtI} Ph_3\overset{+}{P}CH_2CH_3\ I^-$$

Equipment

- Two-necked, round-bottomed flask (500 mL)
- One-necked, round-bottomed flask (500 mL)
- Water-jacketed reflux condenser
- Pressure-equalising dropping funnel (50 mL)
- Magnetic stirrer/hotplate
- Magnetic stirrer bar
- Oil bath

Materials

- Triphenylphosphine (FW 262.3) 40 g, 153 mmol — **irritant, harmful**
- Ethyl iodide (FW 156.0) 15 mL, 188 mmol — **irritant, flammable**
- Chloroform 200 mL — **harmful, cancer suspect agent**
- Propan-2-ol for recrystallisation — **highly flammable**

1. Clean the apparatus and dry in a 120 °C electric oven for at least 4 h. Allow to cool in a desiccator.
2. Add triphenylphosphine (40 g, 153 mmol) to chloroform (200 mL) in the two-necked flask equipped with a reflux condenser, stirrer bar, and a pressure-equalising dropping funnel.
3. Add ethyl iodide (15 mL, 188 mmol) to the solution via the dropping funnel. Heat the mixture slowly until it begins to reflux gently. The reaction is exothermic so care must be taken to ensure that the reaction is not too vigorous, by appropriate cooling.
4. Maintain the mixture at reflux for a further 1 h and leave to cool.[a]
5. Transfer the mixture to a one-necked flask (500 mL) and remove the chloroform under reduced pressure on a rotary evaporator to leave a colourless oil which crystallises on standing.
6. The product may be recrystallised from propan-2-ol to give pure ethyltriphenylphosphonium iodide, m.p. 166–167 °C (lit. m.p. 167 °C).

[a]The displacement follows the normal reactivity trend of an S_N2 reaction. Allyl and benzyl bromides react rapidly at ambient temperatures; in these cases the reaction can be carried out in THF and in most cases the insoluble crystalline phosphonium salt is simply isolated by filtration. Alkyl iodides and bromides react much faster than the corresponding chloride; yields are highest when the reaction is conducted in refluxing toluene. In most cases the phosphonium salts are sufficiently pure for the Wittig reaction. If required they may be recrystallised from higher alcohols or precipitated from strong solutions in ethyl acetate or chloroform with hexane.

Protocol 2.
Preparation of chloromethyltriphenylphosphonium iodide[8]

Caution! All procedures should be carried out in a well-ventilated hood, and disposable vinyl or latex gloves and chemical-resistant safety goggles should be worn.

$$Ph_3P \xrightarrow[\text{THF, reflux}]{ICH_2Cl} Ph_3\overset{+}{P}CH_2Cl \; I^-$$

Equipment

- Three-necked, round-bottomed flask (500 mL)
- Water-jacketed reflux condenser
- Thermometer and thermometer adapter
- Pressure-equalising dropping funnel (50 mL)
- Tubing adapter
- Magnetic stirrer/hotplate
- Magnetic stirrer bar
- Oil bath

Protocol 2. *Continued*

Materials

- Triphenylphosphine (FW 262.3) 315 g, 120 mmol — **irritant, harmful**
- Chloroiodomethane (FW 176.4) 26.5 g, 150 mmol — **irritant, harmful, light sensitive**
- Dry THF 450 mL — **highly flammable, irritant**

1. Clean the apparatus and dry in a 120°C electric oven for at least 4 h. Allow to cool in a desiccator.
2. Dissolve triphenylphosphine (31.5 g, 120 mmol) in THF (200 mL) in the three-necked flask equipped with a magnetic stirrer bar, a pressure-equalising dropping funnel, a thermometer, and a condenser topped with a tubing adapter (see Chapter 1, Fig. 1.13).
3. Flush the apparatus with nitrogen for 10 min.
4. Add chloroiodomethane[9] (26.5 g, 150 mmol) dropwise, via the dropping funnel, stirring the mixture rapidly at room temperature.
5. Once addition is complete, immerse the flask in an oil bath and heat the mixture at reflux for 20 h.
6. Cool the flask and filter the white precipitate and wash thoroughly with THF (5 × 50 mL) under a nitrogen atmosphere.
7. Dry the isolated phosphonium iodide (37 g, 70%) *in vacuo* at 50–60°C for several hours.

Alternatively, phosphonium salts may be made by the alkylation of simpler phosphonium salts. For example, (2-trimethylsilylethyl)triphenylphosphonium bromide and the related tin analogue,[10] useful reagents in the synthesis of allylsilanes and stannanes, are made in this way (see Protocol 3).

Protocol 3.
Alkylation of a phosphorane: synthesis of (2-trimethylsilylethyl) triphenylphosphonium bromide[11]

Caution! All procedures should be carried out in a well-ventilated hood, and disposable vinyl or latex gloves and chemical-resistant safety goggles should be worn.

$$Ph_3\overset{+}{P}{-}Me\ \ Br^- \xrightarrow[\text{2. } ICH_2SiMe_3]{\text{1. n-BuLi}} Ph_3\overset{+}{P}CH_2CH_2SiMe_3\ \ Br^-$$

Equipment

- One-necked, round-bottomed flask (250 mL)
- Glass syringes with needle-lock Luers (20 mL and 5 mL)
- Two six-inch medium-gauge needles
- Magnetic stirrer
- Magnetic stirrer bar
- Ice bath

Materials

- Methyltriphenylphosphonium bromide (FW 357.2) 8.03 g, 22.5 mmol — **irritant, harmful, moisture sensitive**
- Dry THF 40 mL — **highly flammable, irritant**
- *n*-Butyllithium (FW 64.1) 1.66 M in hexane, 15 mL, 25 mmol — **highly flammable, hygroscopic**
- Iodomethyltrimethylsilane (FW 214.1) 2.82 g, 22.5 mmol — **irritant, flammable, light sensitive**

1. Clean the apparatus and dry in a 120 °C electric oven for at least 4 h. Allow to cool in a desiccator.
2. Suspend methyltriphenylphosphonium bromide (8.03 g, 22.5 mmol) in dry THF (40 mL) at 0 °C under a nitrogen atmosphere in the round-bottomed flask stoppered with a rubber septum connected, via a needle, to a positive pressure of dry nitrogen.
3. Add *n*-butyllithium (15 mL, 1.66 M in hexane) dropwise via a syringe (**caution!** see Chapter 1, Protocol 1).
4. Warm the mixture to room temperature and stir for 1 h.
5. Cool the mixture to 0 °C using the ice bath and add iodomethyltrimethylsilane (2.82 g, 22.5 mmol) dropwise via a syringe over 10 min.
6. Warm the mixture to room temperature and stir for 1 h after which time the new phosphonium salt will precipitate and is sufficiently pure to use without purification. The ylide may be generated by the addition of a further equivalent of *n*-butyllithium.

Further examples of the synthesis of common triphenylphosphonium salts are listed in Table 2.1.

The currently accepted mechanism of the Wittig reaction[5] involves the potentially reversible formation of intermediate *erythro*- and *threo*-oxaphosphetanes which undergo stereospecific *syn*-elimination to the (*E*)- and (*Z*)-alkenes respectively. Formation of the oxaphosphetane occurs by an asynchronous cycloaddition and the decomposition to alkene by a *syn*-cycloreversion. The old two-step ionic mechanism seems to have been largely discredited. Stabilised and non-stabilised ylides favour the formation of the *threo*- and *erythro*-oxaphosphetanes respectively. Conditions that favour thermo-

$(R^3)_3P{=}CHR^1$ + R^2CHO ⇌ *threo* oxaphosphetane + *erythro* oxaphosphetane

threo → $R^1CH{=}CHR^2$ (*E*)

erythro → $R^1CH{=}CHR^2$ (*Z*)

Table 2.1 Syntheses of common triphenylphosphonium salts $RPPh_3^+X^-$[12] (see also reference 6)

R	X	From	Ref.
Methyl	Br	CH_3Br, PPh_3, C_6H_6, r.t., 1 day	13
Ethyl	Br	CH_3CH_2Br, $C_6H_5CH_3$, 105 °C, 1 day	14
Propyl	Br	$CH_3(CH_2)_2Br$, PPh_3, C_6H_6, 80 °C, 2 days	15
Isopropyl	I	$(CH_3)_2CHI$ (neat), PPh_3, 100 °C, 20 h	16
Neopentyl	I	$(CH_3)_3CCH_2I$, PPh_3, sulfolane, 160 °C, 24 h	17
Chloromethyl	I	ICH_2Cl, PPh_3, THF, 67 °C, 20 h	18
Bromomethyl	Br	$HOCH_2PPh_3Br$, PCl_5, C_6H_6, 80 °C, 23 h	19
Iodomethyl	I	CH_2I_2, PPh_3, C_6H_6, 50 °C, 16 h	19
Methoxymethyl	I	ICH_2OMe, PPh_3, C_6H_6, 25 °C, 30 min	20
Hydroxymethyl	Br	PPh_3, HBr, CH_2O, r.t., 2 h	19
Cyanomethyl	Br	$BrCH_2CN$, PPh_3, C_6H_6, 20 °C, 14 days	21
Trimethylsilylmethyl	Cl	PPh_3, Me_3SiCH_2Cl (neat), 90 °C, 5 days	22
2-Bromoethyl	Br	$Br(CH_2)_2Br$, PPh_3, xylenes, 140 °C	23
2-Hydroxyethyl	Br	PPh_3, $Br(CH_2)_2OH$, C_6H_6, 80 °C, 18h	24
3-Bromopropyl	Br	PPh_3, $Br(CH_2)_3Br$, xylene, 130 °C, 16 h	25
3-Phenylpropyl	Br	PPh_3, $Ph(CH_2)_3Br$, DMF, 153 °C, 12 h	26
Benzyl	Cl	PPh_3, $PhCH_2Cl$, $CHCl_3$, 61 °C, 6 h	27
p-Cl-benzyl	Cl	PPh_3, *p*-Cl-benzyl chloride, C_6H_6, 80 °C, 3 days	26
p-MeO-benzyl	Br	PPh_3, *p*-Me-benzyl bromide, C_6H_6, 80 °C, 12 h	28
p-Me-benzyl	Br	PPh_3, *p*-Me-benzyl bromide, DMF, 153 °C, 3 h	29
p-NO_2-benzyl	Br	PPh_3, *p*-NO_2-benzyl bromide, C_6H_6, 80 °C, 12 h	28
Allyl	Br	$BrCH_2CH{=}CH_2$, PPh_3, C_6H_6, 80 °C, 1 h	30
Propargyl	Br	PPh_3HBr, $BrCH_2C{\equiv}CH$, 1,4-dioxane, 3 h, r.t.	31
(*E*)2-Butenyl	Br	$BrCH_2CH{=}CHCH_3$, PPh_3, C_6H_6, r.t., 48 h	32
(*E*)2-Cinnamyl	Cl	$PhCH{=}CHCH_2Cl$, PPh_3, xylene, 140 °C, 12 h	29
Carbomethoxymethyl	Br	PPh_3, $BrCH_2CO_2Me$, C_6H_6, 35 °C, 18 h	33
Carboethoxymethyl[34]	Br	PPh_3, $BrCH_2CO_2Et$, C_6H_6, 80 °C, 30 min	35

dynamic equilibration of the *erythro*- ⟶ *threo*-oxaphosphetanes, and hence potential loss of (*Z*)-selectivity, include donor groups (including alkyl groups) on phosphorus, the presence of lithium salts, and sterically encumbered aldehydes and ylides.

If the (*Z*)-alkene is required the conditions detailed by Schlosser *et al.*[36] (Protocol 4) should be followed:

- R_1 and R_2 are alkyl
- R_3 is phenyl
- Lithium salt-free conditions prevail, i.e. use sodium or potassium bases
- A non-protic, polar solvent is used (THF, diethyl ether, DME, *tert*-butyl methyl ether)

Protocol 4.
A general procedure for the Schlosser–Wittig reaction[36]

Caution! All procedures should be carried out in a well-ventilated hood, and disposable vinyl or latex gloves and chemical-resistant safety goggles should be worn.

$$\mathrm{Ph_3\overset{+}{P}CH_2R^1\ X^-} \xrightarrow[\text{2. } \mathrm{HC(O)R^2},\ -78^\circ\mathrm{C} \rightarrow \text{r.t.}]{\text{1. } \mathrm{NaNH_2}, \text{ THF, r.t.}} \mathrm{R^1CH{=}CHR^2}$$

Equipment

- Three-necked, round-bottomed flask (500 mL)
- Water-jacketed reflux condenser
- Pressure-equalising dropping funnel (50 mL)
- Magnetic stirrer
- Magnetic stirrer bar
- Dry-ice/acetone cooling bath
- Septum

Materials

- Alkyltriphenylphosphonium salt 100 mmol — **irritant, harmful, moisture sensitive**
- Sodium amide (FW 39.0) 3.9 g, 100 mmol — **flammable solid, dangerous when wet**
- Aldehyde 100 mmol
- Dry THF 200 mL — **highly flammable, irritant**
- Hexane 300 mL — **highly flammable, harmful**

1. Clean the apparatus and dry in a 120°C electric oven for at least 4 h. Allow to cool in a desiccator.
2. A mixture of alkyltriphenylphosphonium salt (100 mmol) and sodium amide (100 mmol)[a] in dry THF (200 mL) is vigorously stirred at 25°C for 20 min in the three-necked flask equipped with a stirrer bar, pressure-equalising dropping funnel, and a septum with an outlet to a nitrogen line. The solution will turn a dark red colour.
3. Cool the flask with a dry-ice/acetone cooling bath.
4. Charge the dropping funnel with the aldehyde (100 mmol) and add dropwise over 15 min.
5. Allow the flask to warm to room temperature and stir for 15 min.
6. Perform a standard aqueous work-up (see Protocol 6, step 5).
7. The alkene may be purified directly by distillation or chromatography. In the latter case the majority of the triphenylphosphine oxide by-product may be removed by precipitation by adding cold (<−20°C) hexane (300 mL) to the crude reaction residue.

[a]Sodium amide is sometimes supplied as a 50% suspension in toluene. In this case, toluene may be removed by filtration through a glass frit, washed with diethyl ether, dried, and stored under nitrogen.

Schlosser and Christmann[37] also have shown that the oxaphosphetanes may be deprotonated by PhLi and reprotonated to give, highly selectively, the *threo*-isomer thereby providing an exceptionally efficient synthesis of (*E*)-alkenes (Protocol 5).

Protocol 5. Synthesis of (*E*)-oct-2-ene[37]

Caution! All procedures should be carried out in a well-ventilated hood, and disposable vinyl or latex gloves and chemical-resistant safety goggles should be worn.

$Ph_3\overset{+}{P}CH_2CH_3\ Br^-$ → (1. PhLi; 2. n-$C_5H_{11}CHO$; 3. PhLi; 4. HCl, Et_2O) → (*E*)-oct-2-ene

Equipment

- One-necked, round-bottomed flask (250 mL)
- Three glass syringes with needle-lock Luers (50 mL)
- Magnetic stirrer
- Magnetic stirrer bars
- Dry-ice/acetone cooling bath
- Centrifuge and centrifuge tubes

Materials

- Phenyllithium (FW 84.0) 2 M in cyclohexane/ether (70:30), 15 mL, 30 mmol — **flammable, moisture sensitive**
- Ethyltriphenylphosphonium bromide (FW 371.3) 11.1 g, 30 mmol — **irritant, harmful**
- Dry ether 40 mL — **flammable, irritant**
- Hexanal (FW 100.2) 3 g, 30 mmol — **irritant, flammable, air sensitive**
- Hydrogen chloride (FW 36.5) 1 M in ether 33 mL, 33 mmol — **flammable, corrosive**

1. Clean the apparatus and dry in a 120 °C electric oven for at least 4 h. Allow to cool in a desiccator.
2. Add phenyllithium (15 mL, 2.0 M in cyclohexane/ether (70:30)), via a syringe (**caution**! see Chapter 1, Protocol 1), to a suspension of ethyltriphenylphosphonium bromide (11.1 g, 30 mmol) in dry ether (30 mL) in the round-bottomed flask sealed with a rubber septum with an outlet to a nitrogen line (see Chapter 1, Fig. 1.8).
3. Stir the solution for 10 min and cool to −70 °C with a dry-ice/acetone cooling bath.
4. Add hexanal (3 g, 30 mmol) in dry ether (10 mL) dropwise via a syringe and stir vigorously.
5. Allow the solution to warm to −30 °C. As soon as decolorisation is complete add further phenyllithium (30 mmol) in ether, via a syringe. Stir the mixture at −30 °C for 10 min.

6. Add a solution of hydrogen chloride (33 mL, 1.0 m in ether).
7. Stir the solution at room temperature for 2 h.
8. Transfer the mixture to a centrifuge tube, and centrifuge.
9. Pour off the supernatant liquor and wash with water. Dry the ether ($MgSO_4$) layer and evaporate under reduced pressure on a rotary evaporator. The residue may be purified by distillation (b.p. 121°C/740 mmHg) to give (*E*)-oct-2-ene (70%, (*E*):(*Z*) 99:1).

In cases where (*E*)/(*Z*)-selectivity is not important the use of lithium bases is tolerable. For example, methylenation of ketones and aldehydes may be performed with methylenetriphenylphosphorane[38] generated from methyltriphenylphosphonium bromide and LDA (Protocol 6).

Protocol 6.
Preparation of an exomethylene epoxide[39]

Caution! All procedures should be carried out in a well-ventilated hood, and disposable vinyl or latex gloves and chemical-resistant safety goggles should be worn.

$Ph_3\overset{+}{P}$— Br^- —(1. LDA; 2. ketone)→ exomethylene epoxide

Equipment

- One-necked, round-bottomed flasks (100 mL and 250 mL)
- Cannula needle
- Magnetic stirrer
- Magnetic stirrer bars

Materials

• *n*-Butyllithium (FW 64.1) 2.3 M in hexane 6.5 mL, 15 mmol	**highly flammable, hygroscopic**
• Diisopropylamine (FW 101.2) 2.1 mL, 1.5 g, 15 mmol	**highly flammable, corrosive**
• Dry THF	**highly flammable, irritant**
• Ketone 1.26 g, 10 mmol	**assume toxic**
• Ammonium chloride	**harmful, irritant, hygroscopic**
• Dry ether	**highly flammable, irritant**
• Magnesium sulfate	**moisture sensitive**

1. Clean the apparatus and dry in a 120°C electric oven for at least 4 h. Allow to cool in a desiccator.
2. Add dropwise *n*-butyllithium (6.5 mL, 2.3 M in hexane) (**caution!** see Chapter 1,

Protocol 6. *Continued*

Protocol 1) at 0°C to a solution of diisopropylamine (2.1 mL, 1.5 g, 15 mmol) in dry THF (30 mL) in the round-bottomed flask (100 mL), equipped with a stirrer bar and stoppered with a rubber septum connected, via a needle, to a positive pressure of dry nitrogen.

3. Stir the solution at 0°C for 15 min and transfer by cannula (see Chapter 1, Protocol 2) to a suspension of methyltriphenylphosphonium bromide (5.4 g, 15 mmol) in dry THF (30 mL) at 0°C in the round-bottomed flask (250 mL) again stoppered with a rubber septum connected, via a needle, to a positive pressure of dry nitrogen.
4. Stir the mixture for 1.5 h at 0°C. Add the ketone (1.26 g, 10 mmol) in THF (10 mL) via a syringe and stir the mixture for 3 h at room temperature.
5. Add aqueous ammonium chloride solution (50 mL) and then remove the THF under reduced pressure on a rotary evaporator. Extract the aqueous residue with ether (3 × 50 mL). Dry the combined organic phase ($MgSO_4$) and remove the ether under reduced pressure on a rotary evaporator. The crude alkene (91%) is purified by column chromatography on silica gel (petroleum ether:ether 1:1 v/v as eluant).

The Wittig reaction is truly versatile since very nearly any type of functional group may be tolerated α to the phosphonium group, e.g. halides (Protocol 7), heteroatom-containing groups (including OH), nitriles, esters, or ketones. Many of these methods lead to important types of alkene, for example when X = OMe the vinyl ether may be hydrolysed to the one-carbon-extended aldehyde (Protocol 8).

Ph_3P=CHX; RCHO → RCH=CHX

X = H, Protocol 6
X = Cl, Protocol 7
X = OMe, Protocol 8

Protocol 7.
Preparation of chloromethylenecyclohexane[8]

Caution! All procedures should be carried out in a well-ventilated hood, and disposable vinyl or latex gloves and chemical-resistant safety goggles should be worn.

$Ph_3P^+CH_2Cl\ I^-$ → 1. K-*tert*-butoxide; 2. Cyclohexanone → chloromethylenecyclohexane (Cl)

Equipment

- Three-necked, round-bottomed flask (100 mL)
- Thermometer and thermometer adapter
- Water-jacketed reflux condenser
- Powder-dropping funnel
- Pressure-equalising dropping funnel (50 mL)
- Magnetic stirrer/hotplate
- Magnetic stirrer bar
- Oil bath
- Tubing adapter

Materials

- Dry *tert*-butyl alcohol (FW 74.1) 25 mL — **highly flammable, harmful**
- Potassium (FW 39.1) 1.17 g, 30 mmol — **flammable, solid, hygroscopic**
- (Chloromethyl)triphenylphosphonium iodide (FW 438.7) 11.0 g, 25 mmol — **irritant, moisture, sensitive**
- Cycohexanone (FW 98.2) 1.96 g, 2.07 mL, 20 mmol — **flammable, harmful**
- Pentane — **highly flammable, irritant**
- Magnesum sulfate — **irritant**

1. Clean the apparatus and dry in a 120°C electric oven for at least 4 h. Allow to cool in a desiccator.
2. Place dry[a] *tert*-butyl alcohol (25 mL), under a nitrogen atmosphere, in the three-necked flask equipped with a thermometer, stirrer bar, and a condenser topped with a tubing adapter.
3. Add potassium metal (**caution!**) (1.17 g, 30 mmol) and heat the mixture under reflux for 1 h, or until all the metal has completely dissolved, to give a solution of potassium *tert*-butoxide.
4. Cool the flask and quickly attach a powder-dropping funnel containing (chloromethyl)triphenylphosphonium iodide (Protocol 2) (11.0 g, 25 mmol) to the flask. Add the phosphonium salt to the solution portionwise over 5 min. Stir the mixture at room temperature for 1.5 h after which time the mixture will change from white to an orange–red slurry.
5. Replace the powder funnel with a pressure-equalising dropping funnel containing cyclohexanone (1.96 g, 2.07 mL, 20 mmol) in *tert*-butyl alcohol (10 mL). Add the cyclohexanone dropwise over 20 min while retaining the reaction temperature below 25°C. Stir the mixture for 4 h at room temperature.
6. Add water (50 mL) and pentane (50 mL) and separate the phases. Extract the aqueous layer with pentane (3 × 50 mL). Wash the combined organic phase with water (3 × 30 mL) and dry ($MgSO_4$).
7. Remove the solvent under reduced pressure on a rotary evaporator to give the crude product.[b] Distil to give the vinyl chloride (94%).

[a]Dried according to Perrin and Armarego[40] by distillation from CaH_2.
[b]A small amount (3.4%) of dechlorinated alkene is produced (presumably from metal–halogen exchange of the phosphonium salt).

Protocol 8.
Preparation of homologated aldehydes[a] from ketones[41]

Caution! All procedures should be carried out in a well-ventilated hood, and disposable vinyl or latex gloves and chemical-resistant safety goggles should be worn.

$$\mathrm{R_2C{=}O} \xrightarrow[\text{2. } HClO_4]{\text{1. } MeOCH_2PPh_3^+Br^-,\ NaH,\ DMSO} \mathrm{R_2CH{-}CHO}$$

Equipment

- One-necked, round-bottomed flask (100 mL)
- Magnetic stirrer/hotplate
- Magnetic stirrer bar
- Oil bath
- Two glass syringes with needle-lock Luers (20 mL)
- Two six-inch, medium-gauge needles

Materials

- Sodium hydride (50% dispersion in mineral oil) (FW 24.0) 308 mg, 6.4 mmol — **flammable solid, dangerous when wet**
- Dry dimethyl sulfoxide 25 mL — **irritant, hygroscopic**
- (Methoxymethyl)triphenylphosphonium chloride (FW 342.8) 2.26 g, 6.4 mmol — **harmful, irritant, moisture sensitive**
- Ketone 1.2 mmol
- Ethyl acetate — **highly flammable**
- Magnesium sulfate — **hygroscopic**
- Ether — **flammable, irritant**
- Perchloric acid (FW 100.46) — **oxidizing agent, corrosive**

1. Clean the apparatus and dry in a 120 °C electric oven for at least 4 h. Allow to cool in a desiccator.
2. Add NaH (50% dispersion in mineral oil) (308 mg, 6.4 mmol) to dry DMSO[b] (5 mL) in the round-bottomed flask, equipped with a stirrer bar and stoppered with a rubber septum connected, via a needle, to a positive pressure of dry nitrogen.
3. Warm the flask to 55 °C and stir for 40 min during which time gas is evolved.
4. Add dropwise, via syringe, (methoxymethyl)triphenylphosphonium chloride (2.2 g, 6.4 mmol) in dry DMSO (10 mL).
5. Stir the reaction at 55 °C for 30 min.
6. Add the ketone (1.2 mmol) in dry DMSO (10 mL) via syringe.
7. Raise the temperature of the flask to 70 °C and stir for 10 h.
8. Cool the flask and pour the reaction mixture into water (100 mL).
9. Extract the solution with ethyl acetate (3 × 50 mL). Dry the combined

organic extracts ($MgSO_4$) and evaporate under reduced pressure on a rotary evaporator.

10. Add ether (20 mL) and perchloric acid (5 mL of a 70% aqueous solution) to the crude residue.
11. Stir at room temperature for 10 h. Pour the mixture into ice/water (100 mL).
12. Extract the solution with ethyl acetate (3 × 50 mL). Dry the combined organic extracts ($MgSO_4$) and evaporate under reduced pressure on a rotary evaporator. Purify the residue by chromatography or distillation.

[a]If an aldehyde is to be homologated then the reaction time and temperature for step 7 may both be reduced. The excess of phosphorane may also be reduced.
[b]Dried according to Perrin and Armarego[40] by distillation under reduced pressure from CaH_2.

A variety of bases are used in the Wittig reaction including NaH, $NaNH_2$, $NaN(TMS)_2$, NaH/DMSO (Protocol 9), KH, and KO^{t}Bu (Protocol 10), and, in cases where (*E*)/(*Z*)-selectivity is not important, LDA and *n*-BuLi.

Protocol 9.
Preparation of (*Z*)-5-decen-1-ol-1[42]

Caution! All procedures should be carried out in a well-ventilated hood, and disposable vinyl or latex gloves and chemical-resistant safety goggles should be worn.

$\overset{+}{P}Ph_3\ Br^-$ → 1. DMSO/NaH; 2. HO(tetrahydropyran-2-yl, O) → OH

Equipment

- Two one-necked, round-bottomed flasks
- Cannula needle
- Magnetic stirrer/hotplate
- Magnetic stirrer bar
- Oil bath

Materials

- Sodium hydride (50% dispersion in mineral oil) (FW 24.0) 24.0 g, 500 mmol — **flammable solid, dangerous when wet**
- 2-Hydroxytetrahydropyran (FW 102.3) 25.05 g, 116 mmol
- Dry dimethyl sulfoxide 500 mL — **irritant, hygroscopic**
- Pentyltriphenylphosphonium bromide (FW 411.3) 155 g, 375 mmol — **irritant, moisture sensitive**
- Sulfuric acid 10% in water, 100 mL — **highly corrosive, oxidiser**
- Dry pentane — **highly flammable, irritant**

1. Clean the apparatus and dry in a 120°C electric oven for at least 4 h. Allow to cool in a desiccator.

Protocol 9. *Continued*

2. Place NaH (50% dispersion in mineral oil) (24 g, 500 mmol) in the round-bottomed flask, under a nitrogen atmosphere, equipped with a magnetic stirrer bar and stoppered with a rubber septum connected, via a needle, to a positive pressure of nitrogen. Remove the mineral oil by addition of dry pentane (3 × 150 mL); each time stir the mixture, allow to settle, and decant the pentane by syringe.[a]
3. Add anhydrous DMSO (250 mL) to the NaH via a syringe. Stir the mixture efficiently and slowly heat to 75°C keeping hydrogen evolution under control.
4. After cessation of gas evolution heat the mixture at 75°C for a further 10 min and then cool to room temperature.
5. Add the solution of sodium methylsulfinylmethide dropwise,[b] via cannula, to pentyltriphenylphosphonium bromide (155 g, 375 mmol) in dry DMSO (200 mL) in a round-bottomed flask stoppered with a rubber septum and equipped with a magnetic stirrer and an argon inlet. The mixture is stirred at 20°C for a further 30 min to give a deep-red solution of the phosphorane.
6. Add, dropwise, 2-hydroxytetrahydropyran (25.05 g, 116 mmol) in dry DMSO (50 mL) via a syringe.
7. After addition is complete stir the mixture at 20°C for 2 h.
8. Pour the mixture onto crushed ice and extract with hexane (3 × 200 mL). Wash the combined organic extracts successively with sulfuric acid (100 mL of a 10% aqueous solution), saturated sodium bicarbonate solution (100 mL), and water (100 mL), and dry (Na_2SO_4). Evaporate under reduced pressure on a rotary evaporator to give an oil that is distilled to give pure (*Z*)-alkene (31.4 g, 81%) b.p. 109–111°C at 10 mmHg (*(Z):(E)* 93:7).

[a]Care must be taken to destroy any sodium hydride that is removed via syringe by the addition of isopropanol.
[b]An exothermic reaction occurs; the rate of addition is crucial in the control of temperature.

Protocol 10
Preparation of (*Z*)-dec-2′-enyl-1,3-dioxane[43]

Caution! All procedures should be carried out in a well-ventilated hood, and disposable vinyl or latex gloves and chemical-resistant safety goggles should be worn.

$Ph_3\overset{+}{P}$ Br^- (1,3-dioxan-2-yl)ethyl → 1. K–*tert*–butoxide, THF; 2. n-$C_6H_{13}CHO$ → C_6H_{13} (Z)-alkenyl-1,3-dioxane

Equipment

- Two one-necked, round-bottomed flasks (250 mL)
- Cannula needle
- Magnetic stirrer
- Magnetic stirrer bars
- Glass syringe with needle-lock Luer (20 mL)
- Six-inch, medium-gauge needle

Materials

• 2-(1,3–Dioxan-2-yl)-ethyltriphenylphosphonium bromide (FW 457.4) 22.8 g, 50 mmol	**hygroscopic**
• Potassium *tert*-butoxide (FW 112.2) 5.61 g, 50 mmol	**flammable solid, corrosive, moisture sensitive**
• Dry THF 150 mL	**highly flammable, irritant**
• Heptaldehyde (FW 114.9) 11.4 g, 13.9 mL, 100 mmol	**flammable, irritant**
• Ether 150 mL	**flammable, irritant**
• Magnesium sulfate	**irritant, hygroscopic**

1. Clean the apparatus and dry in a 120 °C electric oven for at least 4 h. Allow to cool in a desiccator.
2. Prepare a solution of potassium *tert*-butoxide by adding alcohol-free potassium *tert*-butoxide (5.61 g) to dry THF (100 mL) in the round-bottomed flask equipped with a magnetic stirrer bar and stoppered with a rubber septum connected, via a needle, to a positive pressure of dry nitrogen. Stir the mixture at room temperature for 10 min to complete dissolution of the salt.
3. Add the solution, via a cannula, to a stirred solution of 2-(1,3-dioxan-2-yl)-ethyltriphenylphosphonium bromide in dry THF (50 mL) at room temperature in a round-bottomed flask equipped with a magnetic stirrer bar and stoppered with a rubber septum again connected, via a needle, to a positive pressure of dry nitrogen.
4. Stir the solution at room temperature for 30 min to give the orange phosphorane.
5. Add heptaldehyde (11.4 g, 13.9 mL, 100 mmol) dropwise via a syringe and stir the mixture at room temperature for a further 2 h.
6. Pour the mixture into water (500 mL) and extract with ether (3 × 50 mL). Dry the combined organic extracts ($MgSO_4$) and evaporate under reduced pressure on a rotary evaporator to give an oil that is distilled to give pure (*Z*)-alkene (15.3 g, 72%) b.p. 134 °C at 5 mmHg.

Another important Wittig procedure is the dibromomethylenation of aldehydes with the reagent from carbon tetrabromide, zinc, and triphenylphosphine (Protocol 11).[44] The 1,1-dibromoalkenes are useful compounds since they are readily converted into alkynes by treatment with *n*-butyllithium.

Protocol 11. Dibromomethylenation[a] of isobutyraldehyde[44]

Caution! All procedures should be carried out in a well-ventilated hood, and disposable vinyl or latex gloves and chemical-resistant safety goggles should be worn.

$$\text{(CH}_3)_2\text{CHCHO} \xrightarrow{\text{CBr}_4,\ \text{PPh}_3,\ \text{Zn}} (\text{CH}_3)_2\text{CHCH=CBr}_2$$

Equipment

- Three-necked, round-bottomed flask (1 L)
- Pressure-equalising dropping funnel (100 mL)
- Magnetic stirrer
- Magnetic stirrer bar
- Glass syringe with needle-lock Luer (20 mL)
- Six-inch, medium-gauge needle

Material

• Zinc dust (FW 65.4) 10.5 g, 160 mmol	**flammable solid**
• Triphenylphosphine (FW 262.3) 42.0 g, 160 mmol	**harmful, irritant**
• Dry dichloromethane 410 mL	
• Carbon tetrabromide (FW 331.6) 53.1 g 160 mmol	**harmful, irritant**
• 2–Methylpropanal (isobutyraldehyde) (FW 72.1) 5.76 g, 7.3 mL, 80 mmol	**highly flammable + stench**
• Petroleum ether (2 L)	**highly flammable**

1. Clean the apparatus and dry in a 120°C electric oven for at least 4 h. Allow to cool in a desiccator.
2. Add zinc dust (10.5 g, 160 mmol) to a solution of triphenylphosphine (42 g, 160 mmol) in dry dichloromethane (150 mL) in the round-bottomed flask equipped with a pressure-equalising dropping funnel and stirrer bar, and stoppered with a rubber septum connected, via a needle, to a positive pressure of dry nitrogen.
3. Add carbon tetrabromide (53.1 g, 160 mmol) in dry dichloromethane (50 mL) via the dropping funnel at room temperature. Stir the resulting suspension for 24 h at room temperature.
4. Add 2-methylpropanal (5.76 g, 7.3 mL, 80 mmol) in dry dichloromethane (10 mL) via a syringe and stir the mixture at room temperature for 2 h.
5. Add the mixture to petroleum ether (1 L). Decant the supernatant solution from the oil that separates. Dissolve the oil in dichloromethane (200 mL) and add petroleum ether (1 L) and decant again. Evaporate the combined petroleum ether solutions under reduced pressure on a rotary evaporator. Purify the residue by flash distillation to give 1,1-dibromo-3-methyl-1-butene (10.6 g, 58%).

[a]For an *Organic Syntheses* description of the corresponding dichloromethylenation reaction see reference 45.

3. The Horner–Wadsworth–Emmons reaction of phosphonate anions

Perhaps the most widely used Wittig reaction is the Horner–Wadsworth–Emmons (HWE) reaction of α-phosphonate anions with aldehydes.[4] The (*E*):(*Z*)-selectivity can, in some cases, be tuned by the use of an appropriate R_3 group.

$$R^1CH^{-}P(O)(OR^3)_2 + HC(O)R^2 \rightleftharpoons (R^3O)_2P(O^-)\text{–O (oxaphosphetane, } R^1, R^2) \longrightarrow R^1CH{=}CHR^2$$

The phosphonates are readily obtained by Michaelis–Arbuzov[46] reaction between trialkylphosphites and an organic halide (the bromide is usually used in preference to the less reactive chloride). The two are simply heated in the absence of solvent (Protocol 12). It is most convenient to use trimethyl or triethylphosphite since the by-products methyl and ethyl bromide are volatile. The phosphonate is simply purified by distillation. Table 2.2 lists the reaction conditions for the syntheses of common phosphonates.

$$(MeO)_2P(O)OMe \xrightarrow[\text{heat}]{\text{RBr}} (MeO)_2P(O)R \text{ and } MeBr\uparrow$$

Table 2.2 Syntheses of common phosphonates[47] for the Horner–Wadsworth–Emmons reaction

Phosphonate	From	Ref.
Diethyl methylphosphonate	$[CH_3CH_2O]_3P$, MeI, reflux, 2 h	48
Diisopropyl methylphosphonate	$[(CH_3)_2CHO]_3P$, MeI, reflux, 1 h	49
Diethyl ethyl phosphonate	$[CH_3CH_2O]_3P$, EtI, reflux, 12 h	50
Diethyl trimethylsilylmethylphosphonate	$[CH_3CH_2O]_3P$, Me_3SiCH_2Cl, 170 °C, 6 h	51
Diethyl benzylphosphonate	$[CH_3CH_2O]_3P$, $PhCH_2Cl$, 150 °C, 10 h	50
Diethyl formylmethylphosphonate	$BrCH_2CH(OEt)_2$, $(EtO)_3P$, 160 °C, 3 h	52

Protocol 12.
Preparation of diethyl benzylphosphonate[53]

Caution! All procedures should be carried out in a well-ventilated hood, and disposable vinyl or latex gloves and chemical-resistant safety goggles should be worn.

$$PhCH_2Cl \xrightarrow[\text{heat}]{P(OEt)_3} PhCH_2P(O)(OEt)_2 \text{ and EtCl}$$

Protocol 12. *Continued*

Equipment

- One-necked, round-bottomed flask (250 mL)
- Water-jacketed reflux condenser
- Magnetic stirrer/hotplate
- Magentic stirrer bar
- Oil bath
- Still-head
- Thermometer

Materials

- Triethylphosphite (FW 166.2) 30 g, 31.0 mL, 181 mmol — **flammable, irritant, stench, air sensitive**
- Benzyl chloride (FW 126.6) 22.9 g, 20.8 mL, 181 mmol — **toxic, possible carcinogen, irritant**

1. Clean the apparatus and dry in a 120 °C electric oven for at least 4 h. Allow to cool in a desiccator.
2. Place triethylphosphite (30 g, 31.0 mL, 181 mmol) in the round-bottomed flask equipped with a condenser and magnetic stirrer bar. Add benzyl chloride (22.91 g, 20.8 mL, 181 mmol).
3. Heat the mixture on an oil bath at reflux for 18 h. Carry out the reaction in a fume hood as ethyl chloride is produced.
4. Cool the mixture and distil the diethyl benzylphosphonate under reduced pressure b.p. 106–108 °C/1 mmHg.

The conventional HWE reaction employs NaH as the base, and when the anion is not benzylic the anion is formed prior to the addition of the aldehyde. The anions of benzylic phosphonates are generated in the presence of the aldehyde (Protocol 13), and in these cases it has been found that catalytic 15-crown-5 assists the reaction (Protocol 14). Ethereal solvents such as 1,2-dimethoxyethane (DME) and THF are generally the solvent of choice. The phosphonate anion and aldehyde are generally stirred at room temperature for 24 h.

Protocol 13. Preparation of (*E*)-stilbene[54]

Caution! All procedures should be carried out in a well-ventilated hood, and disposable vinyl or latex gloves and chemical-resistant safety goggles should be worn.

$$\mathrm{PhCH_2P(O)(OEt)_2} \xrightarrow{\text{PhCHO, NaH, 70°C}} \mathrm{PhCH{=}CHPh}$$

Equipment

- One three-necked, round-bottomed flask (250 mL)
- Water-jacketed reflux condenser
- Magnetic stirrer/hotplate
- Magnetic stirrer bar

Materials

- Sodium hydride (50% dispersion in mineral oil) (FW 24.0) 2.4 g, 50 mmol — **flammable solid, dangerous when wet**
- Benzaldehyde (FW 106.1) 5.3 g, 5.08 mL, 50 mmol — **harmful, air sensitive**
- Dry 1,2-dimethoxyethane 100 mL — **highly flammable, harmful toxic, irritant**
- Diethyl benzylphosphonate (FW 228.4) 11.4 g, 10.4 mL, 50 mmol
- Ethanol — **highly flammable, toxic**

1. Clean the apparatus and dry in a 120°C electric oven for at least 4 h. Allow to cool in a desiccator.
2. Place sodium hydride (50% dispersion in mineral oil) (2.4 g, 50 mmol) in the round-bottomed flask, containing dry[a] DME (100 mL), equipped with a condenser and a magnetic stirrer bar and stoppered with a rubber septum connected, via a needle, to a positive pressure of dry nitrogen.
3. Add benzaldehyde (5.3 g, 5.08 mL, 50 mmol) and diethyl benzylphosphonate (11.4 g, 10.4 mL, 50 mmol), replace the septum, and purge the flask with dry nitrogen.
4. Slowly heat the flask with an oil bath. At around 70°C there is a large evolution of gas and a white precipitate forms. Should the reaction become violently exothermic heating should be stopped.
5. Heat the solution under reflux for 30 min.
6. Cool the mixture and pour onto ice/water (300 mL) to give a white precipitate.
7. Filter the solution and recrystallise the solid from a minimum volume of ethanol.

[a]Dried by distillation from calcium hydride and stored over 4 Å molecular sieves.

Protocol 14.
Preparation of (*E*)-3,4-Dimethoxystilbene[55]

Caution! All procedures should be carried out in a well-ventilated hood, and disposable vinyl or latex gloves and chemical-resistant safety goggles should be worn.

MeO, MeO, O, H
1. NaH/THF
15-crown-5
O, P-OEt, OEt
MeO, MeO

Equipment

- One-necked, round-bottomed flask (100 mL)
- Magnetic stirrer
- Magnetic stirrer bar
- Glass syringe with needle-lock Luer (20 mL)
- Six-inch, medium-gauge needle
- Ice bath

Protocol 14. *Continued*

Materials

• 3,4,-Dimethoxybenzaldehyde (FW 166.2) 830 mg, 5 mmol	**irritant, air sensitive**
• Diethyl benzylphosphonate (FW 228.4) 1.14 g, 1.04 mL, 5 mmol	**toxic, irritant**
• Dry THF	**highly flammable, irritant**
• 15-Crown-5 (FW 220.2) 30 mg, 0.14 mmol	**irritant, hygroscopic**
• Sodium hydride (50% dispersion in mineral oil) (FW 24.0) 12.0 mg, 5 mmol	**flammable solid, dangerous when wet**
• Ether	**flammable, irritant**
• Sodium hydrogen sulfite	**irritant, harmful by inhalation**
• Brine	
• Potassium carbonate (dry)	**irritant, moisture sensitive**

1. Clean the apparatus and dry in a 120°C electric oven for at least 4 h. Allow to cool in a desiccator.
2. Add dropwise, via a syringe, a solution of 3,4-dimethoxybenzaldehyde (veratraldehyde) (0.83 g, 5 mmol) and diethyl benzylphosphonate (1.14 g, 1.04 mL, 5 mmol) in dry THF (10 mL) to a slurry of oil-free (Protocol 9, step 2) NaH (0.12 g, 5 mmol) in THF (20 mL) containing 15-crown-5 (30 mg, 0.14 mmol), at 0°C, in the round-bottomed flask equipped with a stirrer bar and stoppered with a rubber septum connected, via a needle, to a positive pressure of dry nitrogen. A rapid evolution of hydrogen is observed and the solution develops a gelatinous orange precipitate.
3. Stir the suspension for 2 h at room temperature and pour into water (100 mL).
4. Separate the aqueous phase and extract with ether (3 × 25 mL). Wash the combined extracts with 10% sodium hydrogen sulfite solution (2 × 20 mL) and brine (2 × 20 mL) and dry (K_2CO_3). Remove the solvent under reduced pressure on a rotary evaporator to give the stilbene as colourless needles (1.19 g, 99%) m.p. 112–113°C (ethanol).

A significant advantage of the HWE procedures over the Wittig reaction of phosphonium ylides is that the phosphorus by-product, the phosphoric acid salt, is water-soluble and can simply be removed from the reaction mixture by aqueous work-up.

The most commonly encountered examples of the HWE reaction employ a stabilising group α to the phosphoryl group, providing an excellent method for the synthesis of α,β-unsaturated carbonyl compounds. Thus, when the methyl phosphonate **1** is used in the reaction with the aldehyde **2** the reaction is mildly (*Z*)-selective. However, by simply using bulky phosphoryl and carboxyl groups the reaction now becomes highly (*E*)-selective (R_1 = ^{i}Pr, R_2 = Et) (Protocol 15).[56] Significantly, the conditions are sufficiently mild that aldehydes bearing an adjacent stereogenic centre are not racemized.

R_1=Me, R_2=Me (*Z*):(*E*) 3:1
R_1=^{i}Pr, R_2=Et (*Z*):(*E*) 5.95

Protocol 15.
Preparation of an unsaturated ester[56]

Caution! All procedures should be carried out in a well-ventilated hood, and disposable vinyl or latex gloves and chemical-resistant safety goggles should be worn.

Equipment

- One necked, round-bottomed flask (100 mL)
- Magnetic stirrer
- Magnetic stirrer bar
- Glass syringe with needle-lock Luer (5 mL)
- Six-inch, medium-gauge needle
- Dry-ice/acetone cooling bath

Materials

- Potassium *tert*-butoxide (FW 112.2) 910 mg, 8.09 mmol — **flammable solid, corrosive, moisture sensitive**
- Diisopropyl (ethoxycarbonylmethyl)phosphonate (FW 252.3) 2.45 g, 2.31 mL, 8.99 mmol — **irritant**
- Dry THF — **highly flammable, irritant**
- Aldehyde **2**[26] (FW 178.2) 400 mg, 2.25 mmol
- Ammonium chloride solution — **harmful, irritant, hygroscopic**
- Dichloromethane — **toxic, irritant**
- Magnesium sulfate (dry) — **irritant**

1. Clean the apparatus and dry in a 120 °C electric oven for at least 4 h. Allow to cool in a desiccator.
2. Add potassium *tert*-butoxide (0.91 g, 8.09 mmol) in dry THF (16 mL) (see Protocol 10, step 2) to diisopropyl (ethoxycarbonylmethyl)phosphonate (2.45 g, 2.31 mL, 8.99 mmol) in THF (10 mL) at 0°C in the round-bottomed flask equipped with a magnetic stirrer bar and stoppered with a rubber septum connected, via a needle, to a positive pressure of dry nitrogen.

Protocol 15. *Continued*

3. Allow the solution to warm to room temperature and stir for 1 h.
4. Cool the flask with a dry-ice/acetone cooling bath.
5. Add the aldehyde **2** (400 mg, 2.25 mmol) in THF (5 mL) dropwise. Stir the mixture for 20 min at −78°C.
6. Allow the mixture to warm to room temperature, pour the mixture into ether (50 mL) and saturated ammonium chloride solution (50 mL), and separate. Extract the aqueous layer with dichloromethane (3 × 30 mL). Dry the combined organic layers ($MgSO_4$) and evaporate under reduced pressure on a rotary evaporator. Purify the alkene by chromatography on silica gel.

An extremely selective method for preparing (*Z*)-α,β-unsaturated esters is provided by the use of the trifluoroethylphosphonate **3** (for preparation see Protocol 16), first used by Still and Gennari (Protocol 17).[57]

$(F_3CCH_2O)_2P(O)CH_2C(O)OMe$ **3**, $KN(TMS)_2$/18-crown-6; aldehyde → CO_2Me

---

Protocol 16.
Preparation of methyl bis(2′,2′,2′-trifluoroethyl)phosphonoacetate 3[57]

Caution! All procedures should be carried out in a well-ventilated hood, and disposable vinyl or latex gloves and chemical-resistant safety goggles should be worn.

$(MeO)_2P(O)CH_2C(O)OMe$ → 1. PCl_5; 2. CF_3CH_2OH → $(F_3CCH_2O)_2P(O)CH_2C(O)OMe$ **3**

Equipment

- One-necked, round-bottomed flask (250 mL)
- Two-necked, round-bottomed flask (500 mL)
- Magnetic stirrer
- Magnetic stirrer bar
- Water-jacketed reflux condenser
- Calcium chloride drying tube
- Still-head
- Pressure-equalising dropping funnel (250 mL)
- Ice bath

Materials

- Phosphorus pentachloride (FW 208.2) 52.7 g, 253 mmol — **corrosive, moisture sensitive**
- Trimethyl phosphonoacetate (FW 182.1) 18.2 g, 16.2 mL, 100 mmol
- 2,2,2-Trifluoroethanol (FW 100.0) 16.7 g, 12.2 mL, 167 mmol — **flammable, toxic, irritant**
- *N,N*-Diisopropylethylamine (FW 129.2) 21.7 g, 29.2 mL, 168 mmol — **highly flammable, irritant**
- Benzene — **flammable, carcinogen**

1. Clean the apparatus and dry in a 120 °C electric oven for at least 4 h. Allow to cool in a desiccator.
2. Add phosphorus pentachloride (52.7 g, 253 mmol) portionwise to trimethyl phosphonoacetate (18.2 g, 16.2 mL, 100 mmol) at 0 °C in the one-necked flask equipped with a magnetic stirrer bar and a reflux condenser topped with a calcium chloride drying tube. An exothermic reaction takes place and methyl chloride is generated.
3. Stir the solution at 25 °C for 1 h and at 75 °C for 3 h.
4. Cool the flask and set up the apparatus for vacuum distillation.
5. Distil the $POCl_3$ by-product (b.p. 40 °C, 20 mmHg) and PCl_5.
6. Distil the product dichlorophosphonoacetate (b.p. 78–80 °C at 0.05 mmHg).
7. Dissolve the dichlorophosphonoacetate in benzene (**caution!**) (100 mL) in the two-necked flask equipped with a magnetic stirrer and a pressure-equalising dropping funnel charged with 2,2,2-trifluoroethanol (16.7 g, 12.2 mL, 167 mmol) and *N*,*N*-diisopropylethylamine (21.7 g, 29.2 mL, 168 mmol) in benzene (150 mL). One neck of the flask is stoppered with a rubber septum connected, via a needle, to a positive pressure of dry nitrogen.
8. Add the trifluoroethanol solution dropwise at 0 °C.
9. Stir the solution at room temperature for 1 h.
10. Evaporate the solvent under reduced pressure on a rotary evaporator in a fume hood.
11. Filter the residue through a four-inch plug of silica gel with ethyl acetate/petroleum ether (7:3 v/v). Evaporate the solvent to give the product.

Protocol 17. Preparation of (*Z,E*)-Methyl octadi-2,4-enoate[57]

Caution! All procedures should be carried out in a well-ventilated hood, and disposable vinyl or latex gloves and chemical-resistant safety goggles should be worn.

$(F_3CCH_2O)_2P(O)CH_2CO_2Me$ **3**; $KN(TMS)_2$/18-crown-6 → CO_2Me

Equipment

- One-necked, round-bottomed flask (100 mL)
- Magnetic stirrer
- Magnetic stirrer bar
- Glass syringe with integral needle (250 μL)
- Glass syringe with needle-lock Luer
- Six-inch, medium-gauge needle

Protocol 17. *Continued*

Materials

- Methyl bis(2′,2′,2′-trifluoroethyl)phosphonoacetate **3** (FW 318.1) 318 mg, 1 mmol — **irritant**
- 18-Crown-6 (FW 264.3) 1.32 g, 5 mmol — **irritant**
- Dry THF — **highly flammable, irritant**
- Potassium bis(trimethylsilyl)amide (FW 199.5) 15% in toluene, 1 mmol — **flammable, corrosive**
- (*E*)-Hex-2-enal (FW 98.1) 98 mg, 116 μL, 1 mmol — **flammable, irritant, keep cold**

1. Clean the apparatus and dry in a 120°C electric oven for at least 4 h. Allow to cool in a desiccator.
2. Place methyl bis(2′,2′,2′-trifluoroethyl)phosphonoacetate **3** (Protocol 16) (318 mg, 1 mmol) and 18-crown-6 (1.32 g, 5 mmol) in the round-bottomed flask containing dry THF (20 mL) and equipped with a magnetic stirrer bar and stoppered with a rubber septum connected, via a needle, to a positive pressure of dry nitrogen.
3. Cool the solution to −78°C.
4. Add potassium bis(trimethylsilyl)amide[a] (1 mmol, as a solution in toluene) dropwise via syringe (see Chapter 1, Protocol 1).
5. Add the (*E*)-hex-2-enal (98 mg, 116 μL, 1 mmol) in THF (1 mL) dropwise via syringe.
6. Stir the mixture at −78°C for 30 min.
7. Add saturated ammonium chloride solution and work-up as in Protocol 6, step 5.

[a]Available from Fluka Chemie.

A convenient way to make the (*E*)-isomer has been provided by Blanchette *et al.*[58] who found that phosphonates bearing a β-carbonyl group can be deprotonated with DBU or diisopropylethylamine and react with aldehydes to give the (*E*)-alkene exclusively (Protocol 18). Since the method is only mildly basic it has proved especially effective where other Wittig methods lead to racemisation of any stereogenic centres adjacent to the carbonyl group.

$(EtO)_2P(O)CH_2C(O)OEt$ —(1. LiCl, DBU; 2. RCHO)→ $RCH{=}CHC(O)OEt$

Protocol 18.
Preparation of (*E*)-ethyl 4-methylpent-2-enoate[58]

Caution! All procedures should be carried out in a well-ventilated hood, and disposable vinyl or latex gloves and chemical-resistant safety goggles should be worn.

$$(EtO)_2P(O)CH_2CO_2Et \xrightarrow[\text{2. } (CH_3)_2CHCHO]{\text{1. LiCl, DBU}} (CH_3)_2CHCH{=}CHCO_2Et$$

Equipment

- One-necked, round bottomed flask (250 mL)
- Glass syringes with needle-lock Luers (two 5 mL and one 1 mL)
- Two six-inch, medium-gauge needles
- Magnetic stirrer
- Magnetic stirrer bar

Materials

- Triethyl phosphonoacetate (FW 224.2) 2.69 g, 2.38 mL, 12 mmol — **irritant**
- 1,8-Diazabicyclo[5.4.0]undec-7-ene (DBU) (FW 152.2) 1.52 g, 1.50 mL, 10 mmol — **corrosive, moisture sensitive**
- Lithium chloride (42.4) 0.51 g, 12 mmol — **irritant, hygroscopic**
- Dry acetonitrile — **highly flammable, lachrymator**
- Isobutyraldehyde (2-methylpropionaldehyde) (FW 72.1) 0.72 g, 0.91 mL, 10 mmol — **highly flammable, stench**
- Ether — **highly flammable**
- Magnesium sulfate (dry) — **irritant**

1. Clean the apparatus and dry in a 120°C electric oven for at least 4 h. Allow to cool in a desiccator.
2. Recrystallise lithium chloride from methanol and dry overnight under vacuum (140°C/0.5 mmHg).
3. Add triethyl phosphonoacetate (2.69 g, 2.38 mL, 12 mmol) and DBU (1.52 g, 1.50 mL, 10 mmol) to a suspension of the dry lithium chloride (0.51 g, 12 mmol) in dry[a] acetonitrile (120 mL) at room temperature in the round-bottomed flask equipped with a magnetic stirrer and stoppered with a rubber septum connected, via a needle, to a positive pressure of dry nitrogen.
4. Add, via a syringe, isobutyraldehyde (0.72 g, 0.91 mL, 10 mmol). Almost all of the salt dissolves and the reaction is complete within 1 h. The progress of the reaction can be monitored by TLC.
5. After the usual work-up (see Protocol 6, step 5) the α,β-unsaturated ester can be isolated in >75% yield ((*E*):(*Z*)>20:1).

[a]Dried according to Perrin and Armarego[40] by distillation under reduced pressure from CaH_2.

4. The Horner–Wittig reaction of phosphine oxide anions

Horner *et al.*[59] described the reaction of phosphine oxides with potassium *tert*-butoxide and aldehydes to give alkenes with moderate to high selectivity. This approach is important since when a lithium base is used in place of the potassium *tert*-butoxide, the intermediate 1,2-phosphinoyl alcohol can be isolated, and hence purified to 100% diastereoisomeric purity.

The synthesis of alkyldiphenylphosphine oxides has been extensively reviewed.[60] The thermal decomposition of alkyltriphenylphosphonium hydroxides represents one of the easiest ways to make phosphine oxides from readily available materials. The reaction involves simply heating the corresponding alkyltriphenylphosphonium halide with 20–40% sodium hydroxide solution (Protocol 19). This procedure is suitable for the synthesis of phosphine oxides when the carbanion of the displaced group is less stable than the phenyl carbanion; the ease of displacement follows the order: allyl, benzyl > phenyl > methyl > 2-phenethyl > ethyl, higher alkyl.[61] Thus benzyl and allyl phosphine oxides must be made by other methods. Normally this will involve the coupling of either an appropriate organometallic nucleophile with diphenylphosphinoyl chloride[62] (Protocol 20) or a metal diphenylphosphide with an alkyl halide.[63] In some cases it may be appropriate to synthesise the phosphine oxide by Michaelis–Arbusov reaction of a diphenylphosphinite ester Ph_2POR and an alkyl halide.[64]

$$Ph{-}\overset{+}{P}(Ph_2)(R)\ Br^- \xrightarrow{NaOH} \left[Ph{-}P(OH)(Ph_2)(R)\right] \longrightarrow R{-}P(=O)(Ph_2) \text{ and } C_6H_6$$

Protocol 19.
Preparation of ethyldiphenylphosphine oxide

Caution! All procedures should be carried out in a well-ventilated hood, and disposable vinyl or latex gloves and chemical-resistant safety goggles should be worn.

$$Ph_3\overset{+}{P}CH_2CH_3\ I^- \xrightarrow[\text{reflux}]{\text{NaOH (aq.)}} Ph_2P(=O)CH_2CH_3$$

Equipment

- One-necked, round bottomed flask (1 L)
- Magnetic stirrer/hotplate
- Water-jacketed reflux condenser
- Still-head and thermometer
- Large magnetic stirrer bar
- Oil bath

Materials

- Ethyltriphenylphosphonium iodide (FW 418.3) 49.17 g, 118 mmol — **harmful, irritant**
- Sodium hydroxide 100 g — **corrosive, toxic**
- Chloroform — **extremely toxic, carcinogen**
- Magnesium sulfate (dry) — **irritant**
- Ethyl acetate — **highly flammable**

1. Clean the apparatus and dry in a 120°C electric oven for at least 4 h. Allow to cool in a desiccator.
2. Place ethyltriphenylphosphonium iodide (Protocol 1) (49.17 g, 118 mmol) in the round-bottomed flask equipped with a reflux condenser and a large magnetic stirrer bar. Add sodium hydroxide (100 g (2.5 mol) in water (170 mL)) and stir the mixture vigorously.
3. Heat the mixture on an oil bath until there is a strong reflux. Drops of benzene (**caution!**) will be seen condensing in the condenser and an oily layer will appear on the surface of the mixture. It is essential that the biphasic mixture be vigorously stirred to ensure complete reaction.
4. Cool the reaction mixture slightly and replace the condenser with still-head and condenser. Azeotropically distil the benzene from the reaction. When the theoretical volume (in this case 10.5 mL) of benzene is collected, cool the mixture remaining in the round-bottomed flask.
5. Extract the mixture with chloroform (3 × 100 mL). Dry the combined extracts ($MgSO_4$) and filter. Evaporate the solvent under reduced pressure on a rotary evaporator to give an off-white solid.
6. Recrystallisation from a minimum amount of boiling ethyl acetate gives the phosphine oxide as colourless needles (20.7 g, 76%), m.p. 123–124°C (lit. m.p. 124.5–125°C).

Protocol 20.
Preparation of phosphine oxides from diphenylphosphinic acid via diphenylphosphinic chloride[65]

Caution! All procedures should be carried out in a well-ventilated hood, and disposable vinyl or latex gloves and chemical-resistant safety goggles should be worn.

$$Ph_2P(=O)-OH \xrightarrow{SOCl_2} Ph_2P(=O)-Cl \xrightarrow[\text{THF, reflux}]{RMgBr} Ph_2P(=O)-R$$

Equipment

- One-necked, round-bottomed flask (1 L)
- Magnetic stirrer/hotplate
- Water-jacketed reflux condenser
- Two one-necked, round-bottomed flasks (250 mL)
- Magnetic stirrer bar

Protocol 20. *Continued*

Materials

- Diphenylphosphinic acid (FW 218.2) 2 g, 9.2 mmol — **irritant**
- Thionyl chloride (FW 119.0) 10 mL, 137 mmol — **lachrymator, corrosive**
- Toluene 50 mL — **highly flammable, harmful**
- Benzene 50 mL — **flammable, carcinogen**
- Hydrochloric acid 10% solution 50 mL — **corrosive**

1. Clean the apparatus and dry in a 120°C electric oven for at least 4 h. Allow to cool in a desiccator.
2. Place diphenylphosphinic acid (2 g, 9.2 mmol) in the round-bottomed flask equipped with a condenser (topped with an outlet to a nitrogen line) and magnetic stirrer bar.
3. Add toluene (50 mL) and thionyl chloride (10 mL, 137 mmol) and heat the mixture under reflux for 1.5 h.
4. Remove the excess thionyl chloride by evaporating the solvent under reduced pressure on rotary evaporator.
5. Dilute the residue with dry benzene (**caution!**) (50 mL).
6. Add the mixture, by cannula dropwise over 1.5 h, to a solution of the Grignard reagent (100 mmol) in ether in the other round-bottomed flask equipped with a magnetic stirrer bar and condenser topped with a tubing adapter attached to a nitrogen line.
7. Heat the mixture under reflux for 2 h.
8. Cool the mixture (0°C) and carefully pour into a mixture of ice (25 g) and hydrochloric acid (50 mL of a 10% solution).
9. Separate the layers. Extract the aqueous layer with ether (3 × 50 mL). Dry the combined ethereal layers ($MgSO_4$) and evaporate the solvent under reduced pressure on a rotary evaporator. The product may be purified by chromatography or recrystallisation as appropriate.

Buss and Warren[66] have shown that the phosphine oxide anion addition to an aldehyde is highly *erythro*-selective (Protocol 21). Treatment of the *erythro*-alcohol with a base such as NaH, KOH, or KO^tBu effects a *syn*-elimination to give the (*Z*)-alkene (Protocols 22 and 23). Some care must be taken when attempting this reaction when the anion of the phosphine oxide from which the *erythro*-alcohol is derived is moderately stabilised. In this case the *erythro*- and *threo*-alcohols are interconvertible and the overall reaction is hence non-stereospecific; this has been well documented for the synthesis of stilbenes.[67]

Protocol 21.
Preparation of an *erythro*-phosphinoyl alcohol[66]

Caution! All procedures should be carried out in a well-ventilated hood, and disposable vinyl or latex gloves and chemical-resistant safety goggles should be worn.

$Ph_2P(O)$–Et → (1. n-Butyllithium; 2. *p*-Methoxybenzaldehyde) → $Ph_2P(O)$–CH(Me)–CH(OH)–C_6H_4–OMe

Equipment

- One-necked, round-bottomed flask (100 mL)
- Magnetic stirrer
- Magnetic stirrer bar
- Ice and dry-ice/acetone cooling baths
- Glass syringes, with needle-lock Luers (5 mL and 1 mL)
- Two six-inch, medium-gauge needles

Materials

- Ethyldiphenylphosphine oxide (FW 230.2) 1.0 g, 4.35 mmol — **irritant**
- Dry THF — **highly flammable, irritant**
- *n*-Butyllithium (FW 64.1) 1.6 M in hexane, 2.7 mL, 4.35 mmol — **highly flammable, hygroscopic**
- 4–Methoxybenzaldehyde (*p*-anisaldehyde) (FW 136.1) 592 mg, 0.53 mL, 4.35 mmol — **harmful, air sensitive**
- Dichloromethane — **toxic, irritant**
- Magnesium sulfate (dry) — **irritant**

1. Clean the apparatus and dry in a 120°C electric oven for at least 4 h. Allow to cool in a desiccator.
2. Dissolve ethyldiphenylphosphine oxide (1.0 g, 4.35 mmol) in anhydrous THF (30 mL) in the round-bottomed flask equipped with a stirrer bar and stoppered with a rubber septum connected, via a needle, to a positive pressure of dry nitrogen. Cool the solution in an ice bath to 0°C.
3. Add *n*-butyllithium (2.7 mL, 1.6 M in hexane, 4.35 mmol) dropwise via a syringe (**caution!** see Chapter 1, Protocol 1) and stir the mixture at 0°C for 30 min.
4. Cool the red solution of the phosphine oxide anion to around −78°C by using a dry-ice/acetone cooling bath.
5. Add neat 4-methoxybenzaldehyde (*p*-anisaldehyde) (592 mg, 0.53 mL, 4.35 mmol) dropwise[a] via a syringe. After the addition is complete the red colour of the anion will have disappeared.
6. Warm the pale-yellow solution to room temperature over 2 h by removing the cooling bath and insulating the flask with cotton wool.

Protocol 21. *Continued*

7. Add water (10 mL) and remove the THF under reduced pressure on a rotary evaporator.
8. Add brine to the aqueous residue and extract with dichloromethane (3 × 30 mL). Dry the combined organic extracts ($MgSO_4$) and evaporate to dryness under reduced pressure on a rotary evaporator to give a mixture of diastereoisomers (*erythro:threo* 87:13, 92%) which can be separated by column chromatography on silica gel (ethyl acetate as eluant) to give the pure *erythro*-alcohol (1.28 g, 80%).

[a]Good *erythro*-selectivity is obtained if the rate of addition of the aldehyde is such that the temperature is maintained at −78°C.

Protocol 22. Preparation of (*Z*)-anethole[66]

Caution! All procedures should be carried out in a well-ventilated hood, and disposable vinyl or latex gloves and chemical-resistant safety goggles should be worn.

$Ph_2P(O)$ … HO … OMe —KOH, DMSO→ OMe

Equipment

- One-necked, round-bottomed flask (50 mL)
- Magnetic stirrer/hotplate
- Magnetic stirrer bar
- Oil bath

Materials

• *erythro*-Phosphinoyl alcohol (from Protocol 21) (FW 366.4) 119 mg, 0.325 mmol	**assume toxic**
• Potassium hydroxide (FW 56.1) 18.2 mg, 0.325 mmol	**corrosive**
• Dry dimethyl sulfoxide 10 mL	**harmful, irritant**
• Brine	
• Ether 60 mL	**flammable, irritant**
• Magnesium sulfate (dry)	**irritant**

1. Clean the apparatus and dry in a 120°C electric oven for at least 4 h. Allow to cool in a desiccator.
2. Add, in one portion, the *erythro*-alcohol (Protocol 21) (119 mg, 0.325 mmol) to a solution of anhydrous potassium hydroxide (18.2 mg, 0.325 mmol) in dry DMSO (10 mL) in the round-bottomed flask equipped with a stirrer bar

and stoppered with a rubber septum connected, via a needle, to a positive pressure of dry nitrogen.

3. Heat the flask for 1 h at 50 °C and cool to room temperature.
4. Add water (15 mL) and brine (10 mL). Extract the mixture with ether (3 × 20 mL).
5. Wash the combined organic extracts with water (3 × 20 mL) (to remove any DMSO present), dry ($MgSO_4$), filter, and evaporate to dryness under reduced pressure. Distil the residue by bulb-to-bulb distillation (Kugelrohr apparatus) to give the (*Z*)-alkene (36 mg, 75%, (*Z*):(*E*) 94:6).

Protocol 23. Elimination of *erythro*-phosphinoyl alcohols with sodium hydride[66]

Caution! All procedures should be carried out in a well-ventilated hood, and disposable vinyl or latex gloves and chemical-resistant safety goggles should be worn.

$Ph_2P(O)$–CH(R^1)–CH(OH)(R^2) —NaH, DMF→ R^1CH=CHR^2 (*Z*)

Equipment

- One-necked, round-bottomed flask (100 mL)
- Magnetic stirrer/hotplate
- Magnetic stirrer bar
- Oil bath

Materials

- Sodium hydride (50% dispersion in mineral oil) (FW 24.0) 96 mg, 2 mmol — **flammable, dangerous when wet**
- *erythro*-Phosphinoyl alcohol (Protocol 21) 1 mmol — **assume toxic**
- Dry *N,N*-dimethylformamide 10 mL — **harmful, irritant**
- Brine
- Ether — **highly flammable, irritant**
- Sodum sulfate (dry) — **irritant**

1. Clean the apparatus and dry in a 120° electric oven for at least 4 h. Allow to cool in a desiccator.
2. Add NaH (2 mmol of an 50% dispersion in oil), in one portion, to a stirred solution of the *erythro*-alcohol (1 mmol) in dry DMF (10 mL) in the round-bottomed flask equipped with a magnetic stirrer and stoppered with a rubber septum connected, via a needle, to a positive pressure of dry nitrogen.
3. Warm the clear solution to 50 °C and stir for 1 h after which time a white precipitate of sodium diphenylphosphinate will have formed.
4. Cool the mixture and carefully add water dropwise to quench any excess

Protocol 23. *Continued*

NaH and add brine (20 mL). Extract with ether (3 × 30 mL). Dry the combined extracts (Na_2SO_4) and evaporate under reduced pressure on a rotary evaporator to give me (*Z*)-alkene.

The (*E*)-alkene may be obtained from the elimination of the *threo*-1,2-phosphinoyl alcohol, which in turn is available from the stereoselective reduction of β-ketophosphine oxides (Protocol 25).[66] Such ketones may be made by either acylation of α-lithiophosphine oxides (Protocol 24) or by oxidation of a 1,2-phosphinoyl alcohol (using PDC as oxidant). In the latter case it is possible to convert the *erythro*-alcohol into the *threo*-isomer thereby providing a method for the synthesis of both (*E*)-and (*Z*)-alkenes from a single intermediate.

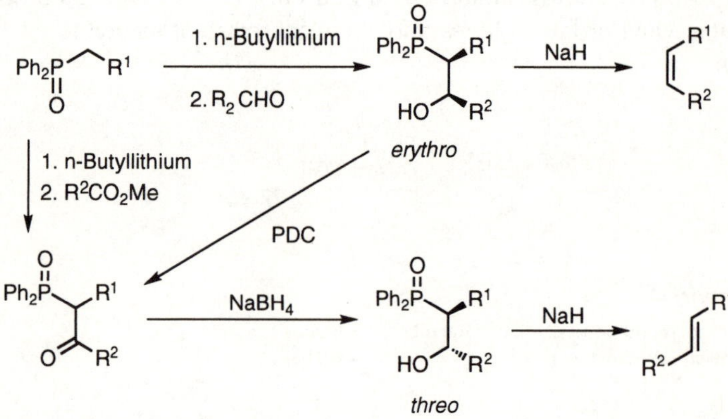

1. n-Butyllithium
2. R_2CHO
NaH
erythro
1. n-Butyllithium
2. R^2CO_2Me
PDC
$NaBH_4$
NaH
threo

Protocol 24. Preparation of a phosphinoyl ketone[66]

Caution! All procedures should be carried out in a well-ventilated hood, and disposable vinyl or latex gloves and chemical-resistant safety goggles should be worn.

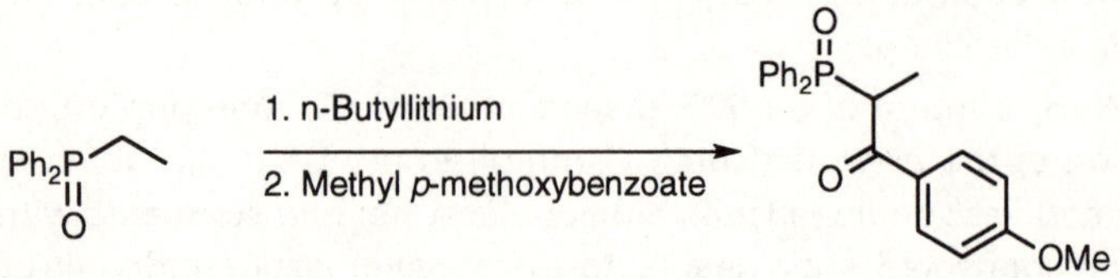

1. n-Butyllithium
2. Methyl *p*-methoxybenzoate

Equipment

- One-necked, round-bottomed flask (100 mL)
- Magnetic stirrer
- Magnetic stirrer bar
- Glass syringe with needle-lock Luer (20 mL)
- Six-inch, medium-gauge needle
- Ice and dry-ice/acetone cooling baths

Materials

- Ethyldiphenylphosphine oxide (FW 230.2) 4.0 g, 17.4 mmol — **irritant**
- Dry THF 35 mL — **highly flammable, irritant**
- *n*-Butyllithium (FW 64.1) 1.5 M in hexane, 11.6 mL, 17.4 mmol — **highly flammable, hygroscopic**
- Methyl *p*-methoxybenzoate (methyl *p*-anisate) (FW 166.2) 2.9 g, 17.4 mmol
- Dichloromethane — **toxic, irritant**
- Brine
- Magnesium sulfate (dry) — **irritant**

1. Clean the apparatus and dry in a 120°C electric oven for at least 4 h. Allow to cool in a desiccator.
2. Dissolve ethyldiphenylphosphine oxide (4.0 g, 17.4 mmol) in dry THF (20 mL) in the round-bottomed flask equipped with a stirrer bar and stoppered with a rubber septum connected, via a needle, to a positive pressure of dry nitrogen.
3. Cool the flask in an ice bath to 0°C. Add dropwise, via a syringe, *n*-butyllithium (11.6 mL, 1.5 M in hexane) and stir the mixture at 0°C for 30 min.
4. Cool the red solution of the phosphine oxide anion to around −78°C by placing the flask in a dry-ice/acetone cooling bath. Add dropwise, via syringe, methyl *p*-methoxybenzoate (methyl *p*-anisate) (2.9 g, 17.4 mmol) in dry THF (15 mL).
5. Warm the pale-yellow solution to room temperature before adding water (20 mL). Remove the THF under reduced pressure on a rotary evaporator. Dilute the aqueous residue with brine (15 mL) and extract with dichloromethane (3 × 30 mL). Dry the combined extracts ($MgSO_4$) and evaporate to dryness to give the ketone (5 g, 79%) as needles.

Protocol 25.
Preparation of a *threo*-phosphinoyl alcohol[66]

Caution! All procedures should be carried out in a well-ventilated hood, and disposable vinyl or latex gloves and chemical-resistant safety goggles should be worn.

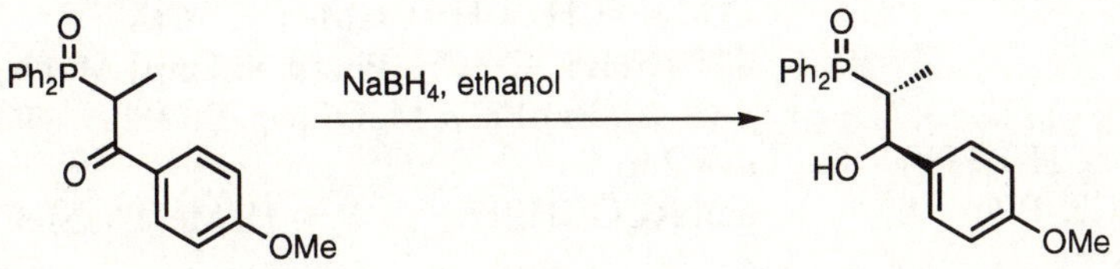

Protocol 25. *Continued*

Equipment

- One-necked, round-bottomed flask (50 mL)
- Water-jacketed reflux condenser
- Isomantle heater

Materials

- Phosphinoyl ketone (Protocol 24) (FW 364.4) 1.5 g, 4.12 mmol
- Sodium borohydride (FW 37.8) 156 mg, 4.12 mmol — **flammable, harmful**
- Ethanol 10 mL — **highly flammable, toxic**
- Saturated aqueous ammonium chloride solution 15 mL — **corrosive, toxic**
- Dichloromethane 150 mL — **toxic, irritant**
- Dilute hydrochloric acid — **corrosive, toxic**
- Anti-bumping granules

1. Clean the apparatus and dry in a 120°C electric oven for at least 4 h. Allow to cool in a desiccator.
2. Dissolve the ketone (1.5 g, 4.12 mmol) in ethanol (10 mL) in the round-bottomed flask. Add sodium borohydride (156 mg, 4.12 mmol).
3. Attach the condenser to the flask, add some anti-bumping granules, and heat the mixture under reflux for 3 h.
4. Cool the mixture to room temperature and add saturated ammonium chloride solution (15 mL). Remove the ethanol under reduced pressure on a rotary evaporator. Add several drops of dilute hydrochloric acid to the aqueous residues.
5. Dilute the residue with brine and extract with dichloromethane (3×50 mL).
6. Dry the combined extracts ($MgSO_4$) and evaporate to dryness to give a solid mixture of diastereoisomeric alcohols (*threo*:*erythro* 90:10, 99%). Chromatography gives the *threo*-isomer (1.35 g, 89%). The *threo*-isomer is converted into the (*E*)-alkene by following Protocol 22.

4.1 Some commercially available Wittig reagents[68]

Phosphonium salts

$Me(CH_2)_nPPh_3^+ X^-$	n = 0 to 13
$RCH_2PPh_3^+ X^-$	R = TMS, $ClCH_2$, $HOCH_2$, $Br(CH_2)_2$, $Br(CH_2)_3$, MeO, MeS, PhS, HO, Cl, CN, CH=CH_2, C(Me)=CH_2, CH=CHMe CH=$C(Me)_2$, CH=CHPh, C≡CH, Ph, (*o*, *m*, or *p*)-MePh), *p*-(F, Cl, Br)-Ph, *p*-MeOPh, *p*-EtOPh, *o*-HOPh
$Ph_3P^+(CH_2)_nPPh_3^+ X_2^-$	n = 2 to 5
$RCOCH_2PPh_3^+ X^-$	and $RCOCHPPh_3$ R = H, Me, Ph, MeO, EtO

Phosphonates

$RCH_2P(O)(OMe)_2$ — R = Me, Et, Pr, $MeOCOCH_2$, $EtOCOCH_2$, $TMSOCOCH_2$, EtOCOC(Me), $(Me)_2CHCH_2$, $MeSCH_2$, ICH_2, $BrCH_2CH_2$, $PhCH_2$, *p*-$MePhCH_2$, *p*-$MeOPhCH_2$, *p*-$ClPhCH_2$

Phosphine oxides

$PH_2P(O)R$ — R = Me, Et, ^{i}Pr, $MeOCH_2$, c-Hx

5. Recent progress in the study of the Wittig reaction

Until recently, very little attention had been paid to chiral Wittig reactions since obviously any chiral centres formed in the course of the reaction are lost in the final elimination to alkene. However, in special circumstances chiral alkenes may be formed from achiral ketones, i.e. when they are disymmetric. Two recent papers from Hanessian and Beaudoin[69] and Denmark and Chen[70] report the use of chiral benzyl phosphonamide **4** and phosphonamidate **5** respectively to synthesise the chiral alkene **6** from 4-*tert*-butylcyclohexanone. Both methods, variants of the HWE reaction, lead to excellent stereochemical induction, giving the alkene in essentially enantiomerically pure form. The reactions proceed via the intermediate β-hydroxyphosphonamide **7** and β-hydroxyphosphonamidate **8** which are both isolable; in both cases equatorial attack of the Wittig reagent has occurred. A similar approach using enantiomerically pure phosphine oxides to effect an asymmetric Horner–Wittig reaction has been disclosed by Warren's group.[71]

4 → (1. n-BuLi; 2. ^{t}Bu-cyclohexanone) → **7** → (AcOH) → **6** 91%, e.e. > 98%

5 → (1. t-BuLi; 2. ^{t}Bu-cyclohexanone) → **8** → ($Ph_3COSO_2CF_3$, 2,6-lutidine) → **6** 65%, e.e. > 99%

Another procedure[72] that uses asymmetric Wittig reagents is the selective HWE reaction of the enantiomerically pure phosphonate **9** derived from 8-phenylmenthol with the *meso*-dialdehyde **10**. The phosphonate **9** reacts selectively with the dialdehyde **10** to give the α,β-unsaturated ester **11** with the expected (*E*)-selectivity.

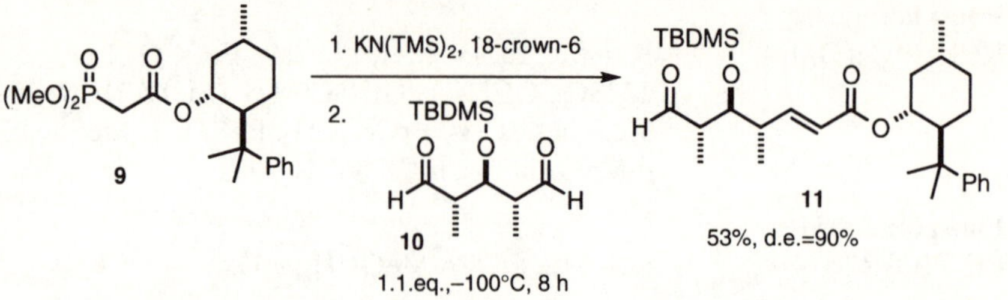

1. KN(TMS)$_2$, 18-crown-6
2. TBDMS
9
10
1.1.eq.,–100°C, 8 h
11
53%, d.e.=90%

The enantiomerically pure phosphonate **9** has also been used to control the exocyclic 5,6 double bond of prostaglandin analogues (the (*E*)-isomers are more biologically active). Two groups[73,74] have studied the asymmetric HWE of **9** with the prostaglandin-like ketones **12**; in both cases the (*E*):(*Z*)-selectivity is 86:14. When the achiral methyl phosphonoacetate is used, the α,β-unsaturated ester **13** is obtained with an (*E*):(*Z*)-selectivity of only 60:40.

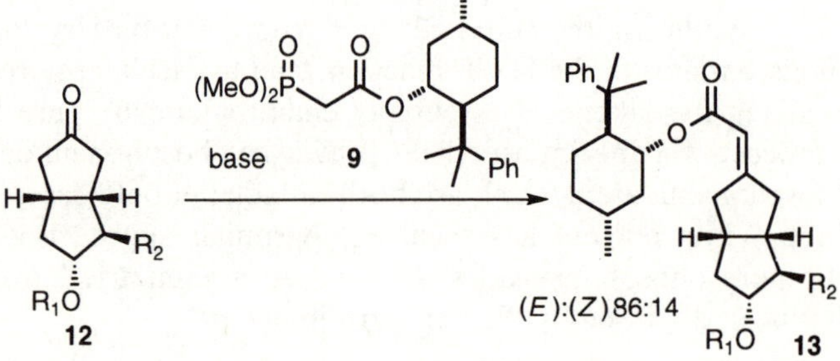

base 9
12
(*E*):(*Z*)86:14
13

Schlosser and co-workers[75–77] have recently introduced new class of Wittig reagents **14**, derived from tris(2-methoxymethoxy)phosphine, with enhanced (*Z*)-selectivity (see Table 2.3). In most cases the selectivity is greater than the corresponding Wittig reagent derived from triphenylphosphine. Stabilised ylids (X = CO_2Me in Table 2.3) show a curious solvent effect; in methanol the reaction is (*Z*)-selective whilst in hexane it is (*E*)-selective.[75] When the phosphorane bears an α-heteroatom the reaction is again impressively (*Z*)-selective, much more so than for the cases of the triphenylphosphine-derived reagents.[76] Tris(2-methoxymethoxy)phosphoniophenylmethanide[77] shows moderate (*Z*)-selectivity in the reaction with benzaldehyde but again is superior to the triphenylphosphine-derived ylide.

Vedejs and co-workers[78–80] have recently shown that the phosphole-derived ethylides **15**[78] and **16**[79] are extremely (*E*)-selective in the reaction with

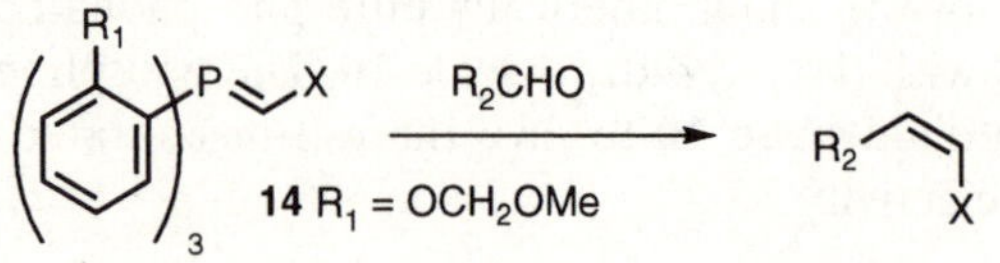

R_2CHO
14 $R_1 = OCH_2OMe$

Table 2.3 Comparison of the selectivity with **14** and triphenylphosphoranes

R_1	R_2	X	**E/Z-selectivity**
H	Me_2CH	CO_2Me	68:32
OCH_2OMe	Me_2CH	CO_2Me	9:91
H	Ph	Cl	23:77
OCH_2OMe	Ph	Cl	2:98
H	Ph	OMe	76:24
OCH_2OMe	Ph	OMe	6:94
H	Ph	Ph	45:55
OCH_2OMe	Ph	Ph	23:77

both aldehydes and ketones.[80] In comparison, the reaction of $Ph_3P{=}CHCH_3$ exhibits (*Z*)-selectivity whilst the reaction with ketones is very much substrate dependent. However, at present, both reagents **15** and **16** suffer drawbacks. The intermediate oxaphosphetane from **15** requires forcing conditions to induce decomposition but **15** is readily available from triphenylphosphine oxide. The decomposition of the oxaphosphetane from **16** is much easier but **16** is expensive.

15 **16**

$Ph_3P{=}CHCH_3$ or **15** or **16**

(*E*):(*Z*) 22:78 with $Ph_3P{=}CHCH_3$
(*E*):(*Z*) 96:4 with **15**
(*E*):(*Z*) 91:9 with **16**

References

1. Wittig, G.; and Geissler, G. *Justus Liebigs Ann. Chem.* **1953,** *580*, 44; Wittig, G.; Schöllkopf, U. *Chem. Ber.* **1954,** *87*, 1318.
2. Kelly, S. E. In *Comprehensive Organic Synthesis*; Pergamon: Oxford, **1991;** Vol. 3, pp. 755–782.
3. Maryanoff, B. E.; Reitz, A. B. *Chem. Rev.* **1989**, *89*, 863.
4. Wadsworth, Jr, W. S. *Org. React. (NY)* **1977,** *73*, 25; Boutagy, J.; Thomas, R. *Chem. Rev.* **1974,** *74*, 87; Walker, B. J. In *Organophosphorus Reagents in Organic Synthesis*; Cadogan, J. I. G., ed., Academic Press: New York, **1979;** p. 155.
5. Maryanoff, B. E.; Reitz, A. B.; Duhl-Emswiler, B. A. *J. Am. Chem. Soc.* **1985,** *107*, 217; Maryanoff, B. E.; Reitz, A. B.; Mutter, M. S.; Inners, R. R.; Almond, Jr.,

H. R. *J. Am. Chem. Soc.* **1985,** *107*, 1068; Maryanoff, B. E.; Reitz, A. B.; Mutter, M. S.; Inners, R. R.; Almond, Jr., H. R.; Whittle, R. R.; Olofson, R. A. *J. Am. Chem. Soc.* **1986,** *108*, 7664; Vedejs, E.; Marth, C. F.; Ruggeri, R. *J. Am. Chem. Soc.* **1988,** *110*, 3940; Vedejs, E.; Marth, C. F. *J. Am. Chem. Soc.* **1988,** *110*, 3948; Maryanoff, B. E.; Reitz, A. B.; Grieden, D. W.; Almond, Jr., H. R. *Tetrahedron Lett.* **1989,** *30*, 1361; Vedejs, E.; Fleck, T. J.; *J. Am. Chem. Soc.* **1989,** *111*, 5861.

6. Maercker, A. *Org. Reac. (NY)* **1965,** *14*, 270.
7. For many common phosphonium salts see: *Dictionary of Organophosphorus Compounds*; Edmundson, R. S. ed.; Chapman and Hall; London, **1988;** see also reference 6.
8. Adapted from: Miyano, S.; Izumi, Y.; Fujii, K.; Ohno, Y.; Hashimoto, H. *Bull. Chem. Soc. Jpn.* **1979,** *52*, 1197.
9. Commercially available or from CH_2Cl_2 and NaI according to: Miyano, S.; Hashimoto, H.; *Bull. Chem. Soc. Jpn.* **1971,** *44*, 2864.
10. Seyferth, D.; Wursthorn, K. R.; Mammarella, R. E. *J. Org. Chem.* **1977,** *42*, 3104.
11. Adapted from: Fleming, I.; Paterson, I. *Synthesis* **1979,** 446.
12. Beck, P. In *Organic Phosphorus Compounds;* Kosolapoff, G. M.; Maier, L., eds; Wiley: New York, **1976;** Vol. 2, Chapter 4, p. 189.
13. Vogel, A. I. *Textbook of Practical Organic Chemistry*, 5th edn; Longman: Harlow **1989;** p. 498.
14. Bergmann, W.; Dusza, J. P. *J. Org. Chem.* **1958,** *23*, 1245.
15. Schlosser, M.; Christmann, K. F. *Justus Leibigs Ann. Chem.* **1967,** *708*, 1.
16. Wittig, G.; Wittenberg, D. *Justus Leibigs Ann. Chem.* **1957,** *606*, 1.
17. Reith, B. A.; Strating, J.; Van Leusen, A. M. *J. Org. Chem.* **1974,** *39*, 2728.
18. See ref 8.
19. Seyferth, D.; Heeren, J. K.; Singh, G.; Grim, S. O.; Hughes, W. B. *J. Organomet. Chem.* **1966,** *5*, 267.
20. Jung, M. E.; Mazurek, M. A.; Lim, R. M. *Synthesis* **1978,** 588.
21. Schiemenz, G. P.; Engelhard, H. *Chem. Ber.* **1961,** *94*, 578.
22. Schmidbaur, H.; Tronich, W. *Chem. Ber.* **1967,** *100*, 1039.
23. Friederich, K.; Henning, H.-G. *Chem. Ber.* **1959,** *92*, 2756.
24. Kitahara, T.; Horiguchi, A.; Mori, K. *Tetrahedron*, **1988,** *44*, 4713.
25. Utimoto, K.; Tamura, M.; Sisido, K. *Tetrahedron* **1973,** *29*, 1169.
26. Bestmann, H. J.; Dornauer, H.; Rostock, K. *Chem. Ber.* **1970,** *103*, 685.
27. Grimshaw, J.; Ramsey, J. S. *J. Chem. Soc. B* **1968,** 63.
28. Johnson, A. W.; Kyllingstad, V. L. *J. Org. Chem.* **1966,** *31*, 334.
29. McDonald, R. N.; Campbell, T. W. *J. Org. Chem.* **1959,** *24*, 1969.
30. Wittig, G.; Schöllkopf, U. *Chem. Ber.* **1954,** *87*, 1318.
31. Schweizer, E. E.; DeVoe Goff, S.; Murray, W. P. *J. Org. Chem.* **1977,** *42*, 200.
32. Bohlmann, F.; Mannhardt, H.-J. *Chem. Ber.* **1956,** *89*, 1307.
33. Isler, O.; Gutman, H.; Montavon, M.; Rüegg, R.; Ryser, G.; Zeller, P. *Helv. Chim. Acta* **1957,** *40*, 1242.
34. For a preparation of the corresponding isolable phosphorane see: Lang, R. W.; Hansen, H. J. *Org. Synth.* **1984,** *62*, 202.
35. Wittig, G.; Haag, W. *Chem. Ber.* **1955,** *88*, 1654.
36. Schlosser, M.; Schaub, B.; de Oliveira-Neto, J.; Jeganathan, S. *Chimia* **1986,** *40*, 244.
37. Schlosser, M.; Christmann, K. F. *Angew. Chem., Int. Ed. Engl.* **1966,** *5*, 126.

38. Leopold, E. J. *Org. Synth.* **1986,** *64*, 164.
39. Adapted from: Marino, J. P.; Abe, H. *Synthesis* **1980,** 872.
40. Perrin, D. D.; Armarego, W. L. F.; *Purification of Laboratory Chemicals* 3rd edn; Pergamon: Oxford, **1988**.
41. Danishefsky, S.; Nagasawa, K.; Wang, N. *J. Org. Chem.* **1975,** *40*, 1989.
42. Adapted from: Ohloff, G.; Vial, C.; Naf, F.; Pawlak, M. *Helv. Chim. Acta* **1977,** *60*, 1161.
43. Adapted from: Stonewell, J. C.; Keith, D. R. *Synthesis* **1979,** 132.
44. Corey, E. J.; Fuchs, P. L. *Tetrahedron Lett.* **1972,** 3769.
45. Speziale, A. J.; Ratts, K. W.; Bissing, D. E. *Org. Synth. Coll.* **1973,** *5*, 361.
46. Crofts, P. *Q. Rev.* **1958,** *12*, 341.
47. Kosolapoff, G. M. *Organophosphorus Compounds;* Wiley: New York, **1950,** Chapter 7; Worms, K. H.; Schmidt-Dunker, M. In *Organic Phosphorus Compounds*; eds. G. M. Kosolapoff L. Maier; Wiley: New York, **1976;** Vol. 7, Chapter 18.
48. Ford-Moore, A. H.; Howarth Williams, J. *J. Chem. Soc.* **1947,** 1465.
49. Ford-Moore, A. H.; Perry, B. J. *Org. Synth. Coll.* **1963,** *4*, 325.
50. Saunders, B. C.; Stacey, G. J.; Wild, F.; Wilding, I. G. E. *J. Chem. Soc.* **1948,** 699.
51. Ginsburg, V. A.; Ya Yakubovich, A. *Zh. Obshch. Khim.* **1958,** *28*, 728; *Chem. Abs.*, **1958,** *52*, 17091h.
52. Nagata, W.; Wakabayashi, T.; Hayase, Y. *Org. Synth. Coll.* **1988,** *6*, 448.
53. Adapted from: Kagan, F; Birkenmeyer, R. D.; Strube, R. E. *J. Am. Chem. Soc.*, **1959,** *81*, 3026; G. M. Kosolapoff, *J. Am. Chem. Soc.* **1945,** *67*, 2259.
54. Adapted from: Wadsworth, Jr. W. S.; Emmons, W. D. *J. Am. Chem. Soc.* **1961,** *83*, 1733.
55. Baker; R.; Sims, R. J. *Synthesis* **1981,** 117.
56. Nagaoka, H.; Kishi, Y. *Tetrahedron* **1981,** *37*, 3873
57. Still, W. C.; Gennari, C. *Tetrahedron Lett.* **1983,** *24*, 4405.
58. Blanchette, M. A.; Choy, W.; Davis, J. T.; Essenfield, A. P.; Masamune, S.; Roush, W. R.; Sakai, T. *Tetrahedron Lett.*, **1984,** *25*, 2183.
59. Horner, L.; Hoffmann, H. M. R.; Wippel, H. G.; Klahre, G. *Chem. Ber.* **1959,** *92* 2499.
60. For an extensive survey of phosphine oxides see: Hays, H. R; Peterson, D. J. In *Organic Phosphorus Compounds*; Kosolapoff, G. M.; Maier, L., eds; Wiley: New York, **1976;** Vol. 3, Chapter 6.
61. Fenton, G. W.; Ingold, C. K. *J. Chem. Soc.* **1929,** 2342; Meisenheimer, J.; Casper, J.;Horing, M.; Lauter, W.; Lichtenstadt, L.; Samuel, W. *Justus Leibigs Ann. Chem.* **1926,** *213*, 449; Zanger, M.; Van der Werf, C. A.; McEwen, W. E. *J. Am. Chem. Soc.* **1959,** *81*, 3806.
62. Morrison, D. C. *J. Am. Chem. Soc.*, **1950,** *72*, 4820.
63. Aguiar, A. M.; Beisler, J.; Mills, A. *J. Org. Chem.* **1962,** *27*, 1001.
64. Arbuzov, A. E.; Nikonorov, K. V. *Zh. Obshch. Khim.* **1948,** *18*, 2008; *Chem. Abstr.*; **1949,** *43*, 3801i.
65. Adapted from: Davidson, A. H; Warren, S. *J. Chem. Soc., Perkin Trans. 1* **1976,** 639.
66. Buss, A. D.; Warren, S. *J. Chem. Soc., Perkin Trans. 1* **1985,** 2307.
67. Buss, A. D.; Warren, S.; Leake, J. S.; Whitham, G. *J. Chem. Soc., Perkin Trans. 1* **1983,** 2215.

68. Available worldwide from Lancaster Synthesis, Aldrich Chemical Company, Fluka Chemie, and others.
69. Hanessian, S.; Beaudoin, S. *Tetrahedron Lett.* **1992,** *33*, 7655.
70. Denmark, S. E.; Chen, C.-T. *J. Am. Chem. Soc.* **1992,** *114*, 10674.
71. Harmat, N. J. S.; Warren S. *Tetrahedron Lett.* **1990,** *31*, 2743.
72. Kann, N.; Rein, T. *J. Org. Chem.* **1993,** *58*, 3802.
73. Gais, H.-J.; Schmiedl, G.; Ball, W. A. *Tetrahedron Lett.* **1993,** *29*, 1773.
74. Rehwinkel, H.; Skupsch, J.; Vorbrüggen, H. *Tetrahedron Lett.* **1993,** *29*, 1775.
75. Patil, V.; Schlosser, M. *Synlett* **1993,** 125.
76. Zhang, X.; Schlosser, M. *Tetrahedron Lett.* **1993,** *34*, 1925.
77. Jeganathan, S.; Tsukamoto, M.; Schlosser, M. *Synthesis* **1990,** 109.
78. Vedejs, E.; Marth, C. F.; *J. Am. Chem. Soc.* **1988,** *110*, 3948.
79. Vedejs, E.; Peterson, M. J. *J. Org. Chem.* **1993,** *58*, 1985.
80. Vedejs, E.; Cabaj, J.; Peterson, M. J. *J. Org. Chem.* **1993,** *58*, 6509.

3

Non-phosphorus stabilised carbanions in alkene synthesis

ALAN ARMSTRONG

1. Introduction

Foremost amongst the methods available to the synthetic chemist for the preparation of alkenes is the reaction of carbanions α- to phosphorus with carbonyl compounds (the Wittig reaction and its variants; Chapter 2). The essential principles of the Wittig reaction—the stabilisation of a carbanion by the phosphorus atom, reaction of this carbanion with a carbonyl compound, and subsequent elimination—can be extended to the use of anions stabilised by other, non-phosphorus heteroatoms (Scheme 3.1). This chapter will consider alkene formation involving carbanions stabilised by trialkylsilyl groups (Peterson olefination), sulfoximines (Johnson methylenation), sulfones (Julia olefination), and boron. These methods can offer useful alternatives to phosphorus-based olefinations, particularly in terms of ease of isolation of the product and, importantly, in the control of alkene stereochemistry.[1]

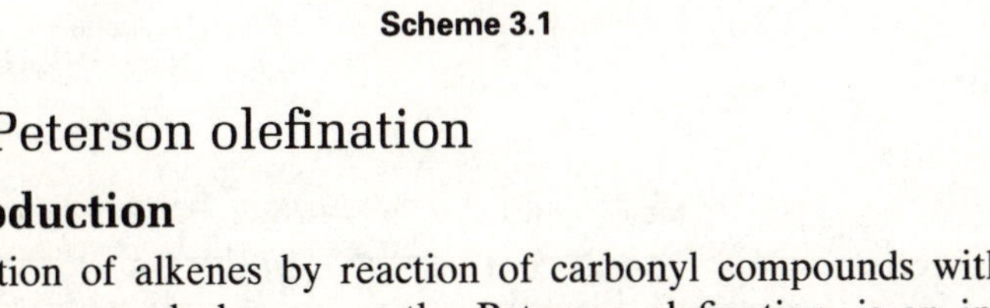

X = Me_3Si — Peterson olefination
$S(O)(N{=}Me)Ph$ — Johnson methylenation
SO_2Ph — Julia olefination
BR_2 — "Boron-Wittig" reaction

Scheme 3.1

2. The Peterson olefination

2.1 Introduction

The formation of alkenes by reaction of carbonyl compounds with α-silyl carbanions, commonly known as the Peterson olefination, is an important method for alkene synthesis (Scheme 3.2).[2] If the metal counterion associated with the α-silyl carbanion forms a relatively ionic intermediate alkoxide (as with Na^+ or K^+), or if R^2 is an anion-stabilising group, then spontaneous elimination of silanol occurs to give the alkene. When the counterion is covalently bound to the alkoxide (for example with Li^+, Mg^{2+}, or Al^{3+}), the alkoxides can be protonated and the β-hydroxysilanes **1** isolated.[3,4] The elimination step can then be performed separately, under either acidic or basic conditions.

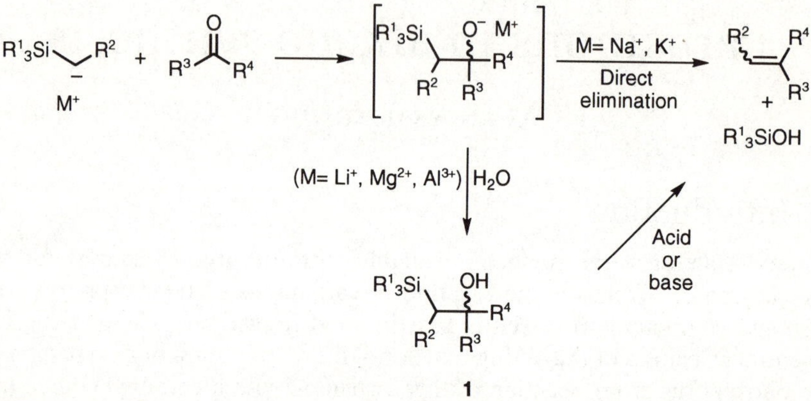

Scheme 3.2

Alkene formation from the β-hydroxysilane can occur by two different mechanisms, depending on whether acidic or basic conditions are employed (Scheme 3.3).[5] Importantly, a given diastereomer of the β-hydroxysilane affords alkenes of opposite geometry depending on which pathway is followed. Under acidic conditions, the observed geometry of the alkene product is consistent with an *anti*-periplanar arrangement **2** for the elimination of the trialkylsilyl and hydroxyl groups. Under basic conditions, a *syn*-periplanar orientation is preferred, presumably leading to formation of an intermediate siloxetane **3**.

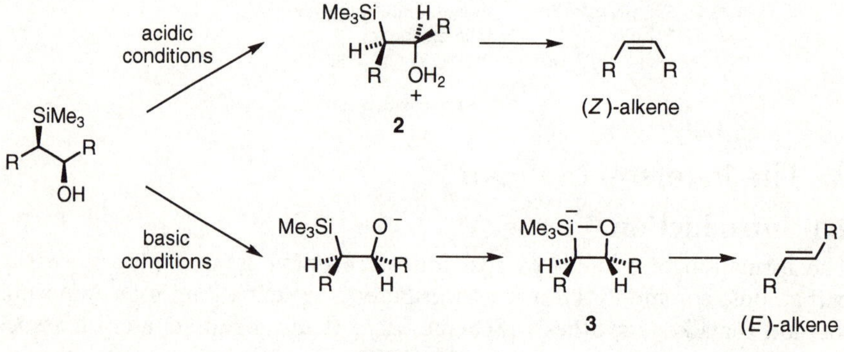

Scheme 3.3

This ability to control the geometry of the product alkene is one of the main advantages of the Peterson olefination sequence, but it is of course necessary to be able to control the stereochemistry of formation of the β-hydroxysilane, or to separate the diastereomers.

2.2 Formation of β-hydroxysilanes

The principal ways of forming the β-hydroxysilanes are:

- Addition of α-metallated silanes to carbonyl compounds
- Addition of hydride sources or organometallic reagents to β-ketosilanes
- Addition of dialkyl cuprates to α,β-epoxysilanes

The last two methods allow the formation of β-hydroxysilanes of high diastereomeric purity, important for stereocontrolled alkene synthesis. Each of these methods will be considered in turn.

2.2.1 Addition of α-metallated silanes to carbonyl compounds

The most obvious way of forming α-metallated silanes is by direct deprotonation; however, since this requires a very strong base and is kinetically very slow, it is generally useful only when the carbon atom α- to silicon bears another anion-stabilising group, for example with benzyltrimethylsilane (Scheme 3.4).[6] Given the strongly basic conditions required for direct deprotonation, this method is rarely used.

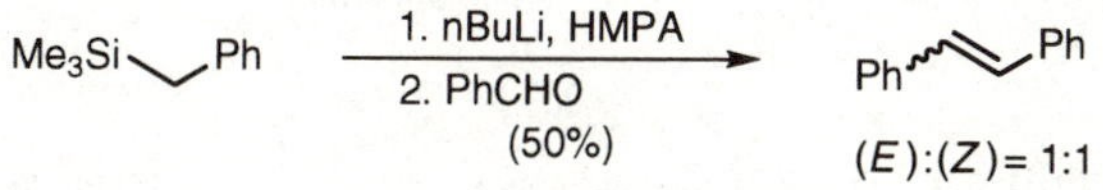

Me_3Si–CH_2–Ph → (1. nBuLi, HMPA; 2. PhCHO; (50%)) → PhCH=CHPh, (*E*):(*Z*) = 1:1

Scheme 3.4

One of the most widely used α-silyl carbanions is trimethylsilyl methylmagnesium chloride **4**, obtained by reaction of readily available chloromethyltrimethylsilane with magnesium in ether (Scheme 3.5). Reagent **4** may be used in a one-pot procedure for the methylenation of ketones (Scheme 3.6),[6] where *in situ* treatment of the intermediate β-alkoxy silane with thionyl chloride or acetyl chloride promotes elimination to give the alkene directly. Trimethylsilyl methylmagnesium chloride seems to be more reactive and less sterically hindered than methylene triphenylphosphorane and has been used successfully for the methylenation of ketones where the Wittig reaction has failed.[7] The relative ease of removal of the hexamethyldisiloxane by-product is another advantage over phosphorus-based olefination procedures.

Protocol 1.
Synthesis of 2,6-dimethyl-1,5-heptadiene 5. Methylenation of a ketone using trimethylsilyl methylmagnesium chloride (adapted from ref. 6)

Caution! All procedures should be carried out in a well-ventilated hood, and disposable vinyl or latex gloves and chemical-resistant safety goggles should be worn.

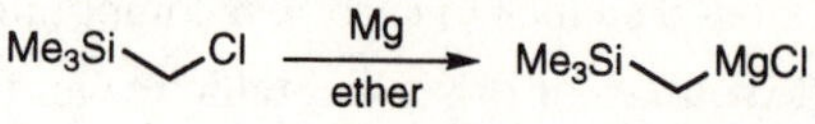

Me_3Si⌒Cl —(Mg, ether)→ Me_3Si⌒MgCl **4**

Scheme 3.5

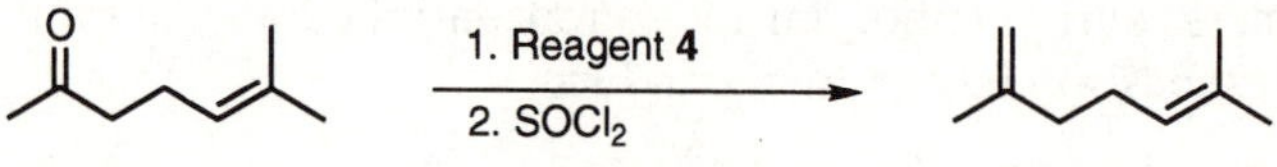

1. Reagent **4**; 2. $SOCl_2$ (57%) → **5**

Scheme 3.6

Equipment

- Magnetic stirrer/hotplate
- Pressure-equalising addition funnel (25 mL)
- Three-necked, round-bottomed flask (100 mL)
- Water-jacketed reflux condenser
- Three-way stopcock
- Septa
- Magnetic stirring bar
- Glass syringes with needle-lock Luers (volume appropriate for the quantity of solution to be transferred)
- Six-inch, medium-gauge needles
- Source of dry argon or nitrogen
- Oil bath or electric heating mantle
- Water-jacketed, semi-micro, short-path distillation apparatus

Materials

• Chloromethyl trimethylsilane (FW 122.7) 2.5 g, 0.02 mol	**highly flammable, irritant**
• Magnesium turnings (FW 24.3) 0.5 g, 0.02 mol	**flammable solid**
• Dry ether 30 mL	**flammable, irritant**
• 6-Methyl-5-hepten-2-one (FW 126.2) 2.5 g, 0.02 mol	**flammable**
• Thionyl chloride (FW 119.0) 1.8 mL, 0.025 mol	**lachrymator, corrosive**
• Saturated aqueous ammonium chloride solution	**corrosive, toxic**
• Technical ether for washing	**flammable, irritant**

1. Clean all glassware, syringes, needles, and stirring bars and dry for at least 4 h in a 120°C electric oven before use. Allow the apparatus to cool in a desiccator.
2. Assemble the three-necked flask, magnetic stirrer bar, condenser, three-way tap, and the pressure-equalising addition funnel. Attach the septum to the third neck of the flask.
3. Place magnesium turnings (0.5 g, 0.02 mol) and the stirrer bar into the reaction flask.

4. Place chloromethyl trimethylsilane (2.5 g, 0.02 mol) into the addition funnel. Close the opening of the addition funnel with a rubber septum.
5. Evacuate the apparatus by turning the three-way tap carefully to a high-vacuum pump. Fill the vessel with dry argon (or nitrogen) by turning the three-way tap to an argon line or a disposable balloon filled with argon. Repeat this evacuation–inert gas filling cycle twice.
6. Assemble the syringe and needle. Flush the syringe with argon or nitrogen (Chapter 1, Fig. 1.6).
7. Charge the flask with dry ether (15 mL) using the syringe, by puncturing the septum. Use the same syringe to add dry ether (10 mL) to the addition funnel, again by puncturing the septum. Swirl gently to ensure mixing.
8. Add the chloromethyl trimethylsilane solution dropwise from the addition funnel to the stirred magnesium turnings in the flask, at such a rate that a gentle reflux of the reaction mixture is maintained.
9. When the addition is complete, place the reaction flask in an oil bath or electric heating mantle and heat the mixture gently to reflux for an additional 1 h.
10. Place a solution of 6-methyl-5-hepten-2-one (2.5 g, 0.02 mol) in dry ether (5 mL) into the addition funnel using a syringe.
11. Add this ketone solution to the Grignard reagent dropwise from the addition funnel, at such a rate that a gentle reflux of the reaction mixture is maintained.
12. When addition is complete, place the reaction flask back into the oil bath or heating mantle and heat the reaction mixture at reflux, maintaining the magnetic stirring.
13. After 3 h, cool the reaction mixture in an ice bath.
14. Add thionyl chloride (1.8 mL, 0.025 mol) to the reaction mixture dropwise by syringe.
15. Remove the ice bath and continue stirring the mixture at room temperature.
16. After 1 h, add saturated aqueous ammonium chloride solution dropwise by syringe until a coagulated solid results.
17. Filter the mixture under suction, washing the solid with ether. Concentrate the filtrate under reduced pressure using a rotary evaporator.
18. Distil the residue; the product has b.p. 135–136 °C. This gives 2,6-dimethyl-1,5-heptadiene **5** (1.4 g, 57%) as a clear, colourless liquid.

Addition of alkyllithium reagents to vinylsilanes is an another important method for the preparation of α-metallated silanes for reaction with carbonyl compounds (Scheme 3.7).[5] Grignard reagents may not be used in place of organolithiums here.

Protocol 2.
Synthesis of 5-trimethylsilyl-4-octanol 6. Addition of alkyllithium reagents to vinylsilanes (adapted from ref. 5)

Caution! All procedures should be carried out in a well-ventilated hood, and disposable vinyl or latex gloves and chemical-resistant safety goggles should be worn.

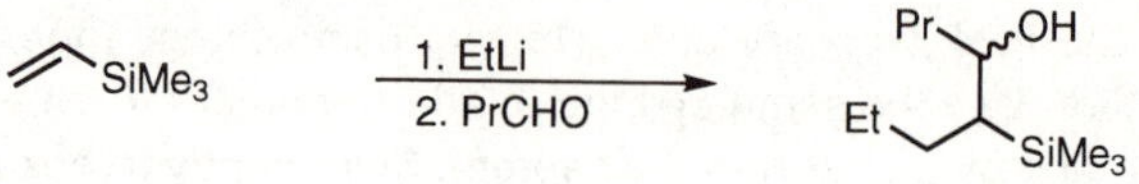

$$\text{CH}_2\text{=CH-SiMe}_3 \xrightarrow[\text{2. PrCHO}]{\text{1. EtLi}} \mathbf{6}$$

Scheme 3.7

Equipment

- Magnetic stirrer
- Two-necked, round-bottomed flask (50 mL)
- Septum
- Glass syringes with needle-lock Luers (volume appropriate for the quantity of solution to be transferred)
- Magnetic stirring bar
- Six-inch, medium-gauge needle
- Disposable argon balloon
- Conical flask (100 mL)
- Water-jacketed, semi-micro, short-path distillation apparatus

Materials

• Dry THF (15 mL)	**flammable, irritant**
• Vinyltrimethylsilane (FW 100.2) 1.05 mL, 6.80 mmol	**highly flammable, irritant**
• Ethyllithium[a] (FW 36.0) 7.65 mL, 1.15 M in ether, 8.8 mmol	**flammable, moisture sensitive**
• *n*-Butyraldehyde[b] (FW 72.1) 0.66 mL, 0.54 g, 7.5 mmol	**highly flammable corrosive**
• Saturated aqueous ammonium chloride solution	**corrosive, toxic**
• Ether for extraction	**flammable, irritant**
• Anhydrous magnesium sulfate	**hygroscopic**

1. Clean all glassware, syringes, needles, and stirring bars and dry for at least 4 h in a 120°C electric oven before use. Allow the apparatus to cool in a desiccator.
2. Assemble the flask and the stopcock and place the magnetic stirrer bar inside the flask. Introduce an atmosphere of argon as described in Chapter 1.
3. Put dry, freshly distilled THF (15 mL) into the flask using a syringe, by puncturing the septum.
4. Add vinyltrimethylsilane (1.05 mL, 6.80 mmol) to the flask by syringe.
5. Cool the flask to −78°C in a dry-ice/acetone bath.
6. Add ethyllithium (7.65 mL, 1.15 M in ether) via syringe (**caution!** see Chapter 1, Protocol 1).
7. Stir the reaction mixture for 2 h, keeping the reaction flask in the −78°C cooling bath.

8. Warm the reaction mixture gradually over a period of 1 h to −30 °C. This can be done by allowing the dry ice in the cooling bath to evaporate, or by adding acetone (at room temperature) to the cooling bath, as appropriate.
9. Recool the reaction mixture to −78 °C, by adding more dry ice to the cooling bath.
10. Add *n*-butyraldehyde (0.66 mL, 0.54 g, 7.5 mmol) to the reaction flask, dropwise by syringe.
11. Allow the reaction mixture to warm to room temperature over a period of 1 h. Stir for a further 2 h at room temperature.
12. Pour the reaction mixture into a separatory funnel containing saturated aqueous sodium chloride solution (30 mL) and ether (50 mL). Shake the separatory funnel cautiously, releasing any build-up of pressure.
13. Transfer the top, ethereal layer to a conical flask (100 mL). Dry ($MgSO_4$), filter, and concentrate the filtrate under reduced pressure using a rotary evaporator.
14. Distil the residue on a Kugelrohr apparatus (oven temperature 120 °C). This gives the 5-trimethylsilyl-4-octanol **6** (1.272 g, 93%), which displays the appropriate spectroscopic properties.

[a]Titrate ethyllithium in ether with a 1.00 M solution of *s*-butanol in xylene using 1,10-phenanthroline as the indicator just prior to use.

[b]Purify *n*-butyraldehyde (b.p. 75 °C) by distillation from $CaCl_2$ or $CaSO_4$ under an inert atmosphere (nitrogen or argon) prior to use.

2.2.2 Nucleophilic addition to β-ketosilanes

The hydroxysilanes **6** prepared as in Protocol 2 are obtained as a mixture of diastereomers. As mentioned earlier, diastereomerically pure compounds are needed if geometrically pure alkenes are desired. One solution to this problem is to oxidise **6** to the β-ketosilane **7**, and then effect addition of a nucleophile such as a hydride source or an organometallic reagent. This addition can occur with high selectivity (Scheme 3.8),[5] giving the isomer **8** predicted by the Felkin–Anh model for nucleophilic addition to carbonyl groups.

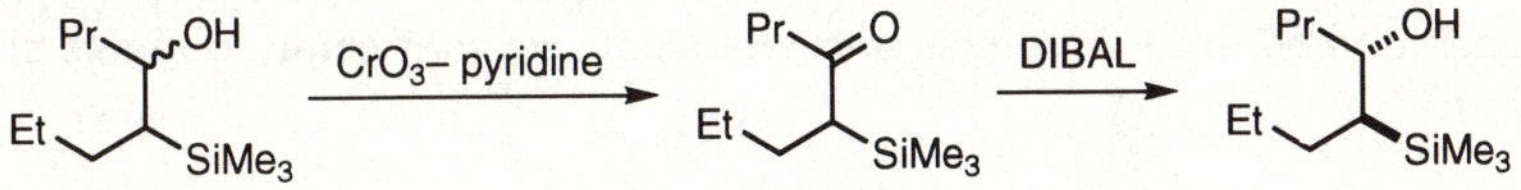

Pr, OH, Et, $SiMe_3$ **6** —(CrO_3– pyridine)→ Pr, O, Et, $SiMe_3$ **7** —(DIBAL)→ Pr, OH, Et, $SiMe_3$ **8**

Scheme 3.8

2.2.3 Opening of α,β-epoxysilanes with dialkylcuprates

Reaction of α,β-epoxysilanes **9** with dialkylcuprates results in regio- and stereospecific opening at the end of the epoxide α- to the silicon atom, affording the β-hydroxysilane (Scheme 3.9).[8] The α,β-epoxysilanes themselves are readily obtained by peracid oxidation of vinylsilanes. This procedure is therefore useful for the formation of β-hydroxysilanes with defined stereochemistry.

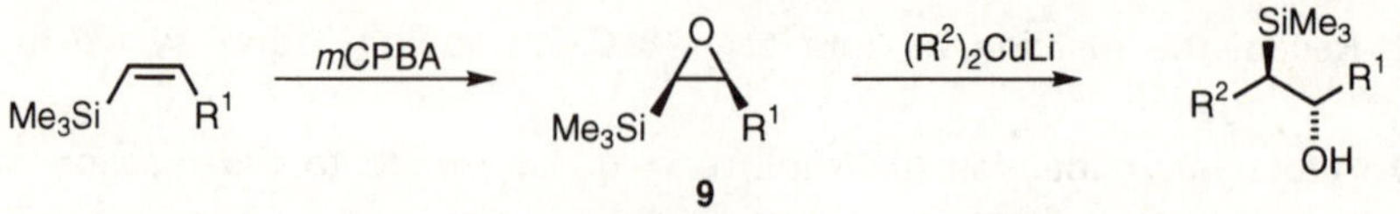

Scheme 3.9

2.3 Alkene formation from β-hydroxysilanes

5-Trimethylsilyl-4-octanol will be used here to exemplify methods for alkene formation from β-hydroxysilanes. While the material prepared in Protocol 2 is a mixture of diastereomers **6**, the experimental procedures given here utilise diastereomerically pure material **8** in order to illustrate how alkenes of opposite configuration can be prepared by appropriate choice of elimination conditions. The single diastereomer **8** can be obtained either by the oxidation–reduction procedure of Scheme 3.8[5] or, in principle, by separation of the isomers.

2.3.1 Elimination under acidic conditions

Either protic acids (for example sulfuric acid in THF, or sodium acetate in acetic acid) or Lewis acids (such as boron trifluoride etherate in dichloromethane) may be used to effect alkene formation from β-hydroxysilanes in good yield.[5] In the case of the β-hydroxysilane **8**, elimination in an *anti*-periplanar fashion results in formation of the (*Z*)-isomer (Scheme 3.10).

Protocol 3.
Synthesis of (*Z*)-4-octene 10. Elimination under acidic conditions (adapted from ref. 5)

Caution! All procedures should be carried out in a well-ventilated hood, and disposable vinyl or latex gloves and chemical resistant safety goggles should be worn.

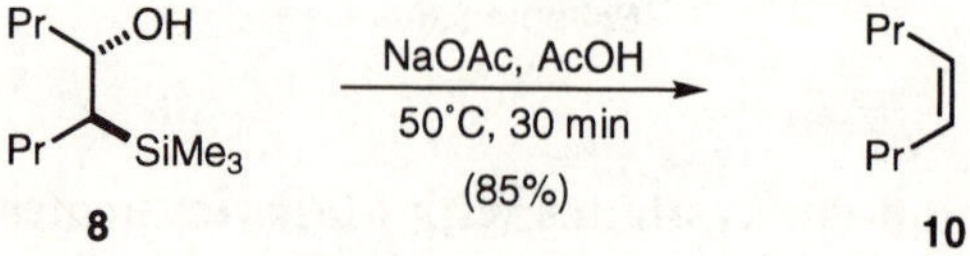

Scheme 3.10

Materials

- Magnetic stirrer/hotplate
- Two-necked, round-bottomed flask (50 mL)
- Magnetic stirring bar
- Water-jacketed reflux condenser
- Three-way stopcock
- Source of argon
- Oil bath and thermometer

- 5-Trimethylsilyl-4-octanol **8** (FW 202.4) 98.1 mg, 0.485 mmol — **assume toxic**
- Glacial acetic acid 15 mL — **corrosive**
- Sodium acetate — **irritant, hygroscopic**
- Saturated aqueous sodium bicarbonate solution 30 mL — **corrosive, toxic**
- Pentane 30 mL — **flammable, irritant**
- Magnesium sulfate (dry) — **hygroscopic**

1. Clean all glassware and stirring bars and dry for at least 4 h in a 120 °C electric oven before use. Allow to cool in a desiccator.
2. Place 5-trimethylsilyl-4-octanol **8** (98.1 mg, 0.485 mmol) and the stirrer bar into the two-necked flask.
3. Add glacial acetic acid (15 mL), previously saturated with sodium acetate at 50 °C.
4. Assemble the condenser and stopcock and attach the argon source.
5. Place the flask into an oil bath at 50 °C and stir for 30 min.
6. Allow the mixture to cool to room temperature.
7. Pour the mixture into a separatory funnel containing saturated aqueous sodium bicarbonate solution (30 mL) overlaid with pentane (30 mL). Shake the separatory funnel, releasing any build-up of pressure cautiously.
8. Run off the lower, aqueous layer. Wash the organic layer with saturated aqueous sodium bicarbonate solution (10 mL), then run it into a conical flask (100 mL). Dry ($MgSO_4$) and filter. Analysis of this solution by gc (with *n*-butylbenzene as internal standard) showed a 98:2 ratio of (*Z*)- and (*E*)-4-octene formed in 85% yield.

2.3.2 Elimination under basic conditions

If the alkene product of stereochemistry corresponding to *syn*-elimination of silanol is required, use of basic reaction conditions is necessary. Elimination using sodium hydride is very slow at room temperature in THF, proceeding at a synthetically useful rate only when the solvent is changed to the more highly solvating (but highly toxic!) HMPA.[5] More satisfactory reaction conditions involve using potassium hydride in THF; in the case of the β-hydroxysilane **8**, the *syn*-elimination process affords the (*E*)-isomer **11** (Scheme 3.11).[5]

Protocol 4.
Synthesis of (*E*)-4-octene 11. Elimination under basic conditions (adapted from ref. 5)

Caution! All procedures should be carried out in a well-ventilated hood, and disposable vinyl or latex gloves and chemical-resistant safety goggles should be worn.

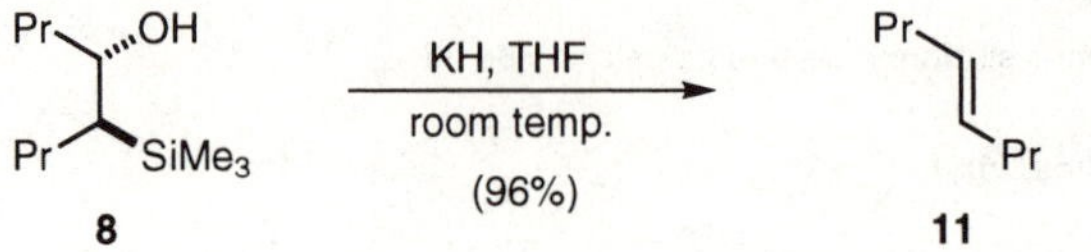

Pr, OH, Pr, $SiMe_3$ **8** → KH, THF, room temp. (96%) → Pr, Pr **11** (*Z*):(*E*)=5:95

Scheme 3.11

Equipment

- Magnetic stirrer
- Two-necked, round-bottom flask (25 mL)
- Magnetic stirring bar
- Three-way stopcock
- Septum
- Glass syringes with a needle-lock Luers (volume appropriate for the quantity of solution to be transferred)
- Six-inch, medium-gauge needles

Materials

- Potassium hydride (FW 40.1) 50% dispersion in oil, 100 mg, 1.25 mmol — **flammable solid, hygroscopic**
- Dry pentane[a] — **flammable, irritant**
- 5-Trimethylsilyl-4-octanol **8** (FW 202.4) 76.5 mg, 0.378 mmol — **assume toxic**
- Dry THF — **flammable, irritant**
- Saturated aqueous ammonium chloride solution 10 mL — **corrosive, toxic**
- Ether for extraction 10 mL — **flammable, irritant**
- Magnesium sulfate (dry) — **hygroscopic**

1. Clean all glassware, syringes, needles, and stirring bars and dry for at least 4 h in a 120 °C electric oven before use. Allow to cool in a desiccator.
2. Place the potassium hydride (as a 50% dispersion in mineral oil) (100 mg) and the magnetic stirrer bar into the two-necked flask.
3. Assemble the flask and stopcock and introduce an atmosphere of argon (see Chapter 1, Figs 1.1.–1.5).
4. Wash the potassium hydride dispersion with dry pentane.
   (a) Add dry pentane (4 mL) by syringe by piercing the septum, and slurry the mixture. Allow to settle.
   (b) Draw off the pentane carefully by syringe.
   (c) Repeat this procedure twice more.
5. Using a syringe, add to the flask a solution of 5-trimethylsilyl-4-octanol **8** (76.5 mg, 0.378 mmol) in dry THF (5 mL).

6. Stir the reaction mixture at room temperature for 1 h.
7. Carefully quench the reaction by cautious, dropwise addition of saturated aqueous ammonium chloride solution (10 mL).
8. Carefully pour the reaction mixture into a separatory funnel containing saturated aqueous ammonium chloride solution (10 mL) and ether (10 mL). Shake the separatory funnel cautiously.
9. Run the upper, organic layer into a conical flask (25 mL). Dry ($MgSO_4$) and filter. Analysis of this solution by gc (with *n*-butylbenzene as internal standard) showed a 5:95 ratio of (*Z*)- and (*E*)-4-octene formed in 96% yield.

[a]Pentane may be dried by distillation from sodium hydride under an inert atmosphere (nitrogen or argon).

3. The Johnson methylenation

α-Carbanions derived from sulfoximines react with aldehydes or ketones to give β-hydroxysulfoximines. Reduction with aluminium amalgam then results in formation of the alkene (Scheme 3.12). The method is most often carried out using *N*-methylphenylsulfonimidoylmethyllithium, derived from sulfoximine **12**, a procedure known as the Johnson methylenation. The reaction sequence can be carried out in one pot, without isolation of the intermediate β-hydroxysulfoximines.[9] *N*-methylphenylsulfonimidoylmethyllithium appears to be more nucleophilic than methylene triphenylphosphorane, and there are several examples of successful Johnson methylenation where the Wittig reaction has failed.[10]

Protocol 5.
Preparation of 4-*t*-butylmethylenecyclohexane 13. Johnson methylenation of a ketone (adapted from ref. 9)

Caution! All procedures should be carried out in a well-ventilated hood, and disposable vinyl or latex gloves and chemical-resistant safety goggles should be worn.

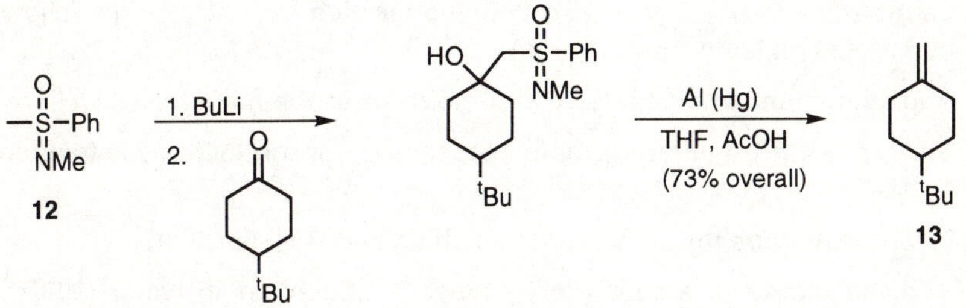

O
—S—Ph
NMe
12

1. BuLi
2. (4-tBu-cyclohexanone)

HO, S(O)(NMe)Ph, tBu

Al (Hg)
THF, AcOH
(73% overall)

tBu
13

Scheme 3.12

Protocol 5. *Continued*

Equipment

- Magnetic stirrer
- Two-necked, round-bottomed flask (250 mL)
- Magnetic stirring bar
- Three-way stopcock
- Septum
- Source of dry argon or nitrogen
- Glass syringes with needle-lock Luers (volume appropriate for the quantity of solution to be transferred)
- Six-inch, medium-gauge needles
- Water-jacketed, semi-micro, short-path distillation apparatus

Materials

- *N,S*-Dimethyl-*S*-phenylsulfoximine[a] (FW 169.2) 1.69 g, 0.01 mol — **moisture sensitive**
- Triphenylmethane (FW 244.3) 15 mg
- Butyllithium[b] (FW 64.1) 1.5 M in hexane, 6.67 mL, 0.01 mol — **flammable, moisture sensitive**
- 4-*t*-Butylcyclohexanone (FW 154.2) 1.54 g, 0.01 mol
- Dry THF 40 ml — **flammable, irritant**
- THF 65 ml — **flammable, irritant**
- Glacial acetic acid 35 ml — **corrosive, flammable**
- Aluminium amalgam[c] — **toxic**
- Celite® for filtration

1. Clean all glassware, syringes, needles, and stirring bars and dry for at least 4 h in a 120°C electric oven before use. Allow to cool in a desiccator.
2. Place the *N,S*-dimethyl-*S*-phenylsulfoximine (1.69 g, 0.01 mol), triphenylmethane (15 mg), and the stirrer bar into the flask.
3. Place the three-way stopcock into one neck of the flask, and the septum on the other, and introduce an atmosphere of argon (see Chapter 1, Figs 1.1–1.5).
4. Add dry THF (25 ml) by syringe, as described in Protocol 1, steps 6 and 7.
5. Cool the flask to 0°C by placing it in an ice bath. Start the magnetic stirring.
6. Add butyllithium (1.5 M in hexane) dropwise by syringe until an orange colour persists. This should require *c.* 6.7 mL of the butyllithium solution (**caution!** see Chapter 1, Protocol 1).
7. Add the 4-*t*-butylcyclohexanone (1.54 g, 0.01 mol) in THF (15 mL) by syringe over a period of 5 min.
8. Stir the mixture at 0°C for 30 min, then at room temperature for 1 h.
9. Cautiously add acetic acid (35 mL) to the reaction flask by syringe, followed by THF (35 mL) and water (35 mL).
10. Add aluminium amalgam (0.16 mol)[c] to the reaction mixture.
11. Stir for 1–4 h (monitoring the disappearance of the hydroxysulfoximines by TLC).
12. Filter the mixture through Celite®, washing with THF (*c.* 30 mL).
13. Add the filtrate to a separatory funnel (1 L) containing water (300 mL). Extract with pentane (2 × 300 mL).

14. Wash the pentane extracts with 20% aqueous sodium hydroxide (2 × 100 mL), then water (100 mL).

15. Dry the organic layer ($MgSO_4$), filter, and remove pentane using a rotary evaporator.

16. Distil the product 4-*t*-butylmethylenecyclohexane **13**, b.p. 72–73°C/30 mmHg (73% yield).

[a]*N,S*-Dimethyl-*S*-phenylsulfoximine is prepared by the methylation (CH_2O/HCO_2H) of *S*-methyl-*S*-phenylsulfoximine.[11]

[b]Commercial solutions of butyllithium in hexane may be used directly since triphenylmethane is present as indicator in the actual reaction mixture.

[c]Aluminium amalgam may be prepared by stirring granular aluminium (60 mesh; 4.32 g, 0.16 mol) in 2% aqueous mercuric chloride (100 mL) for 2 min, followed by filtration and successive washing with water and ethanol. This material may be used directly.

Di- and trisubstituted alkenes can be prepared from aldehydes and ketones, respectively, by using α-substituted sulfoximines, but mixtures of (*E*)- and(*Z*)-isomers are obtained.[9] Synthesis of tetrasubstituted alkenes is not possible as the reduced nucleophilicity of an α-disubstituted sulfoximine causes it simply to act as a strong base.[9]

Methylenation of aromatic aldehydes is complicated by the rapid reaction of the styrene products with thiophenol produced during the reductive elimination.[9] Use of smaller quantities of aluminium amalgam to try to avoid formation of thiophenol succeeds only in lowering the yield of alkene, so alternative olefination methods are to be preferred in these cases.

One great advantage of this olefination procedure is that the sulfoximine sulfur atom is an asymmetric centre. If an enantiomerically pure sulfoximine is used, the diastereomeric β-hydroxysulfoximines obtained upon addition to a chiral ketone can be separated and converted to the enantiomerically pure alkenes (Scheme 3.13).[12] Alternatively, the starting ketone can be regenerated by thermolysis of the diastereomerically pure β-hydroxysulfoximine, the procedure thus providing a method for the resolution of ketones.[13]

4. The Julia olefination

The Julia olefination[14,15] is a valuable connective, regiospecific method for alkene synthesis; in the synthesis of 1,2-disubstituted alkenes, it generally affords products with high (*E*)-selectivity. Up to three distinct steps may be involved (Scheme 3.14). First, deprotonation α- to a sulfone group is effected using a strong base such as butyllithium or LDA, to give an anion that then reacts with an aldehyde or ketone to give a mixture of diastereomeric β-hydroxysulfones. Next, the hydroxyl group is usually converted into a better leaving group (acetate, benzoate, or mesylate). The acetates or benzoates can be prepared directly by quenching the addition reaction with acetyl chloride or benzoyl chloride, respectively, but higher yields are often obtained if the

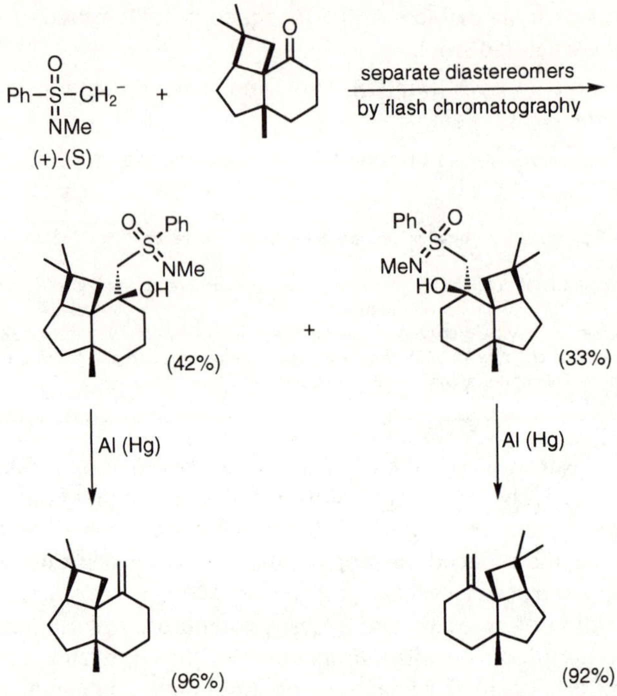

Scheme 3.13

β-hydroxysulfones are isolated before derivatisation of the hydroxyl group. The final step, reductive elimination to give the alkene, is usually effected by treatment with 6% sodium amalgam in a solvent such as THF or ethyl acetate, in the presence of methanol as proton source. The reaction is carried out at low temperature (−20 °C) in order to avoid reaction between sodium and the alcohol and formation of undesirable sodium methoxide.[16] Na_2HPO_4 may be added as a heterogeneous base scavenger.[17] The β-hydroxysulfones have sometimes been used directly in this reductive elimination step,[18] without functionalisation of the hydroxyl group, but retro-addition or simple reductive desulfonylation[19] sometimes compete. The reductive elimination is particularly rapid if the sulfone occupies an allylic position, and the method has often been used for the synthesis of conjugated dienes and trienes.

The high (*E*)-selectivity in the reaction was explained by Kocienski *et al.*,[16] who also showed that the stereochemistry of the alkene product is independent of the stereochemistry of the intermediate β-oxygenated sulfones. They considered that the two diastereomers must form a common intermediate radical or anion **14**, which has sufficient lifetime to undergo rotation about the central bond, adopting a conformation where the two R-groups are as far apart as possible before *anti*-periplanar elimination to give the (*E*)-alkene.

The Julia olefination has proved to be useful during the synthesis of many complex natural products[15] (e.g. Scheme 3.15),[20] but the length of the sequence is one of its main disadvantages, and each of the stages has proved problematic on occasions. The initial addition of the α-sulfonyl carbanion

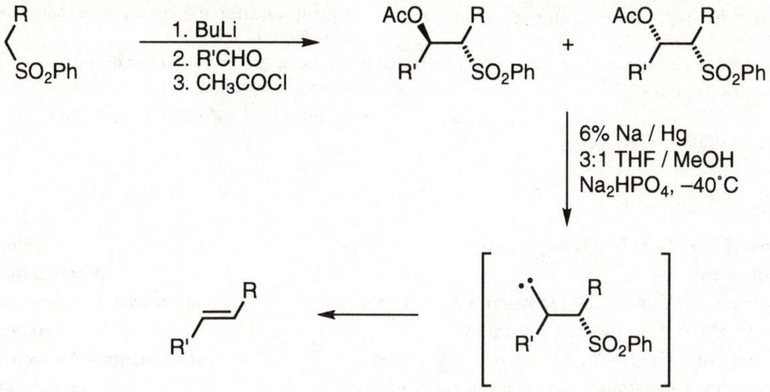

Scheme 3.14

to the carbonyl compound is reversible, and best results are obtained with aldehydes rather than ketones. If the carbonyl compound is enolisable, use of the magnesio-sulfone (generally prepared by reaction of the sulfone with EtMgBr) rather than the lithio-sulfone can prove advantageous. Use of Lewis acid additives such as diisobutyl aluminium methoxide,[21] boron trifluoride etherate,[22] or magnesium bromide[23] in the addition step have also been used to improve the yields of β-hydroxysulfones on occasion.

Protocol 6.
Preparation of alkene 17. Synthesis of a 1,2-disubstituted alkene from an aldehyde by Julia olefination (adapted from ref. 20)

Caution! All procedures should be carried out in a well-ventilated hood, and disposable vinyl or latex gloves and chemical-resistant safety goggles should be worn.

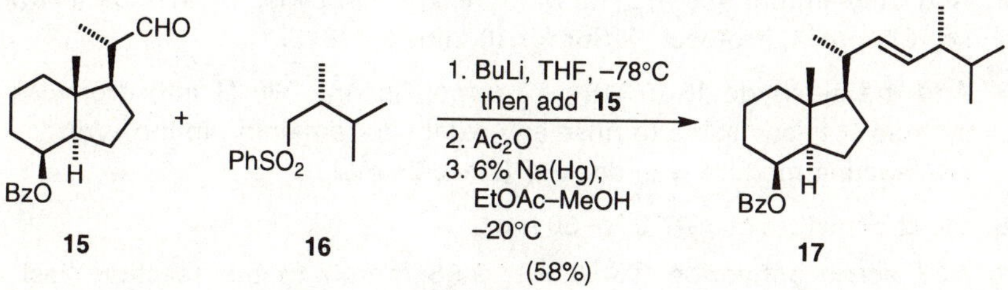

Scheme 3.15

Protocol 6. *Continued*

Equipment

- Magnetic stirrer
- Two two-necked, round-bottomed flasks (20 mL)
- Magnetic stirring bars
- Three-way stopcock
- Septa
- Source of dry argon or nitrogen
- Glass syringes with needle-lock Luers (volume appropriate for the quantity of solution to be transferred)
- Six-inch, medium-gauge needles
- Mortar and pestle
- Column for flash chromatography

Materials

- Sulfone **16**[20] (FW 226.3) 0.226 g, 1 mmol — **assume toxic**
- Dry THF (7 ml) — **flammable, irritant**
- Butyllithium[a] (FW 64.1) 1.3 M in hexane 0.8 mL, 1.04 mmol — **flammable, moisture sensitive**
- Aldehyde **15**[20] (FW 314.4) 0.31 g, 1 mmol — **assume toxic**
- Acetic anhydride[b] (FW 102.1) 0.185 mL, 1.96 mmol — **corrosive, flammable, lachrymator**
- Saturated aqueous ammonium chloride solution 5 mL — **corrosive, toxic**
- Ether for extraction 135 mL — **flammable, irritant**
- Magnesium sulfate (dry) — **hygroscopic**
- Dry methanol[c] 4 mL — **highly flammable, toxic**
- Ethyl acetate[d] 2 mL — **flammable, irritant**
- 6% Sodium amalgam[e] 1.0 g — **toxic**
- Silica gel for flash chromatography — **irritant dust**
- Ethyl acetate for flash chromatography — **flammable, irritant**
- Toluene for flash chromatography — **highly flammable, harmful**

1. Clean all glassware, syringes, needles, and stirring bars and dry for at least 4 in a 120 °C electric oven before use. Allow to cool in a desiccator.
2. Place the sulfone **16** (0.226 g, 1 mmol) and the stirrer bar into one of the two-necked flasks.
3. Place a three-way stopcock into one neck of the flask, and a septum on the other. Introduce an atmosphere of argon (see Chapter 1, Figs 1.1–1.5).
4. Add dry THF (5 mL) by syringe, as described in Protocol 1, steps 6 and 7.
5. Cool the flask to −78 °C by placing it in a dry-ice/acetone bath. Start the magnetic stirring.
6. Add butyllithium (0.8 mL, 1.3 M in hexane) dropwise by syringe (**caution**! see Chapter 1, Protocol 1). Stir for 10 min at −78 °C.
7. Add the aldehyde **15** (0.316 g, 1 mmol) in dry THF (1 mL) dropwise by syringe. It is advisable to rinse any aldehyde remaining in the syringe into the reaction mixture with dry THF (2 × 0.25 mL).
8. Stir the mixture at −78 °C for 30 min.
9. Add acetic anhydride (0.185 mL, 1.96 mmol) to the reaction flask by syringe.

10. Stir the mixture at −78°C for 4 h, then allow to warm to room temperature. Stir for a further 1 h at room temperature.

11. Quench the reaction by cautiously adding saturated aqueous ammonium chloride solution (5 mL) to the reaction flask, dropwise by syringe.

12. Transfer the mixture to a separatory funnel (100 mL). Extract with ether (3 × 15 mL).

13. Wash the combined organic extracts with water (20 mL), dry the organic layer ($MgSO_4$), filter, and remove the solvents using a rotary evaporator to give the oily, crude acetoxy sulfones (550 mg).

14. Transfer the acetoxy sulfones into the other two-necked flask. Place a stirrer bar in the flask and provide an inert atmosphere for the reaction as before.

15. Add dry methanol (4 mL) and ethyl acetate (2 mL) to the flask by syringe and cool the flask to −20°C. This temperature can be achieved using a cooling bath containing an ice–salt mixture, or a slush made from carbon tetrachloride and dry ice.

16. Crush 6% sodium amalgam (1.0 g, prepared as in Protocol 7) in a mortar and pestle and add it to the reaction flask as quickly as possible, by briefly removing the septum.

17. Stir the mixture at −20°C for 8 h.

18. Pour the mixture into a separatory funnel (100 mL) containing water (25 mL). Extract with ether (3 × 30 mL).

19. Dry the organic layers ($MgSO_4$), filter, and remove the solvents using a rotary evaporator to give crude alkene.

20. Apply the crude alkene to a flash column packed with silica gel. Elute with a mixture of 4% ethyl acetate–toluene to give the alkene **17** as an amorphous solid (220 mg, 58%), which displays the appropriate spectroscopic data.

[a]Titrate butyllithium with a 1.00 M solution of *s*-butanol in xylene using 1,10-phenanthroline as the indicator just prior to use.
[b]Acetic anhydride should be distilled from P_2O_5 under an inert atmosphere (nitrogen or argon) prior to use.
[c]Methanol should be distilled under an inert atmosphere (nitrogen or argon) from magnesium metal activated by iodine prior to use.
[d]Ethyl acetate may be dried over P_2O_5 before distillation under an inert atmosphere (nitrogen or argon) prior to use.
[e]6% Sodium amalgam can be prepared by the procedure of Protocol 7.

Protocol 7.
Preparation of 6% sodium amalgam

Caution! All procedures should be carried out in a well-ventilated hood, and disposable vinyl or latex gloves and chemical-resistant safety goggles should be worn. The reaction between sodium and mercury may be violent, and should be performed behind a protective screen. Strong gloves should be worn whilst adding sodium to mercury.

Equipment

- Conical flask (100 mL)
- Glass funnel (*c.* 10 cm diameter)
- Source of dry argon or nitrogen
- Tweezers
- Hammer
- Paper towel
- Glass jar for storage of amalgam

Materials

- Sodium (FW 23.0) 3 g, 130 mmol — **flammable solid, dangerous when wet**
- Mercury (FW 200.6) 47 g, 234 mmol — **corrosive, toxic**
- Dry hexanes for washing sodium — **flammable, irritant**

1. Charge the conical flask with mercury (47 g).
2. Support the conical flask using a clamp and a stand with a heavy base. Above the neck of the conical flask, place an inverted glass funnel through which is flowing argon or nitrogen.
3. Weigh a piece of sodium (3 g) with freshly cut edges and wash it by immersing it in a small beaker of hexanes. Cut the sodium into pieces (*c.* 0.5 cm $\times$ 0.5 cm $\times$ 0.5 cm) and store these under hexanes until required.
4. Using tweezers, add a piece of sodium to the mercury. **Caution!** The reaction is **extremely** vigorous; fumes are evolved violently. If reaction does not take place, **cautiously** swirl the conical flask, holding it by the clamp, to initiate this process. Strong gloves **must** be worn when performing this operation.
5. While the mixture is still hot and molten, carefully add the remainder of the sodium, piece by piece. **Caution!** Addition of each piece may be accompanied by vigorous reaction.
6. When addition of sodium is complete, allow the amalgam to cool with nitrogen or argon flowing through the inverted funnel.
7. Wrap the conical flask in paper towel and break the glass with a hammer. Carefully remove the large lump(s) of solid amalgam from the glass and transfer it quickly to a glass jar that has been flushed with nitrogen or argon. If desired, the amalgam can be analysed for sodium before use by titration of a sample with 0.1 M sulfuric or hydrochloric acid.

In order to avoid the use of 6% sodium amalgam, the use of magnesium with catalytic mercuric chloride[24] and also samarium iodide[25] have been recently reported as alternative procedures.

In contrast to the high (*E*)-selectivity observed in the synthesis of 1,2-disubstituted alkenes, application of the Julia olefination to the synthesis of trisubstituted alkenes can result in mixtures of isomers. (Scheme 3.16).[26]

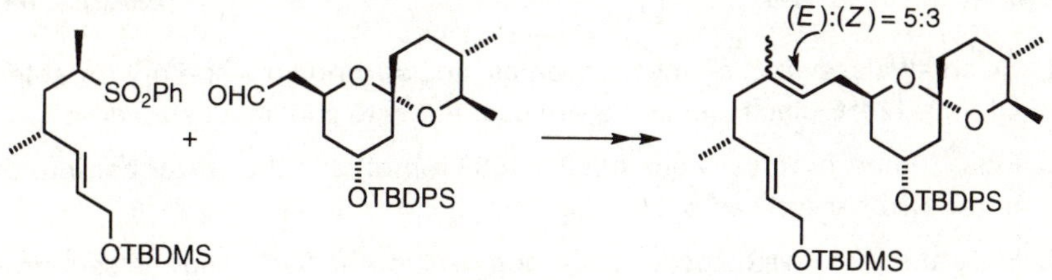

(*E*):(*Z*) = 5:3

SO_2Ph OHC + OTBDMS OTBDPS → OTBDPS OTBDMS

Scheme 3.16

5. Olefination using boron-stabilised anions

Dimesitylboron-stabilised carbanions react with diaryl ketones to give the corresponding alkenes in good yield (Scheme 3.17);[27] the reaction may be considered to be the boron analogue of the Wittig reaction. The method has rarely been used in synthesis to date.

Protocol 8.
Synthesis of 1,1-diphenylethene 18 from benzophenone. Boron–Wittig reaction with an aromatic ketone (adapted from ref. 27)

Caution! All procedures should be carried out in a well-ventilated hood, and disposable vinyl or latex gloves and chemical-resistant safety goggles should be worn.

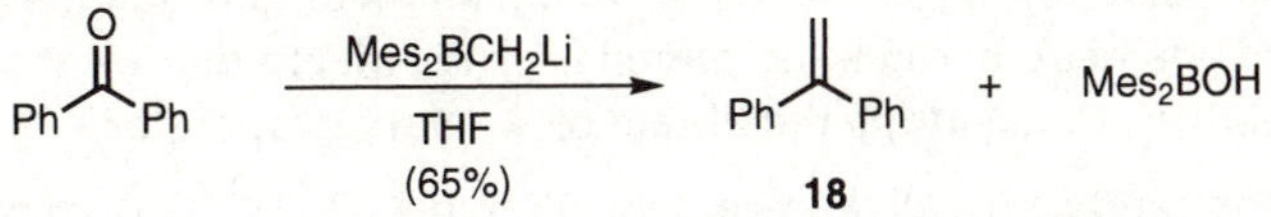

$$Ph_2C{=}O \xrightarrow[\text{THF (65\%)}]{Mes_2BCH_2Li} Ph_2C{=}CH_2\ (\mathbf{18}) + Mes_2BOH$$

Equipment

- Magnetic stirrer
- Two two-necked, round-bottomed flasks (25 mL)
- Magnetic stirring bar
- Three-way stopcock
- Septum
- Source of dry argon or nitrogen
- Six-inch, medium-gauge needles
- Glass syringes with needle-lock Luers (volume appropriate for the quantity of solution to be transferred)
- Glass sinter
- One-necked, round-bottomed flask (50 mL)
- Water-jacketed, semi-micro, short-path distillation apparatus

Protocol 8. *Continued*

Materials

- Bromomesitylene[a] (FW 199.1) 0.56 g, 2.83 mmol
- Dry THF (16.6 mL) — **flammable, irritant**
- *t*-Butyllithium[b] (FW 64.1) 2.83 mL, 2.0 M in hexane, 5.66 mmol — **flammable, moisture sensitive**
- *B*-Methyldimesitylborane[c] (FW 264.2) 0.69 g, 2.58 mmol — **assume toxic**
- Benzophenone (FW 182.2) 0.474 g, 2.6 mmol
- Light petroleum (b.p. 40–60°C) 15 mL — **flammable, irritant**

1. Clean all glassware, syringes, needles, and stirring bars and dry for at least 4 h in a 120°C electric oven before use. Allow to cool in a desiccator.
2. Place the bromomesitylene (0.56 g, 2.83 mmol) and the stirrer bar into one of the two-necked flasks (25 mL).
3. Place the three-way stopcock into one neck of the flask, and the septum on the other, and introduce an atmosphere of argon (see Chapter 1, Figs 1.1–1.5).
4. Add dry THF (6 mL) by syringe, as described in Protocol 1, steps 6 and 7.
5. Cool the flask to −78°C by placing it in a dry-ice/acetone bath and stir.
6. Add *t*-butyllithium (2.83 mL, 2.0 M in hexane) dropwise by syringe (**caution!** see Chapter 1, Protocol 1). The solution should become pale yellow and a white solid should precipitate.
7. Stir the mixture at −78°C for 15 min, then place the reaction flask in a bath at 25°C for 15 min. The precipitate should dissolve.
8. Transfer this mesityllithium solution by cannula or by syringe into a solution of *B*-methyldimesitylborane (0.69 g, 2.58 mmol) in THF (8 mL) in the other two-necked flask at 25°C. Stir the resulting mixture at 25°C for 1 h.
9. Add the benzophenone (0.474 g, 2.6 mmol) in THF (2.6 mL) by syringe. Stir the resulting pink solution at room temperature for 24 h.
10. Cautiously quench the reaction by adding water to the flask by syringe.
11. Remove the solvents using a rotary evaporator.
12. Add light petroleum (b.p. 40–60°C, 15 ml) and filter through the glass sinter into the one-necked, round-bottomed flask (50 mL) to remove the precipitate of Mes_2BOH. Concentrate the filtrate on a rotary evaporator.
13. Distil the product 1,1-diphenylethene **15**, b.p. 137–139°C/12 mmHg (0.31 g, 65% yield).

[a]Distil bromomesitylene at reduced pressure under an inert atmosphere (nitrogen or argon) immediately before use.
[b]Titrate butyllithium with a 1.00 M solution of *s*-butanol in xylene using 1,10-phenanthroline as the indicator just prior to use.
[c]*B*-Methyldimesitylborane can be prepared by sequential reaction of boron trifluoride etherate with mesitylmagnesium bromide (2 equivalents) and methylmagnesium bromide according to ref. 26, and should be dried in a drying pistol at 35°C/2 mmHg for 2 h prior to use.

Boron–Wittig reactions using aromatic aldehydes are more complex, but control of alkene stereochemistry can be achieved by choice of appropriate reaction conditions.[27] Condensation of the boron-stabilised carbanion with the aromatic aldehyde, quenching with chlorotrimethylsilane followed by aqueous HF gives (*E*)-alkenes in high yields and with good stereoselectivity (Scheme 3.18).[27] Alternatively, quenching with trifluoroacetic anhydride leads to formation of the (*Z*)-alkene, except where the aromatic aldehyde has an electron-withdrawing group in the *para*-position.[27]

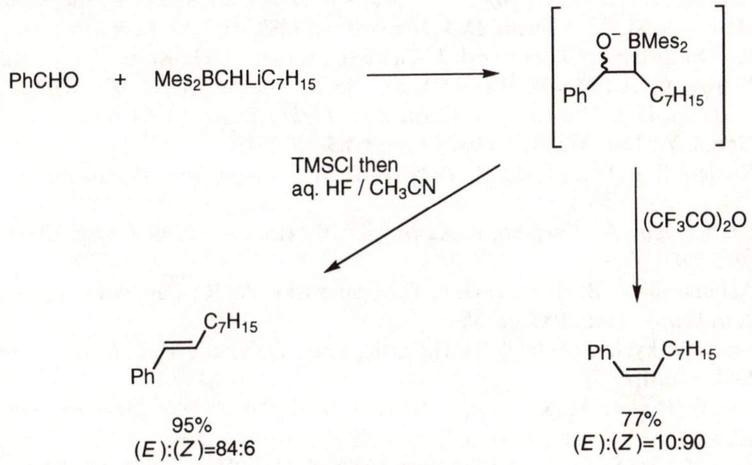

PhCHO + $Mes_2BCHLiC_7H_{15}$ → [O^- $BMes_2$ / Ph, C_7H_{15}]

TMSCl then aq. HF / CH_3CN → C_7H_{15}/Ph (*E*)-alkene: 95%, (*E*):(*Z*)=84:6

$(CF_3CO)_2O$ → Ph/C_7H_{15} (*Z*)-alkene: 77%, (*E*):(*Z*)=10:90

Scheme 3.18

References

1. For a review of methods for forming alkenes from carbonyl compounds and carbanions, covering all the methods included in this chapter, see: Kelly, S. E. In *Comprehensive Organic Synthesis*; Trost, B. M., ed.; Pergamon: Oxford, **1990**; Vol. 1, p. 729.
2. For a review of the Peterson olefination, see: Ager, D. J. *Org. React.* **1990,** *38*, 1.
3. Peterson. D. J. *J. Org. Chem.* **1968,** *33*, 780.
4. Hudrlik, P. F.; Peterson, D. J. *Tetrahedron Lett.* **1974,** *15*, 1133.
5. Hudrlik, P. F.; Peterson, D. *J. Am. Chem. Soc.* **1975,** *97*, 1464.
6. Chan, T. H.; Chang, E. *J. Org. Chem.* **1974,** *39*, 3264.
7. Boeckman, Jr, R. K.; Silver, S. M. *Tetrahedron Lett.* **1973,** *14*, 3497.
8. Hudrlik, P. F.; Peterson, D.; Rona, R. J. *J. Org. Chem.* **1975,** *40*, 2263.
9. Johnson, C. R.; Kirchhoff, R. A. *J. Am. Chem. Soc.* **1979,** *101*, 3602.
10. Boeckman, Jr, R. K.; Blum, D. M.; Arthur, S. D. *J. Am. Chem. Soc.* **1979,** *101*, 5060. Niwa, H.; Wakamatsu, K.; Hida, T.; Niiyama, K.; Kigoshi, H.; Yamada, M.; Nagase, H.; Suzuki, M.; Yamada, K. *J. Am. Chem. Soc.* **1984,** *106*, 4547. Morton, Jr, J. R.; Brokaw, F. C. *J. Org. Chem.* **1979,** *44*, 2880.
11. Johnson, C. R.; Schroeck, C. W.; Shanklin, J. R. *J. Am. Chem. Soc.* **1973,** *95*, 7424.
12. Johnson, C. R.; Meanwell, N. A. *J. Am. Chem. Soc.* **1981,** *103*, 7667.

13. Johnson, C. R.; Zeller, J. R. *Tetrahedron* **1984,** *40*, 1225.
14. Julia, M.; Paris, J.-M. *Tetrahedron Lett.* **1973,** *14*, 4833.
15. For reviews of the Julia olefination, see: Kocienski, P. J. *Phosphorus Sulfur* **1985,** *24*, 97; Kocienski, P. J. In *Comprehensive Organic Synthesis*; Trost, B. M., ed.; Pergamon: Oxford, **1990;** Vol. 6, p. 975.
16. Kocienski, P. J.; Lythgoe, B.; Ruston, S. *J. Chem. Soc., Perkin Trans. 1* **1978,** 829.
17. Trost, B. M.; Arndt, H. C.; Strege, P. E.; Verhoeven, T. R. *Tetrahedron Lett.* **1976,** *17*, 3477.
18. For example: Ley, S. V.; Anthony, N. J.; Armstrong, A.; Brasca, M. G.; Clarke, T.; Culshaw, D.; Greck, C.; Grice, P.; Jones, A. B.; Lygo, B.; Madin, A.; Sheppard, R. N.; Slawin, A. M. Z.; Williams, D. J. *Tetrahedron.* **1989,** *45*, 7161; Ley, S. V.; Armstrong, A.; Díez-Martín, D.; Ford, M. J.; Grice, P.; Knight, J. G.; Kolb, H. C.; Madin, A.; Marby, C. A.; Mukherjee, S.; Shaw, A. N.; Slawin, A. M. Z.; White, A. D.; Williams, D. J.; Woods, M. *J. Chem. Soc., Perkin Trans. 1* **1991,** 667.
19. Gaoni, Y.; Tomazic, A. *J. Org. Chem.* **1985,** *50*, 2948.
20. Kocienski, P. J.; Lythgoe, B.; Roberts, D. A. *J. Chem. Soc., Perkin Trans. 1* **1978,** 834.
21. Spaltenstein, A.; Carpino, P. A.; Miyake, F.; Hopkins, P. B. *J. Org. Chem.* **1987,** *52*, 3759.
22. Achmatowicz, B.; Baranowska, E.; Daniewski, A. R.; Panowski, J.; Wicha, J. *Tetrahedron Lett.* **1985,** *26*, 5597.
23. Danishefsky, S. J.; Selnick, H. G.; Zelle, R. E.; DeNinno, M. P. *J. Am. Chem. Soc.* **1988,** *110*, 4368.
24. Lee, G. H.; Lee, H. K.; Choi, E. B.; Kim, B. T.; Pak, C. S. *Tetrahedron Lett.* **1995,** *60*, 3194.
25. Keck, G. E.; Savin, K. A.; Weglarz, M. A. *J. Org. Chem.* **1995,** *60*, 3194. Marko, I. E.; Murphy, F.; Dolan, S. *Tetrahedron Lett.* **1996,** *37*, 2089.
26. Barrett, A. G. M.; Carr, R. A. E.; Atwood, S. V.; Richardson, G.; Walshe, N. D. A. *J. Org. Chem.* **1986,** *51*, 4840.
27. Pelter, A.; Buss, D.; Colclough, E.; Singaram, B. *Tetrahedron* **1993,** *49*, 7077.
28. Pelter, A.; Singaram, B; Warren, L.; Wilson, J. W. *Tetrahedron* **1993,** *49*, 2965.

4

Chromium- and titanium-mediated synthesis of alkenes from carbonyl compounds

DAVID M. HODGSON and LEE T. BOULTON

1. Introduction

This chapter contains organometallic-based one-step procedures for:

- Reductive coupling of aldehydes or ketones to produce alkenes (the McMurry reaction)
- Carbonyl methylenation using titanium-based reagents (the Oshima–Lombardo reaction)
- Conversion of esters and amides into enol ethers and enamines
- Homologation of aldehydes to (*E*)-1,2-difunctionalised alkenes using *gem*-dichromium reagents

The procedures in this chapters are distinct from the organometallic methods described in Chapter 7, since here the alkene functional group is constructed from two components. The two individual components are coupled via organometallic intermediates such that a new C=C bond is constructed.

2. Reductive coupling of aldehydes or ketones to produce alkenes (the McMurry reaction)

The reductive coupling of carbonyl compounds directly to produce alkenes by low-valent transition metals constitutes an important method of C=C bond formation.[1] The most widely applied procedure is the McMurry reaction which uses $TiCl_3$ and a Zn–Cu couple.[2] The reaction usually provides good yields of alkenes from both aromatic and aliphatic aldehydes and ketones, as exemplified in Protocol 1—the synthesis of cyclohexylidenecyclohexane **2** from cyclohexanone **1**.[3]

$\xrightarrow[\text{Zn-Cu, 86\%}]{TiCl_3(DME)_{1.5}}$

1 → 2

The McMurry reaction is particularly useful in the synthesis of strained alkenes, such as adamantylideneadamantane **4** from 2-adamantone **3**,[4] and in the synthesis of cyclic alkenes (ring sizes from 3 to 20 have been prepared), such as the diterpene, kempene-2 **6**, from the diketoaldehyde **5**.[5] The synthesis of medium and large ring alkenes requires slow (syringe pump) addition of the dicarbonyl compound to the low-valent titanium reagent in order to create the high dilution conditions necessary to avoid intermolecular coupling. Non-symmetrical alkenes can be prepared from a mixture of two ketones, if one is used in excess. Where applicable, (*E*)-alkene geometry is usually favoured in the McMurry reaction, particularly when the difference in energy between the (*E*)- and (*Z*)-isomers exceeds 4–5 kcal mol^{-1}.

$\xrightarrow[\text{Li, 76\%}^{4}]{TiCl_3}$

3 → 4

CHO, OAc, H, O

$\xrightarrow[\text{Zn-Cu, 32\%}^{5}]{TiCl_3(DME)_{1.5}}$

5 → 6

Functional groups that are generally compatible in the McMurry reaction include alcohols (however, allylic alcohols couple to give 1,5-dienes), alkenes (retention of position and geometry), amines, ethers, halides, and sulfides. Functional groups that generally survive a few hours' exposure to typical McMurry reaction conditions include alkynes, amides, esters, ketones (e.g. the diketoaldehyde **5**), nitriles, and tosylates.[2] In practice this means that these groups should survive most intermolecular reactions and those intramolecular reactions that form five- or six-membered rings. Functional groups which are reduced competitively include epoxides, nitro groups, oximes, and sulfoxides. Mechanistically, the reaction is believed to start with pinacolate formation on the surface of a small zero-valent titanium particle, followed by stepwise cleavage of the C–O bonds which results in formation of the alkene and an oxide-coated surface.

Ti surface Ti surface Ti surface Ti surface

From a practical aspect, most problems in the McMurry reaction arise in the preparation of the active low-valent titanium species, for which a variety of methods are available.[3] McMurry has reported an optimised procedure for titanium-induced carbonyl coupling using a $TiCl_3(DME)_{1.5}$ complex and Zn–Cu couple, on which Protocol 1 is based.[3a]

Protocol 1.
Synthesis of cyclohexylidenecyclohexane 2[3a]

Caution! All procedures should be carried out in a well-ventilated hood, and disposable vinyl or latex gloves and chemical-resistant safety goggles should be worn.

$TiCl_3(DME)_{1.5}$, Zn-Cu, 86%

1 → 2

Equipment

- Two-necked, round-bottomed flask (250 mL)
- Water-jacketed reflux condenser
- Septum
- Magnetic stirring bar
- Magnetic stirrer
- Glove bag (supplied, for example, by Philip Harris Scientific)
- Portable electric balance
- Oil bath
- Glass syringe with a needle-lock Luer (volume appropriate for quantity of solution to be transferred)
- Six-inch, medium-gauge needle
- Source of dry argon

Materials

- Dry DME[a] 80 mL — **flammable, irritant**
- Titanium(III) chloride–DME complex[b] (FW 289.4) 3.8 g, 13.1 mmol — **flammable, corrosive, moisture sensitive**
- Zinc–copper couple[b] 3.6 g, 51 mmol — **flammable, irritant, moisture sensitive**
- Cyclohexanone **1** (FW 98.1) 0.32 g, 3.3 mmol — **flammable, harmful**
- Pentane 100 mL — **flammable, irritant**
- Florisil — **irritant**

1. Clean all glassware, syringes, needles, and stirring bar and dry for at least 4 h in a 120 °C electric oven before use.
2. Assemble the flask with stirring bar and reflux condenser whilst still hot and allow to cool under argon, sealed with septa.

Protocol 1. *Continued*

3. Transfer the sealed flask with attached condenser to a glove bag[c] containing two spatulas, bottles of titanium(III) chloride–DME complex and the zinc–copper couple, and a portable electric balance.
4. Flush the glove bag thoroughly with argon, then seal and put the titanium(III) chloride–DME complex (3.8 g, 13.1 mmol) and zinc–copper couple (3.6 g, 51 mmol) into the flask.
5. Transfer the sealed flask and condenser to a stand with a heavy base and support with two clamps. Reconnect via a needle to the argon supply.
6. Assemble the syringe and needle while hot and allow the assembled syringe to cool to room temperature in a desiccator. Flush the syringe with argon (see Chapter 1, Fig. 1.6).
7. Stir the contents of the flask vigorously and charge the flask with dry DME (70 mL) via syringe by puncturing the septum on the reaction flask.
8. Heat the stirring reaction mixture to reflux for 3 h to form the active titanium coupling agent as a black suspension.
9. Remove the oil bath and as soon as the solvent has ceased to reflux add the cyclohexanone **1** (0.32 g, 3.3 mmol) in dry[a] DME (10 mL) rapidly to the reaction mixture and then continue heating to reflux for a further 8 h.
10. Cool the reaction mixture to ambient temperature and dilute with pentane (80 mL). Filter the reaction slurry through a small (5 cm × 10 cm) pad of Florisil to remove metal salts, and rinse the flask and Florisil with additional pentane (20 mL).
11. Concentrate the filtrate under reduced pressure by means of a rotary evaporator (25°C/40 mmHg) to afford cyclohexylidenecyclohexane **2** (0.23 g, 86% yield) as white crystals, which display the appropriate spectroscopic data.

[a]Distil DME from sodium benzophenone ketyl under an argon atmosphere and use immediately.
[b]Titanium(III) chloride–DME complex and zinc–copper couple can both be purchased from Aldrich, or prepared according to McMurry.[3a]
[c]Use of an inflatable glove bag is recommended for convenience; it functions as an inexpensive, flexible, and portable glove box and provides a dry inert atmosphere on connection to an inert gas supply.

3. Carbonyl methylenation using titanium-based reagents (the Oshima–Lombardo reaction)

3.1 Methylenation of ketones and aldehydes

The methylenation of aldehydes and ketones can generally be accomplished using $Ph_3P{=}CH_2$ (Chapter 2). A range of titanium reagents has been developed to effect the same transformation.[6] These are useful in cases where the carbonyl group to be methylenated is hindered, or contains a potentially epimerisable centre or other base-sensitive functionality. For these cases, the Tebbe reagent **7**, dimethyltitanocene **8** (see Section 3.2), or the Oshima–Lombardo reagent **9** should be considered.

$Cp_2Ti(\mu\text{-}CH_2)(\mu\text{-}Cl)AlMe_2$ **7**, 77%[6]

The Oshima–Lombardo reagent **9** has had the broadest range of applications with ketones and aldehydes, since it does not react with esters and many other functional groups (THP ethers, TBDMS ethers, acetals, carboxylic acids, alcohols, and lactones).

Protocol 2.
Synthesis of (+)-3-methylene-*cis*-*p*-menthane 11[7]

Caution! All procedures should be carried out in a well-ventilated hood, and disposable vinyl or latex gloves and chemical-resistant safety goggles should be worn.

10 → **11**: $Zn\text{-}CH_2Br_2\text{-}TiCl_4$ **9**, 95%

Equipment

- Overhead mechanical stirrer
- Two-necked, round-bottomed flask (500 mL)
- Septum
- Pressure-equalising dropping funnel
- Glass syringe with a needle-lock Luer (volume appropriate for quantity of solution to be transferred)
- Six-inch, medium-gauge needle
- Erlenmeyer flask (1 L)
- Distillation column (20 cm × 1 cm) packed with glass helices
- Source of dry nitrogen or argon

Materials

- Dry THF 110 mL — **flammable, irritant**
- Activated zinc powder[a] (FW 65.4) 12.5 g, 0.19 mol — **flammable, moisture sensitive**
- Dibromomethane (FW 173.8) 4.39 mL, 0.063 mol — **harmful**
- Titanium(IV) chloride[b] (FW 189.7) 5 mL, 0.045 mol — **harmful, moisture sensitive**
- (+)-Isomenthone **10** (FW 154.2) 7.26 g, 0.047 mol — **harmful**
- Dry dichloromethane 40 mL — **harmful**
- Pentane for extraction 200 mL — **flammable, irritant**

Protocol 2. *Continued*

1. Clean all glassware, syringes, and needles and dry for at least 4 h in a 120 °C electric oven before use.
2. Assemble the flask with overhead stirrer and pressure-equalising dropping funnel whilst still hot and allow to cool under nitrogen (see Chapter 1, Figs 1.12 and 1.13).
3. Put activated zinc powder (12.5 g, 0.19 mol) into the flask followed by dry THF (110 mL) and dibromomethane (4.39 mL, 0.063 mol).
4. Cool the mixture with stirring to −40 °C by means of a dry-ice/acetonitrile cooling bath. Add $TiCl_4$ (5 mL, 0.045 mol) to the dropping funnel via syringe (Chapter 1, Protocol 1) and then drip into the reaction over 30 min (**caution**! extremely vigorous reaction). Unless the syringe and needle have been carefully dried, the needle will become blocked with white solids.
5. Remove the cooling bath and stir the reaction mixture for 3 days at 0°C by means of, for example, a Dewar containing cold water.
6. Cool the dark-grey slurry to 0°C by means of an ice/water bath and add dry dichloromethane (20 mL), followed by a solution of (+)-isomenthone **10** (7.26 g, 0.047 mol) in dry dichloromethane (20 mL) over 10 min.
7. Remove the cooling bath and stir the reaction mixture at room temperature (20 °C) for 1.5 h before adding pentane (100 mL), followed by cautious addition of a slurry of $NaHCO_3$ (50 g) in water (30 mL) over 1 h.
8. Pour off the clear organic layer into an Erlenmeyer flask (1 L) and wash the residue with pentane (3 × 20 mL). Combine the organic layers and dry over a mixture of Na_2SO_4 (50 g) and $NaHCO_3$ (10 g). Filter the solution through a sintered glass funnel (no. 2), and wash the solid desiccant thoroughly with pentane.
9. Remove the solvent at atmospheric pressure by flash distillation through a column (20 cm × 1 cm) packed with glass helices.
10. Distil the liquid residue under reduced pressure to obtain (+)-3-methylene-*cis-p*-menthane **11** (b.p. 105–107 °C/90 mmHg, 6.81 g, 95% yield) as a clear colourless oil which displays the appropriate spectroscopic data.

[a]Wash zinc powder using sequentially 5% hydrochloric acid, water, ethanol, and finally diethyl ether and dry before use.[8]

[b]Titanium(IV) chloride can be purchased from Aldrich. Material from bottles which have already been opened should be distilled before using.

Two useful modifications of the Oshima–Lombardo reagent involve (a) substituting $TiCl_4$ by $Ti(OPr^i)_4$ for aldehyde-selective addition in the presence of a ketone and (b) *in situ* pre-complexation of an aldehyde with $Ti(NEt_2)_4$ for ketone-selective addition in the presence of an aldehyde.[6]

3.2 Conversions of esters and amides into enol ethers and enamines

The conversion of esters and amides to enol ethers and enamines cannot be achieved using conventional Wittig-type chemistry. However, the methylenation of esters and amides can be achieved using Tebbe's reagent **7**,[6,9] or dimethyltitanocene **8**.[10] The alkylidenation of ester carbonyl groups (including trimethylsilyl esters) requires the use of 1,1-dibromoalkanes, Zn, catalytic $PbCl_2$, and $TiCl_4$ in the presence of TMEDA.[6d]

TBDMSO–(N-CO₂Me piperidin-2-one) —[Cp_2TiMe_2 **8**, 82%[10b]]→ TBDMSO–(N-CO₂Me 2-methylenepiperidine)

Dimethyltitanocene **8** has seen increasing application in organic synthesis since its introduction as a methylenating agent in 1990[10a] and is often preferred over the Tebbe reagent because of its comparative ease of use. The methylenation of dihydrocoumarin **12** (Protocol 3) illustrates the use of dimethyltitanocene **8**, prepared *in situ* for additional experimental convenience.

Protocol 3.
Synthesis of 3,4-dihydro-2-methylene-2*H*-1-benzopyran 13[9,10a]

Caution! All procedures should be carried out in a well-ventilated hood, and disposable vinyl or latex gloves and chemical-resistant safety goggles should be worn.

12 —[Cp_2TiMe_2 **8**, 66%]→ **13** ($=CH_2$)

Equipment

- Magnetic stirrer
- Oil bath
- Two-necked, round-bottomed flask (100 mL)
- Water-jacketed reflux condenser
- Septum
- Tubing adapter incorporating a two-way stopcock (Chapter 1, Fig. 1.3)
- Magnetic stirring bar
- Portable, electric balance
- Glove bag (supplied, for example, by Philip Harris Scientific)
- Glass syringe with a needle-lock Luer (volume appropriate for quantity of solution to be transferred)
- Six-inch, medium-gauge needle
- Aluminium foil
- Source of dry argon

Protocol 3. *Continued*

Materials

- Ethanol 1 L **flammable**
- Concentrated hydrochloric acid 20 mL **corrosive, harmful**
- Potassium hydroxide 7 g **corrosive, irritant**
- Dry ether 10 mL **flammable, irritant**
- Titanocene dichloride (FW 249.0) 1.5 g, 6.0 mmol **corrosive, moisture sensitive**
- Methyllithium, 1.4 M in Et_2O, 8.7 mL, 12.2 mmol **flammable, moisture sensitive**
- Dry THF 15 mL **flammable, irritant**
- Dihydrocoumarin **12** (FW 148.2) 0.44 g, 3.0 mmol **harmful, irritant**
- Activated basic alumina (Brockmann grade I) 35 g **toxic**
- Light petroleum (b.p. 40–60 °C) for chromatography **flammable, irritant**

1. Add concentrated hydrochloric acid (20 mL) cautiously to ethanol (500 mL), with stirring, to provide cleaning solution A.
2. Add potassium hydroxide pellets (7 g) to ethanol (500 mL), with stirring, to provide cleaning solution B.
3. Clean all glassware, syringes, needles, and stirring bar using sequentially solution A, deionised water, solution B, deionised water, and finally acetone and then dry for at least 4 h in a 120 °C electric oven before use.
4. Assemble the flask with the stirring bar and the tubing adapter whilst still hot and allow to cool under argon by introducing argon via the tubing adapter.
5. Close the stopcock on the tubing adapter and transfer the sealed flask to a glove bag[a] containing a spatula, the bottle of titanocene dichloride, and a portable electric balance.
6. Flush the glove bag thoroughly with argon, then seal and put the titanocene dichloride (1.5 g, 6 mmol) into the flask.
7. Transfer the sealed flask to a stand with a heavy base and support with a clamp. Reconnect to the argon supply, via a needle.
8. Assemble the syringe and needle while hot and allow the assembled syringe to cool to room temperature in a desiccator. Flush the syringe with argon (see Chapter 1, Fig. 1.6).
9. Stir the contents of the flask and then charge the flask with dry ether (10 mL) by puncturing a septum on the tubing adapter and opening the two-way stopcock.
10. Wrap the flask in aluminium foil (to keep the flask dark) and place in a cold water bath (10 °C).
11. Add the methyllithium (8.7 mL 1.4 m in ether) dropwise by syringe to the stirring suspension (see Chapter 1). On completion of addition allow the reaction mixture to stir for 30 min at room temperature.
12. Apply water aspirator reduced pressure via a cold trap to the tubing

adapter and remove the majority of the solvent to leave a light-sensitive yellow slurry. Remove the vacuum and then add dry THF (15 mL) followed by dihydrocoumarin **12** (0.44 g, 3.0 mmol) to the flask.

13. Replace the stopper on the flask with the reflux condenser. Use the oil bath to heat the stirring reaction mixture under reflux for 8 h.
14. Allow the reaction mixture to cool to room temperature and dilute with ether (50 mL) and then remove the solvents by means of a rotary evaporator (35°C/15 mmHg) to yield a red–brown slurry.
15. Apply the oily residue to a flash column packed with alumina. Elute with a mixed solvent of ether–light petroleum (5:95) to obtain 0.285 g of 3,4-dihydro-2-methylene-2*H*-1-benzopyran **13** (66%) as a colourless oil, which displays the appropriate spectroscopic data.

[a]For details see Protocol 1.

4. Homologation of aldehydes to (*E*)-1,2-difunctionalised alkenes using *gem*-dichromium reagents

The homologation of aldehydes to (*E*)-1,2-difunctionalised alkenes can often be accomplished using Wittig-type chemistry (Chapters 2 and 3). However, certain substitution patterns are impossible to prepare using this strategy; the stereoselectivity obtained may not be satisfactory, or functional groups elsewhere in the starting aldehyde may not be tolerated in the coupling step. For these cases, *gem*-dichromium reagents usually derived from *gem*-dihalides have provided some useful solutions.[11]

(*E*)-1,2-Dialkyl-substituted alkenes, alkenyl halides, sulfides, silanes, and

$$\mathrm{RCHO} \xrightarrow[\mathrm{CrCl_2}]{\mathrm{Hal_2CHX}} \mathrm{R{\diagdown}\!\!{=}\!\!{\diagdown}X} \qquad \mathrm{X = R', Hal, SiMe_3, SPh, SnBu_3}$$

stannanes can all be prepared using this chemistry, usually with good to excellent (*E*)-stereoselectivity. A further characteristic of organochromium-mediated reactions with aldehydes is the high chemoselectivity usually observed. Other electrophilic sites in the aldehyde seldom compromise the alkylidenation step. Esters, cyano groups, and ketonic functionality are tolerated and, in addition, sites sensitive to epimerisation or elimination are usually unaffected. Whilst 1,2-addition is observed with α,β-unsaturated aldehydes, the stereoselectivity for the (*E*)-isomer is usually slightly lower than that seen with aliphatic and aromatic aldehydes. The reactions have found widespread use in natural product synthesis.

The (*E*)-1,2-dialkyl-substituted alkenes prepared using this chemistry require *gem*-diiodides[12] or 1-acetoxy-1-bromides[13] to partner the aldehyde. The reactions are generally run in THF; small quantities of DMF are often added to facilitate reductive removal of the halogen atoms and sonication has also been used in difficult cases. Yields are usually excellent (80–95%). (*E*:*Z*)-Stereoselection is in the region of 90:10 to 95:5.

The conditions used to prepare alkenyl halides depend upon the specific halide required.[14] Alkenyl halide formation is the most popular of the chromium-mediated olefination procedures and is most often used to prepare alkenyl iodides because of the greater reactivity of the carbon–iodine bond for subsequent reactions. Alkenyl iodides are prepared in good yields (70–90%, with stereoselectivity (85:15–95:5) dependent on the nature of the aldehyde) using iodoform with $CrCl_2$ in THF. Stereoselectivity can be enhanced using 1,4-dioxane as a co-solvent.[15]

Alkenyl silanes are produced exclusively as the (*E*)-isomers from aldehydes and $Br_2CHSiMe_3$ in THF in good yields (70–85%).[16] Alkenyl stannane formation uses $CrCl_2$ with $Br_2CHSnBu_3$ and requires LiI and additional DMF as a co-solvent with the THF.[17] Mechanistically, the reactions are envisaged to proceed via two successive halogen atom transfers[18] to $CrCl_2$ in which the intermediate radicals are immediately reduced to give ultimately a *gem*-dichromium species which adds to the aldehyde and then eliminates to provide the alkene.

$$\mathrm{Hal_2CHX} \xrightarrow{\mathrm{CrCl_2}} \mathrm{Hal\dot{C}HX} \xrightarrow{\mathrm{CrCl_2}} \mathrm{(Cl_2Cr)(Hal)CHX} \longrightarrow \mathrm{(Cl_2Cr)_2CHX} \xrightarrow{\mathrm{RCHO}} \mathrm{RCH(OCrCl_2)CH(CrCl_2)X} \longrightarrow \mathrm{RCH{=}CHX}$$

From a practical standpoint, the crucial aspects are minimising exposure of the chromium(II) chloride to air and using dry, oxygen-free solvents.

Protocol 4.
Synthesis of β-iodostyrene 15[14]

Caution! All procedures should be carried out in a well-ventilated hood, and disposable vinyl or latex gloves and chemical-resistant safety goggles should be worn.

$$\underset{\mathbf{14}}{\mathrm{PhCHO}} \xrightarrow[\text{THF, } 87\%^{14}]{\mathrm{CHI_3,\ CrCl_2}} \underset{\mathbf{15}\ (E):(Z)\ 94:6}{\mathrm{PhCH{=}CHI}}$$

Equipment

- Magnetic stirrer
- Two-necked, round-bottomed flask (50 mL)
- Septa
- Magnetic stirring bar
- Glove bag (supplied, for example, by Philip Harris Scientific)
- Portable electric balance
- Glass syringe with a needle-lock Luer (volume appropriate for quantity of solution to be transferred)
- Six-inch, medium-gauge needle
- Source of dry argon

Materials

- Dry THF[a] 15 mL **flammable, irritant**
- Chromium(II) chloride[b] 95% w/w pure (FW 122.9) 0.78 g, 6 mmol **harmful, moisture sensitive**
- CHI_3 (FW 393.7) 0.79 g, 2 mmol **harmful**
- Benzaldehyde **14**[c] (FW 106.1) 0.11 g, 1 mmol **harmful**
- Ether for extraction 30 mL **flammable, irritant**
- Silica gel for flash chromatography **irritant**
- Light petroleum (b.p. 40–60 °C) **flammable, irritant**

1. Clean all glassware, syringes, needles, and stirring bar and dry for at least 4 h in a 120 °C electric oven before use.
2. Assemble the two-necked flask with stirring bar whilst still hot and allow to cool under argon.
3. Transfer the sealed flask to a glove bag[d] containing a spatula, a bottle of chromium(II) chloride, and a portable electric balance.
4. Flush the glove bag thoroughly with argon, then seal and put the chromium(II) chloride (0.78 g, 95% w/w pure, 6 mmol) into the flask.
5. Transfer the sealed flask to a stand with a heavy base and support with a clamp. Reconnect to the argon supply, via a needle.
6. Assemble the syringe and needle while hot and allow the assembled syringe to cool to room temperature in a desiccator. Flush the syringe with argon (see Chapter 1, Fig. 1.6).
7. Stir the contents of the flask slowly and charge the flask with dry THF (10 mL) by puncturing the septum on the reaction flask.
8. Cool the flask to 0°C by means of an ice/water bath. Add dropwise a mixture of CHI_3 (0.79 g, 2 mmol) and benzaldehyde **14** (0.11 g, 1 mmol) in dry THF (5 mL) to the reaction mixture and allow to stir at 0°C for 3 h.
9. Dilute the reaction mixture with water (25 mL) and transfer the mixture to a separatory funnel. Extract the water layer with ether (3 × 10 mL).
10. Transfer the combined ethereal layers to a 50 mL flask. Dry (Na_2SO_4) and filter through a filter paper. Concentrate the filtrate under reduced pressure by means of a rotary evaporator (25 °C/15 mmHg).
11. Apply the residue to a flash column packed with silica gel. Elute with light petroleum (b.p. 40–60 °C) to obtain 0.20 g of β-iodostyrene **15** (87%, (*E*):(*Z*) 94:6) as a colourless oil, which displays the appropriate spectroscopic data.

[a]Bubble argon through for 15 min prior to use.
[b]Chromium(II) chloride can be purchased from Aldrich.
[c]Distil benzaldehyde before use.
[d]For details see Protocol 1.

Protocol 5. Synthesis of (*E*)-1-tributylstannyl-1-decene 17[17]

Caution! All procedures should be carried out in a well-ventilated hood, and disposable vinyl or latex gloves and chemical-resistant safety goggles should be worn.

$$n\text{-}C_8H_{17}CHO \ (\mathbf{16}) \xrightarrow[\text{LiI, THF, DMF}\ \ 60\%^{17}]{Br_2CHSnBu_3,\ CrCl_2} n\text{-}C_8H_{17}CH{=}CHSnBu_3 \ (\mathbf{17})$$

Equipment

- Magnetic stirrer
- Two-necked, round-bottomed flask (50 mL)
- Septa
- Magnetic stirring bar
- Glove bag (supplied, for example, by Philip Harris Scientific)
- Portable electric balance
- Glass syringe with a needle-lock Luer (volume appropriate for quantity of solution to be transferred)
- Six-inch, medium-gauge needle
- Source of dry argon

Materials

• Dry THF[a] 24 mL	**flammable, irritant**
• Dry DMF[b] 0.78 mL	**harmful**
• Chromium(II) chloride[c] 95% w/w pure (FW 122.9) 1.33 g, 10 mmol	**harmful, moisture sensitive**
• Anhydrous lithium iodide[c] (FW 133.8), 0.535 g, 4 mmol	**harmful, moisture sensitive**
• $Bu_3SnCHBr_2$[17] (FW 289.4) 0.926 g, 2 mmol	**harmful**
• Nonanal **16**[d] (FW 142.2) 0.144 g, 1 mmol	**flammable, harmful**
• Light petroleum (b.p. 40–60°C) 40 mL	**flammable, irritant**
• Reversed-phase silica gel for flash chromatography[e]	**irritant**
• Acetonitrile for flash chromatography	**flammable, harmful**
• Dichloromethane for flash chromatography	**harmful**

1. As steps **1** to **7** in Protocol 4.
2. Repeat steps **6** and **7** in Protocol 4 using DMF (0.78 mL).
3. Add a mixture of $Br_2CHSnBu_3$ (0.926 g, 2 mmol) and nonanal **16** (0.144 g, 1 mmol) in dry THF (4 mL) to the reaction mixture, followed by a solution of anhydrous LiI (0.535 g, 4 mmol) in THF (4 mL). Cover the flask in aluminium foil to exclude light and allow to stir at room temperature for 24 h.
4. Dilute the reaction mixture with water (30 mL) and light petroleum (20 mL). Transfer the mixture to a separatory funnel and separate the two layers. Extract the water layer with light petroleum (20 mL). Combine the light petroleum layers and wash with water (20 mL) then brine (20 mL).
5. Dry the organic layer ($MgSO_4$), filter, and concentrate the filtrate under reduced pressure by means of a rotary evaporator (25°C/15 mmHg).

6. Apply the oily residue to a flash column packed with reversed-phase silica gel. Elute with a mixed solvent of dichloromethane–acetonitrile (3:7) to obtain 0.26 g of (*E*)-1-tributylstannyl-1-decene **17** (60%) as a colourless oil, which displays the appropriate spectroscopic data.

[a]For details see Protocol 4.
[b]Distil DMF under reduced pressure from calcium hydride and store over 4 Å molecular sieves.
[c]Chromium(II) chloride and anhydrous lithium iodide can be purchased from Aldrich.
[d]Distil nonanal before use.
[e]Reversed-phase silica gel (Preparative C18 125 Å bulk packing material, 55–105 μm) can be purchased from Millipore.

References

1. (a) Robertson, G. M. In *Comprehensive Organic Synthesis*; Trost, B. M.; Fleming I., eds; Pergamon Press: Oxford, **1991**, Vol. 2, p. 563. (b) Dushin, R. G. In *Comprehensive Organometallic Chemistry II*; Abel, E. W.; Stone, F. G. A.; Wilkinson, G., eds; Pergamon Press: Oxford, **1995**, Vol. 12, p. 1071.
2. McMurry, J. E. *Chem. Rev.* **1989**, *89*, 1513.
3. (a) McMurry, J. E.; Lectka, T.; Rico, J. G. *J. Org. Chem.* **1989**, *54*, 3748. (b) Fürstner, A.; Hupperts, A. *J. Am. Chem. Soc.* **1995**, *117*, 4468.
4. Fleming, M. P.; McMurry, J. E. *Org. Synth. Coll.* **1990**, *7*, 1.
5. Dauben, W. G.; Farkas, I.; Bridon, D. P.; Chuang, C.-P. *J. Am. Chem. Soc.* **1991**, *113*, 5883.
6. (a) Pine, S. H. *Org. React.* **1993**, *43*, 1. (b) Kelly, S. E. In *Comprehensive Organic Synthesis*; Trost, B. M.; Fleming I., eds; Pergamon Press: Oxford, **1991**; Vol. 1, p. 729. (c) Stille, J. R. In *Comprehensive Organometallic Chemistry II*; Abel, E. W.; Stone, F. G. A.; Wilkinson, G., eds; Pergamon Press: Oxford, **1995**, Vol. 12, p. 577. (d) Takai, K.; Kataoka, Y.; Miyai, J.; Okazoe, T.; Oshima, K.; Utimoto, K. *Org. Synth.*, **1995**, *73*, 73.
7. Lombardo, L. *Org. Synth. Coll.* **1993**, *8*, 386.
8. Fieser, L. F.; Fieser, M. *Reagents for Organic Synthesis*; Wiley: New York, **1967**; Vol. I, p. 1276.
9. Pine, S. H.; Kim, G.; Lee, V. *Org. Synth. Coll.* **1993**, *8*, 512.
10. (a) Petasis, N. A.; Bzowej, E. I. *J. Am. Chem. Soc.* **1990**, *112*, 6392. (b) Herdeis, C.; Heller, E. *Tetrahedron: Asymmetry* **1993**, *4*, 2085.
11. Hodgson, D. M. *J. Organomet. Chem.* **1994**, *476*, 1.
12. Okazoe, T.; Takai, K.; Utimoto, K. *J. Am. Chem. Soc.* **1987**, *109*, 951.
13. Knecht, M.; Boland, W. *Synlett* **1993**, 1234.
14. Takai, K.; Nitta, K.; Utimoto, K. *J. Am. Chem. Soc.* **1986**, *108*, 7408.
15. Evans, D. A.; Cameron Black, W. *J. Am. Chem. Soc.* **1993**, *115*, 4497.
16. Takai, K.; Kataoka, Y.; Okazoe, Y; Utimoto, K. *Tetrahedron Lett.* **1987**, *28*, 1443.
17. Hodgson, D. M.; Boulton L. T.; Maw, G. N. *Tetrahedron* **1995**, *51*, 3713.
18. Kochi, J. K. *Organometallic Mechanisms and Catalysis*; Academic Press: New York, **1978**; p. 138.

5

Elimination and addition–elimination reactions

ANDREW D. WESTWELL and JONATHAN M. J. WILLIAMS

1. Introduction

When two groups are lost from adjacent atoms to form an alkene the process is known as β-elimination. This type of reaction can be readily divided into two major categories, depending on the species eliminated. Thus, alkenes may be formed by the elimination of H–X (type 1) or X–X (type 2) as shown in Scheme 5.1. Type 1 and type 2 reactions will be treated separately, and a final section will be concerned with the synthesis of alkenes via addition of enolates to carbonyl compounds followed by dehydration.

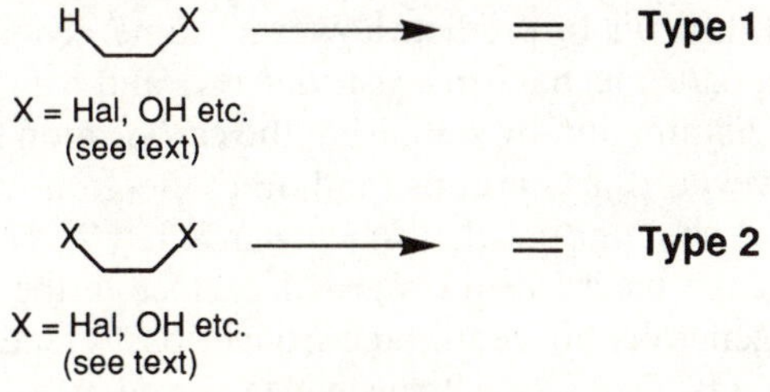

Scheme 5.1

The mechanism of β-elimination reactions can be classified into three basic types, namely E2, E1, and E1cB (Scheme 5.2).

(1) In the E2 mechanism (bimolecular elimination) the two groups depart simultaneously, with the proton being pulled off by a base. The mechanism is second order and is analogous to the S_N2 mechanism with which it can sometimes compete. An important feature of this mechanism is that it is often stereospecific since the H and X groups which leave must usually be *anti*-periplanar.

(2) In the two-step E1 mechanism (unimolecular elimination) the rate-determining step is ionisation of the substrate to give a carbocation which rapidly loses a β-proton to give an alkene, i.e. X leaves before H.

(3) In contrast, the two-step E1cB mechanism (unimolecular elimination, conjugate base) involves rate-determining loss of a proton (to give a carbanion) followed by loss of X.

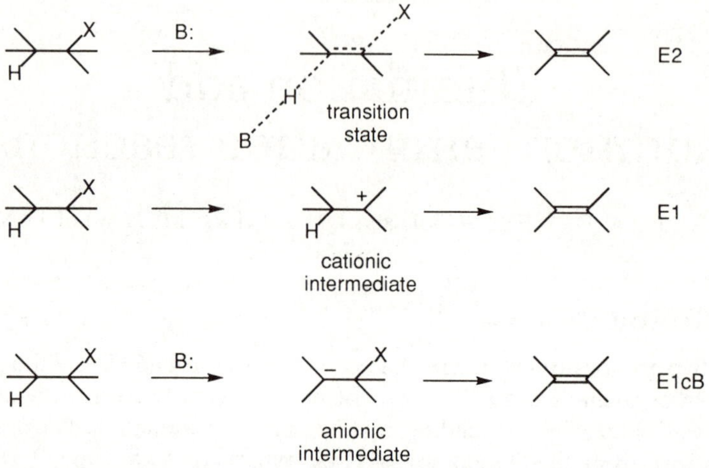

Scheme 5.2

The extent to which one elimination mechanism predominates over another is often difficult to predict; however, some general guidelines concerning the dominant mechanism exist. α-Alkyl and α-aryl groups increase the extent of E1 elimination by stabilising the carbocation nature of the transition state. However, β-aryl groups (and other electron-withdrawing groups in the β-position) shift the mechanism towards E1cB. The effect of an external base shifts the mechanism towards E2, since in the E1 mechanism an external base is generally not required. Stronger bases tend to shift the mechanism towards E1cB. The E1 mechanism is favoured by substrates containing better leaving groups.

Where elimination of H–X can lead to two or more different alkenes, some general rules can be applied in order to predict the major isomer in each case.

For eliminations proceeding via E1 mechanisms the leaving group is gone before the choice of which alkene predominates is made and the crucial factor is the relative stability of the various possible products. In such cases Zaitsev's rule applies, which states that the double bond goes mainly towards the most highly substituted carbon (Scheme 5.3). For the E2 mechanism an *anti* β-proton is necessary. If *anti* β-protons are present on two or three carbons, the orientation of the resulting double bond is dependent on substrate structure and the nature of the leaving group. Compounds containing uncharged leaving groups usually follow Zaitsev's rule, as for E1 eliminations. However, acyclic substrates containing charged leaving groups, e.g. NR_3^+, usually follow Hofmann's rule which states that the double bond formed goes mainly towards the least substituted carbon (Scheme 5.3).

If there is a double bond (C–C or C–O) already present in the substrate, the conjugated product usually predominates. This is the case for alkenes formed by nucleophilic addition of enolates to carbonyl species followed by dehydration, as discussed in Section 4 of this chapter.

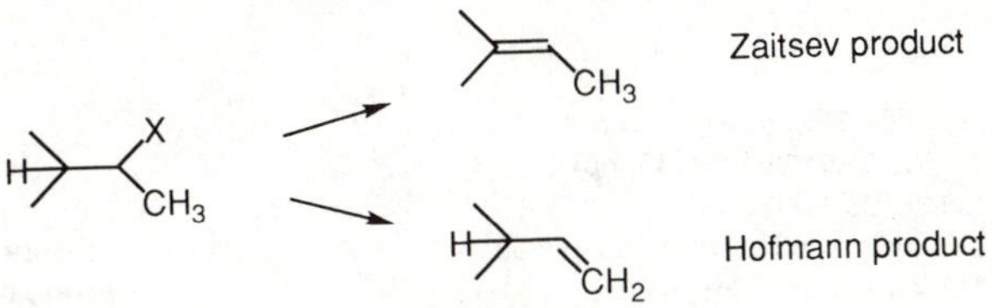

Zaitsev product

Hofmann product

Scheme 5.3

2. Elimination of H–X to give alkenes

2.1 Alkenes from alkyl halides

The formation of alkenes by the elimination of a hydrogen halide is well-established methodology. A variety of bases have been used to carry out this transformation; typical examples include DBN,[1] DBU,[2] LiF/Li_2CO_3/HMPA,[3] KO^{t}Bu,[4] and NEt_3.[5]

A common application of this methodology involves the synthesis of α,β-unsaturated carbonyl compounds by a bromination–dehydrobromination sequence; an approach exemplified by the synthesis of γ-crotonolactone from γ-butyrolactone (Protocol 1).[5]

Protocol 1.
Synthesis of γ-crotonolactone 3[5]

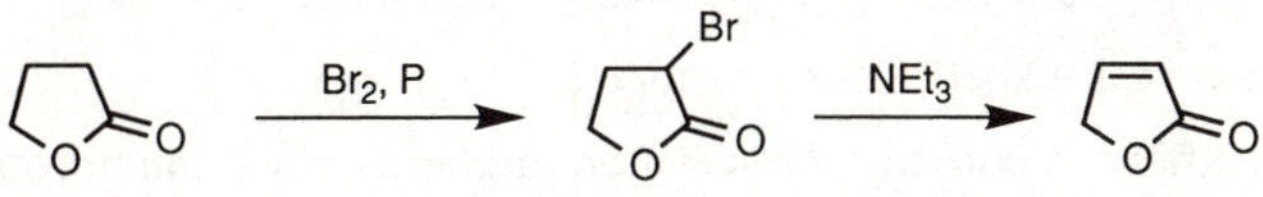

Br_2, P; NEt_3

1 2 3

Caution! All procedures should be carried out in a well-ventilated hood, and disposable vinyl or latex gloves and chemical-resistant safety goggles should be worn.

Equipment

- Magnetic stirrer/hotplate
- Ice/water bath
- Oil bath
- Three-necked, round-bottomed flask (1 L)

Protocol 1. *Continued*

- Three-necked, round-bottomed flask (500 mL)
- Water-jacketed reflux condenser
- Dry-ice/acetone trap for bromine–hydrogen bromide vapours
- Septum
- Pressure-equalising dropping funnel (250 mL)
- Magnetic stirrer bar
- Calcium chloride drying tubes
- Tubing adapter
- Source of dry and oxygen-free nitrogen
- Source of compressed air
- Glass inlet tube and inlet tube adapter

Materials

- γ-Butyrolactone[a] **1** (FW 86.1) 100 g, 1.16 mol — **harmful, irritant**
- Red phosphorus (FW 31.0) 13.4 g, 0.43 mol — **flammable, poison**
- Bromine (FW 159.8) 133.0 ml, 2.44 mol — **corrosive, poison**
- α-Bromo-γ-butyrolactone **2** (FW 165.0) 83 g, 0.50 mol — **severe irritant, vesicant**
- Triethylamine[b] (FW 101.2) 84.5 mL, 0.60 mol — **flammable, toxic, irritant**
- Dry ether 270 mL — **highly flammable, irritant**
- γ-Crotonolactone (product) — **irritant**

(A) Preparation of α-bromo-γ-butyrolactone **2**

1. Clean all of the glassware and the magnetic stirrer bar and dry in a 120°C electric oven for at least 4 h before use.
2. Assemble the three-necked flask (1 L), stirrer bar, condenser with drying tube, and pressure-equalising dropping funnel whilst still hot and allow to cool under nitrogen (by attaching the tubing adapter directly to the three-necked flask).
3. Add redistilled γ-butyrolactone (100 g, 1.16 mol) and red phosphorus (13.4 g, 0.43 mol) to the flask, and replace the tubing adapter with a glass stopper.
4. Add bromine (66.5 mL, 1.22 mol) to the pressure-equalising dropping funnel.
5. Cool the flask by means of an ice/water bath and add the bromine slowly over a 30 min period with stirring.
6. Heat the mixture to 70°C (oil bath) and add additional bromine (66.5 mL, 1.22 mol) via the dropping funnel over a 30 min period.
7. Heat the mixture at 80°C for a further 3 h.
8. Allow the flask to cool to room temperature and detach the dropping funnel and condenser. Then blow air into the reaction mixture for 1 h to remove excess bromine and hydrogen bromide (using a cold trap to catch the vapours). The air should be delivered from a glass inlet tube immersed under the surface of the reaction mixture, and attached through an inlet tube adapter (which is the same as a thermometer adapter). Stopper one of the necks of the round-bottomed flask and to the other attach a tubing adapter with a tube connected to a round-bottomed flask immersed in a dry-ice/acetone bath and equipped with a dry-ice/acetone condenser (to condense the bromine and hydrogen bromide vapours).

9. Re-attach the dropping funnel and reflux condenser and heat the mixture to 80 °C then cautiously add water (25 mL) via the dropping funnel with stirring (**caution**! vigorous reaction). Upon cessation of the reaction add additional water (300 mL).
10. Heat the reaction mixture (two layers and some solid residue) under reflux for 4 h then allow to cool.
11. Extract the product using ether (2 × 200 mL) and dry ($MgSO_4$). Filter the solution and remove the solvent by means of a rotary evaporator (35°C/15 mmHg).
12. Distil the crude product under reduced pressure to obtain α-bromo-γ-butyrolactone **2** (b.p. 107–109 °C/13 mmHg, 105 g, 55% yield) as a colourless oil which displays the appropriate spectroscopic data.

(B) Preparation of γ-crotonolactone **3**

1. Clean all of the glassware and the magnetic stirrer bar and dry in a 120 °C electric oven for at least 4 h before use.
2. Assemble the three-necked flask (500 mL), stirrer bar, condenser with drying tube, and dropping funnel whilst still hot and allow to cool under nitrogen (by attaching the tubing adapter directly to the three-necked flask).
3. Place α-bromo-γ-butyrolactone (83 g, 0.50 mol) and dry diethyl ether (200 mL) in the flask, replace the tubing adapter with a glass stopper and heat to reflux with stirring.
4. Add a solution of distilled triethylamine (84.5 mL, 0.60 mol) in dry ether (70 ml) slowly via the dropping funnel to the solution over 5 h with stirring, then continue to maintain reflux for a further 24 h.
5. Allow the mixture to cool and then remove the brown precipitate (40 g) by suction filtration (the precipitate is predominantly triethylamine hydrobromide).
6. Remove the solvent from the filtrate by means of a rotary evaporator (35 °C/15 mmHg) and distil the liquid residue under reduced pressure to obtain γ-crotonolactone **3** (b.p. 107–109 °C/24 mmHg, 25 g, 60% yield) as a colourless oil which displays the appropriate spectroscopic data.

[a]Distil γ-butyrolactone (b.p. 204–206 °C) before use.
[b]Triethylamine should be dried over KOH pellets then distilled (b.p. 89–90 °C) before use.

2.2 Alkenes from alcohols (directly)

There are numerous reagents available for the dehydration of alcohols. As with other eliminations, tertiary alcohols dehydrate most easily and elimination usually gives the Zaitsev alkene product. Listed below are selected reagents which have been used on representative substrates, often under mild conditions.

Entry	Conversion	Reagent	Ref.
1		$KHSO_4$, heat, 78% yield	6
2		DMSO, 160°C, 81% yield	7
3		$CuSO_4$ on silica gel, 83% yield	8
4		$Cu(OTf)_2$, 71% yield	9
5	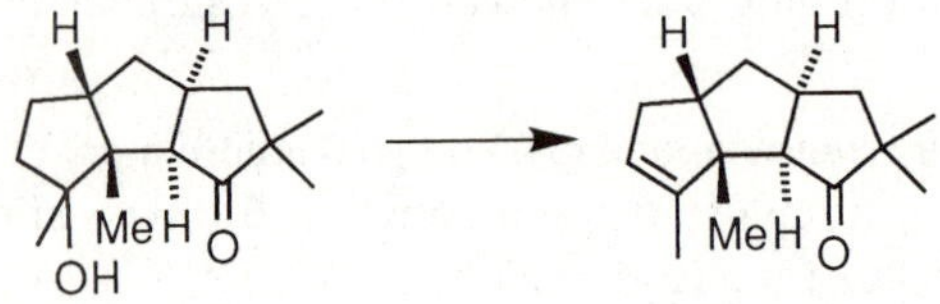	$POCl_3$-py, 75% yield	10
6	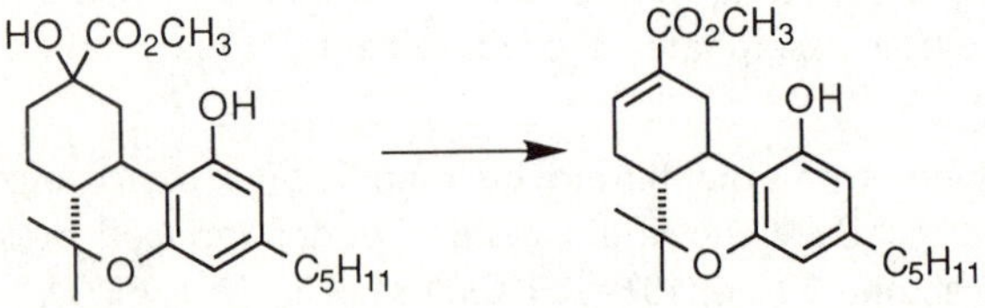	$SOCl_2$, py, 73% yield	11
7	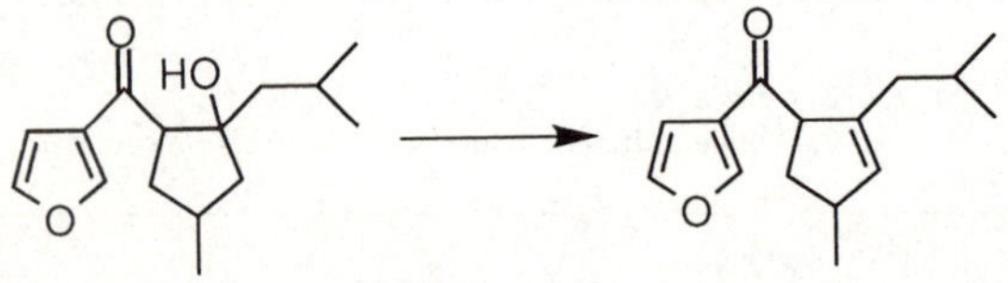	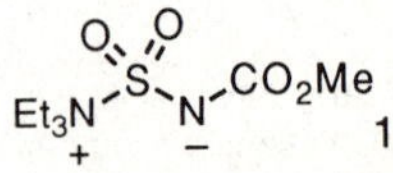 $Et_3\overset{+}{N}SO_2\overset{-}{N}CO_2Me$, 50% yield	12

Many alcohol dehydration methods convert the alcohol into a better leaving group *in situ* prior to elimination. However, it is often more convenient to make an isolable intermediate and effect elimination later. This approach is discussed in the next section.

2.3 Alkenes from alcohols (indirectly)

Conversion of the hydroxyl group to better leaving groups such as tosylate[13,14] or mesylate[15] is a commonly employed strategy in the synthesis of alkenes. This methodology is demonstrated below for the synthesis of the cyclohexene **6** from cyclohexanol **4** via the tosylate **5** in two steps (Protocol 2).[14]

Protocol 2.
Synthesis of cyclohexene 6[14]

OH (4) —TsCl, pyridine→ OTs (5) —^{t}BuOK, DMSO→ (6)

Caution! All procedures should be carried out in a well-ventilated hood, and disposable vinyl or latex gloves and chemical-resistant goggles should be worn.

Equipment

- Magnetic stirrer
- Ice/water bath
- Two-necked, round-bottomed flask (500 mL)
- Septum
- Thermometer (−10 to 110°C)
- Thermometer adapter
- Magnetic stirrer bar
- Three-necked, round-bottomed flask (1 L)
- Pressure-equalising dropping funnel (500 mL)
- Water-jacketed reflux condenser
- Tubing adapter
- Calcium chloride drying tube
- Source of dry nitrogen

Materials.

• Cyclohexanol[a] **4** (FW 100.16) 45.1 g, 0.45 mol	**harmful, irritant**
• *p*-Toluenesulfonyl chloride[b] (FW 190.65) 89.8 g, 0.47 mol	**corrosive, moisture sensitive**
• Pyridine[c] (FW 79.10) 200 mL, 2.48 mol	**flammable, irritant**
• Hydrochloric acid (concentrated) 260 mL	**corrosive, irritant**
• Ether (900 mL)	**highly flammable, irritant**
• Cyclohexyl tosylate **5** (FW 254.56) 104.4 g, 0.41 mol	**unknown, assume toxic**
• Potassium *tert*-butoxide (FW 112.2) 92.0 g, 0.82 mol	**toxic, irritant**
• Dimethyl sulfoxide[d] 450 mL	**teratogen, toxic, irritant**
• Petroleum ether (40–60) 600 mL	**flammable, irritant**
• Cyclohexene (product)	**flammable, harmful**

(A) Preparation of cyclohexyl tosylate **5**

1. Clean all of the glassware and the magnetic stirrer bar and dry in a 120°C electric oven for at least 4 h before use.

Protocol 2. *Continued*

2. Assemble the two-necked flask (500 mL) and stirrer bar whilst still hot and allow to cool under nitrogen, by attaching the tubing adapter directly to the flask.
3. Fit the thermometer and charge the flask with pyridine (200 mL, 2.48 mol) and cyclohexanol (45.1 g, 0.45 mol) with stirring.
4. Cool the flask to 0°C and add the *p*-toluenesulfonyl chloride (89.8 g, 0.47 mol) in portions over a 20 min period (briefly remove the tubing adapter to allow for each addition).
5. Allow the mixture to warm to room temperature then stir for a further 16 h.
6. Re-cool the solution to 0°C and pour into a mixture of concentrated hydrochloric acid (260 mL) in ice/water (800 mL) in a large conical flask.
7. Extract the mixture using ether (3 × 300 mL) and dry the organic extracts ($MgSO_4$). Filter, and remove the solvent by means of a rotary evaporator (35°C/15 mmHg). Apply high vacuum (0.03 mmHg) to the residue for 3 h in order to remove volatiles, leaving cyclohexyl tosylate **5** (104.4 g, 0.41 mol, 92% yield) as a pale yellow oil.

(B) Preparation of cyclohexene **6**

1. Clean all glassware and the magnetic stirrer bar and dry in a 120°C electric oven for at least 4 h before use.
2. Assemble the three-necked flask (1 L), dropping funnel, condenser, calcium chloride drying tube, and stirrer bar whilst still hot and allow to cool under nitrogen.
3. Charge the flask with potassium *tert*-butoxide (92.0 g, 0.82 mol) and dry DMSO (100 mL) with stirring.
4. Place a solution of cyclohexyl tosylate **5** (104.4 g, 0.41 mol) in dry DMSO (350 mL) in the dropping funnel.
5. Stir the potassium *tert*-butoxide suspension vigorously and add the cyclohexyl tosylate solution dropwise over a period of 10 min, using an ice bath to maintain a reaction temperature of 20–25°C.
6. Stir the reaction mixture at room temperature for a further 30 min.
7. Attach a short-path distillation apparatus connected to a receiver flask which is immersed as fully as possible in a dry-ice/acetone bath. Ideally, the exit from the receiver flask should be connected to another receiver flask also immersed in a dry-ice/acetone bath.
8. Distil the cyclohexene from the reaction mixture at room temperature by attaching a vacuum (*c.* 10–15 mmHg). Do not heat. The temperature in the still-head should not exceed room temperature.

9. The crude distillate may contain small quantities of DMSO and *tert*-butanol, and hence a normal pressure distillation is recommended for further purification (b.p. 83–83 °C) 22.4g, 0.27 mol, 66% yield.

[a]Distil cyclohexanol (b.p. 160–161 °C) before use.
[b]Recrystallise *p*-toluenesulfonyl chloride from hexane before use.
[c]Distil pyridine (b.p. 114–115 °C) and store over potassium hydroxide.
[d]Distil dimethyl sulfoxide (b.p. 189 °C) from calcium hydride.

The conversion of mainly secondary and tertiary alcohols into the corresponding xanthate followed by pyrolytic *syn*-elimination[16,17] (the Chugaev reaction) is a well-established technique for the synthesis of alkenes, although regioselectivity can be a problem in the elimination step due to the rather harsh conditions employed. The synthesis of (*E*)-2,2-dimethyl-3-hexene **9** from 2,2-dimethyl-3-hexanol **7** via the intermediate **8** using this method is illustrated below.[16] The xanthate intermediate need not be isolated.

OH

(i) NaH, ether reflux

(ii) CS_2, MeI

O SMe

S

(iii) 200°C

7 **8** **9**

82% yield

2.4 The Hofmann elimination

The Hofmann elimination reaction is a classical alkene synthesis. In this process a primary, secondary, or tertiary amine is treated with enough methyl iodide to convert it into the quaternary ammonium iodide, then the iodide is converted into the corresponding hydroxide salt and heated to effect elimination and the formation of an alkene. The reaction is not, however, an important synthetic tool. On the question of regioselectivity, the Hofmann rule applies when there is a choice of elimination products.

The synthesis of diphenylmethylvinyl ether **13** from 2-(diphenylmethoxy)-*N,N*-dimethylethylamine **10** via the intermediates **11** and **12** illustrates the application of this method.[18]

2.5 The Cope reaction

An alternative to the Hofmann reaction is the Cope reaction, which involves the cleavage of amine oxides, usually formed *in situ* from the corresponding amine. The reaction conditions are quite mild and there are few side-reactions. The reaction is thus useful for the preparation of many alkenes, although one notable exception is that the reaction fails for six-membered rings containing

$Ph_2CH{-}O{-}CH_2CH_2{-}N(CH_3)_2$ **10** $\xrightarrow{MeI}$ $Ph_2CH{-}O{-}CH_2CH_2{-}N^+(CH_3)_3I^-$ **11** $\xrightarrow{\text{Anion-exchange resin, methanol}}$ $Ph_2CH{-}O{-}CH_2CH_2{-}N^+(CH_3)_3OH^-$ **12** $\xrightarrow{\text{heat}}$ $Ph_2CH{-}O{-}CH{=}CH_2$ **13**

nitrogen. An example is provided by the synthesis of methylenecyclohexane **16** from *N,N*-dimethylcyclohexanemethylamine **14** via the *N*-oxide **15**.[19]

Protocol 3.
Synthesis of methylenecyclohexane 16[19]

$C_6H_{11}CH_2NMe_2$ **14** $\xrightarrow{H_2O_2}$ $C_6H_{11}CH_2N^+Me_2{-}O^-$ **15** $\xrightarrow{\text{heat}}$ methylenecyclohexane **16**

Caution! All procedures should be carried out in a well-ventilated hood, and disposable vinyl or latex gloves and chemical-resistant safety goggles should be worn.

Equipment

- Erlenmeyer flask (500 mL)
- Watch glass
- Magnetic stirrer/hotplate
- Oil bath
- Single-necked, round-bottomed flask (500 mL)
- Magnetic stirrer bar
- Glass distillation column (20 cm)
- Trap cooled in dry-ice/acetone
- Pipette (50 mL)

Materials

- *N,N*-Dimethylcyclohexanemethylamine **14** (FW 141.3) 49.4 g, 0.35 mol, — **corrosive, flammable**
- 30% Hydrogen peroxide 118.5 mL, 1.05 mol — **oxidiser, irritant, explosion hazard**
- Methanol 45 mL — **flammable, toxic, irritant**
- Platinum black — **tumorigen**
- 10% Hydrochloric acid solution 10 mL — **corrosive, irritant**
- Saturated sodium bicarbonate solution 5 mL — **hygroscopic, irritant**
- Sodium (small piece) — **flammable when exposed to heat or moisture**
- Methylenecyclohexane **16** (product) — **highly flammable**

1. To the carefully cleaned Erlenmeyer flask, covered with a watch glass, add *N,N*-dimethylcyclohexanemethylamine **14** (49.4 g, 0.35 mol), 30% hydrogen

peroxide (39.5 mL, 0.35 mol), and methanol (45 mL). After swirling, allow the homogeneous solution to stand at room temperature.

2. After 2 h and 5 h add further 30% hydrogen peroxide (39.5 mL, 0.35 mol portions each time) and swirl. Allow the solution to stand at room temperature for a total of 36 h.
3. Destroy the excess hydrogen peroxide by stirring the mixture with a small amount of platinum black (aqueous suspension) until the evolution of gas ceases.
4. Filter the solution into the round-bottomed flask and concentrate by means of a rotary evaporator (50–60 °C/15 mmHg) initially and finally under high vacuum, until the amine oxide hydrate solidifies. (**Caution!** Peroxides are potentially explosive. Check for their absence with peroxide test paper.)
5. Add the stirrer bar to the flask then connect via a 20-cm distillation column to a trap cooled in dry-ice/acetone. This may be conveniently achieved by using a conventional distillation apparatus and immersing the receiver flask in a dry-ice/acetone bath.
6. Evacuate the apparatus to a pressure of *c.* 10 mmHg and heat the flask to 90–100 °C (oil bath) with stirring of the liquefied amine oxide hydrate.
7. When the contents of the flask re-solidify, raise the temperature of the oil bath to 160 °C and continue heating at this temperature for 2 h.
8. Add water (100 mL) to the contents of the trap (receiver flask).
9. Remove the alkene layer using a pipette and wash with water (2 × 5 mL), ice-cold 10% hydrochloric acid[a] (2 × 5 ml) and 5% sodium bicarbonate solution (5 mL).
10. Cool the alkene in a dry-ice/acetone bath and filter through glass wool.
11. Distil the crude product over a small piece of sodium to give methylenecyclohexane **16** (b.p. 100–102 °C, 30.7–32.7 g, 0.32–0.34 mol, 90–96% yield) as a clear colourless liquid which displays the appropriate spectroscopic data.

[a]Methylenecyclohexane does not rearrange to 1-methylcyclohexene under these conditions. In preparations of more acid-sensitive alkenes, washing with acid should be omitted.

2.6 Elimination of epoxides to give allylic alcohols

Epoxides can be converted into allylic alcohols upon treatment with strong bases such as lithium diethylamide. The reaction is quite general and often proceeds with good regio- and stereochemical control.[20] For example, the base-induced elimination reaction of 2,3-epoxy pinane **17** gives specifically *trans*-pinocarveol **18** (Protocol 4).[21]

Protocol 4.
Synthesis of *trans*-pinocarveol 18[21]

17 → ($LiNEt_2$, Et_2O, reflux) → 18

Caution! All procedures should be carried out in a well-ventilated hood, and disposable vinyl or latex gloves and chemical-resistant safety goggles should be worn.

Equipment

- Magnetic stirrer/hotplate
- Two-necked, round-bottomed flask (300 mL)
- Water-jacketed reflux condenser
- Pressure-equalising dropping funnel (50 mL)
- Tubing adapter
- Septum
- Magnetic stirrer bar
- Ice/water bath
- Glass syringes with needle-lock Luers
- Six-inch, medium-gauge needles
- Source of dry nitrogen
- Water-jacketed, semi-micro, distillation apparatus

Materials

• 2,3-Epoxy pinane[a] **17** (FW 152.2) 5.00 g, 0.033 mol	**flammable**
• Diethylamine[b] (FW 73.1) 2.40 g, 0.034 mol	**flammable, toxic, irritant**
• *n*-Butyllithium (FW 64.1) 1.4 M in hexanes, 25 mL, 0.035 mol	**toxic, flammable, air and moisture sensitive**
• Dry ether 220 mL	**flammable, irritant**
• Hydrochloric acid (1 M) (100 mL)	**corrosive, irritant**
• Saturated aqueous sodium bicarbonate 100 mL	**irritant**
• *trans*-Pinocarveol **18** (product)	**unknown, assume toxic**

1. Clean all of the glassware and the magnetic stirrer bar and dry in a 120°C electric oven for at least 4 h before use.
2. Assemble the three-necked flask, pressure-equalising dropping funnel, condenser, and stirrer bar whilst still hot and allow to cool under nitrogen by attaching the tubing adapter to the top of the condenser.
3. Flush the flask with nitrogen then charge with diethylamine (2.40 g, 0.034 mol) and dry ether (100 mL).
4. Immerse the flask in an ice/water bath then slowly add butyllithium (1.4 M in hexanes, 25 mL) by means of a syringe with stirring (**caution!** see Chapter 1, Protocol 1).
5. After stirring for 10 min remove the ice bath and add a solution of 2,3-epoxy pinane (5.00 g, 0.033 mol) in dry ether (20 mL) via the stoppered dropping funnel over a 10 min period.

6. Heat the resulting mixture to reflux for 6 h, then allow to cool.
7. Cool the mixture in an ice/water bath then add water (100 mL) whilst stirring vigorously.
8. Separate the organic phase and wash successively with 1 M hydrochloric acid (100 mL), water (100 mL), saturated aqueous sodium bicarbonate (100 mL), and water (100 mL).
9. Dry the organic phase ($MgSO_4$), filter the solution, and remove the solvent by means of a rotary evaporator (35 °C/15 mmHg).
11. Distil the liquid residue under reduced pressure through a short-path distillation apparatus to obtain *trans*-pinocarveol **18** (b.p. 92–93 °C/8 mmHg, 4.50–4.75 g, 0.029–0.031 mol, 90–95% yield) as a clear colourless oil which displays the appropriate spectroscopic data.

[a] Prepare 2,3-epoxy pinane by *m*CPBA oxidation of α-pinene.
[b] Distilled from calcium hydride (b.p. 55–58 °C) before use.

2.7 Elimination of selenoxides

Selenoxides[22] undergo elimination reactions, often at room temperature. The mildness of the reaction conditions means that this method is very popular for the synthesis of alkenes. The method has been used frequently as a means of converting ketones, aldehydes, and carboxylic esters to their α,β-unsaturated derivatives, as illustrated below by the conversion of cyclohexenone **19** to its α,β-unsaturated derivative **21**[23] by initial deprotonation followed by formation of the α-selenyl derivative **20** then oxidation to the corresponding selenoxide and elimination under mild conditions.

OTs; LDA, PhSeCl, THF, −78°C, 60% yield; PhSe; OTs; H_2O_2, THF, 76% yield; OTs; O; **19**; **20**; **21**

This method provides an alternative (especially on a reasonably small scale) to the preparation of α,β-unsaturated carbonyl compounds by α-bromination followed by dehydrobromination described earlier. An analogous elimination reaction can also be carried out using the corresponding sulfoxides[24] or sulfinate esters.[25]

2.8 Alkenes from toluene-*p*-sulfonylhydrazones (the Shapiro reaction)

The treatment of toluene-*p*-sulfonylhydrazones (derived from the corresponding aldehyde or ketone) with two or more equivalents of a strong base, such as methyllithium, results in decomposition of the hydrazone, yielding an alkene.[26] This procedure, known as the Shapiro reaction, usually gives good yields of alkenes without side-reactions and gives the less highly substituted alkene where a choice is possible. For example, treatment of the toluene-*p*-sulfonylhydrazone **22** (derived from camphor) with methyllithium in ether followed by aqueous work-up gives the 2-bornene **23** in quantitative yield.[23]

More recent advances in the synthetic utility of the Shapiro reaction have involved quenching the vinyllithium precursor to **23** with electrophiles other than the proton quench involved in formation of **23**. For example, using benzaldehyde as the electrophile affords the corresponding allylic alcohol.[28]

NNHTs — MeLi (2eq.), Et_2O, quantitative yield →

22 **23** (—E)

3. Elimination of X–X to give alkenes

3.1 Alkenes from vicinal dihalides

Numerous reagents have been used to effect dehalogenation, the most common being zinc in ethanol or acetic acid.[29] Although the reaction usually gives good yields, its synthetic utility is limited since the best way to prepare *vic*-dihalides is often by the addition of a halogen to a double bond! For example, debromination of *meso*-stilbene dibromide **24** by zinc in ethanol gives (*E*)-stilbene **25** in 97% yield (Protocol 5).[29] The elimination of OR and a halogen from β-halo ethers using zinc is also well known (the Boord reaction).[30]

Protocol 5.
Synthesis of (*E*)-stilbene 25

Ph Ph Br Br — Zn, EtOH, reflux → Ph Ph

24 **25**

Caution! All procedures should be carried out in a well-ventilated hood, and disposable vinyl or latex gloves and chemical-resistant safety goggles should be worn.

Equipment

- Magnetic stirrer/hotplate
- Oil bath
- One-necked, round-bottomed flask (100 mL)
- Anti-bumping granules
- Water-jacketed reflux condenser
- Thermometer (−10 to 110°C)

Materials

- *meso*-Stilbene dibromide **24**[a] (FW 340.1) 40 mg, 0.18 mmol — **lachrymator**
- Powdered zinc[b] (FW 65.4) 100 mg, 1.5 mmol — **flammable solid, dangerous when wet, hygroscopic**
- 95% Ethanol 60 mL — **flammable, irritant**
- Dichloromethane 30 mL — **toxic, irritant**
- (*E*)-Stilbene (product) **25**

1. Clean all of the glassware carefully before starting the experiment.
2. Add *meso*-stilbene dibromide **24** (40 mg, 0.18 mmol) to the round-bottomed flask and add 95% ethanol (50 mL) along with a few anti-bumping granules.
3. Add powdered zinc (100 mg)[b] to the flask and heat the reaction mixture under reflux for 15 h.
4. Allow the flask to cool then filter the solution under suction, using further 95% ethanol (10 mL) to ensure complete transfer of material from the flask. The solid residue should be destroyed in hydrochloric acid 1 M.
5. Concentrate the filtrate by means of a rotary evaporator (35 °C/15 mmHg) until most of the ethanol is removed.
6. Add brine (10 mL) to the filtrate then extract the aqueous layer using dichloromethane (3 × 10 mL).
7. Dry the organic phase ($MgSO_4$), filter the solution, and remove the solvent by means of a rotary evaporator (35 °C/15 mmHg) to obtain (*E*)-stilbene **25** (31 mg, 0.17 mmol, 97% yield) as a solid which displays the appropriate spectroscopic data.

[a] *meso*-1,2-dibromo-1,2-diphenylethane (Aldrich).

[b] Zinc can be activated by washing in a Büchner flask with hydrochloric acid 1 M (3 × 30 mL) then ethanol (30 mL) then ether (30 mL). The activated zinc is then dried *in vacuo* and used soon afterwards.

3.2 The Corey–Winter reaction

vic-Diols can be converted to cyclic thionocarbonates by treatment with thiophosgene and 4-dimethylaminopyridine (DMAP),[31] which can be converted into alkenes by heating with trivalent phosphorus compounds such as trimethyl phosphite[32] (the Corey–Winter reaction). This conversion is particularly useful for carbohydrates and related compounds and is illustrated by the conversion of the diol **26** to the corresponding 5,6-alkene **28** in 75% yield via the intermediate cyclic thionocarbonate **27**.[32]

HO, HO, O, O, O — **26** —($Cl_2C{=}S$, DMAP)→ S, O, O, O, O, O — **27** —($P(OMe)_3$, 150°C, 75% yield)→ O, O, O — **28**

4. Addition of enolates to carbonyl compounds followed by dehydration

4.1 The aldol reaction and subsequent dehydration

Some of the most widely used alkene-forming reactions involve processes related to the aldol condensation. The reaction of an enolate with a carbonyl compound affords a β-hydroxy carbonyl (aldol) adduct, which can undergo elimination to an alkene (Scheme 5.4). The elimination step often occurs *in situ*. The reaction is most commonly carried out under base catalysis although both basic and acidic catalysts have been used.

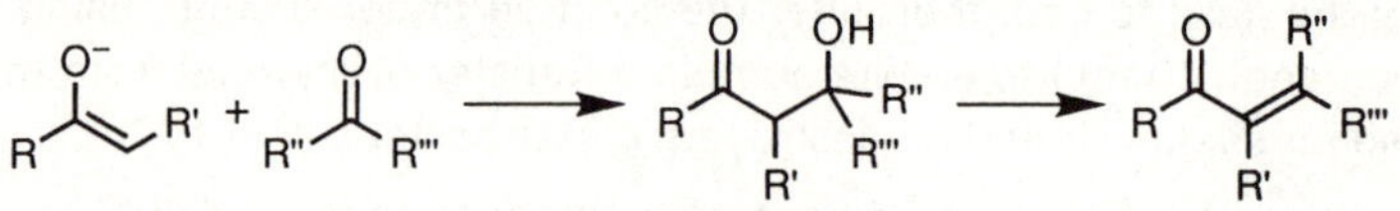

Scheme 5.4

A particularly well-known example of the aldol reaction–dehydration sequence is the synthesis of chalcone **29** from acetophenone and benzaldehyde in the presence of base.[33] Because benzaldehyde is the more electrophilic of the carbonyl compounds, and acetophenone is the only one which is enolisable, the aldol condensation proceeds smoothly to afford one product. However, when two aldehydes or two ketones are employed, a mixture of aldol products may be obtained.

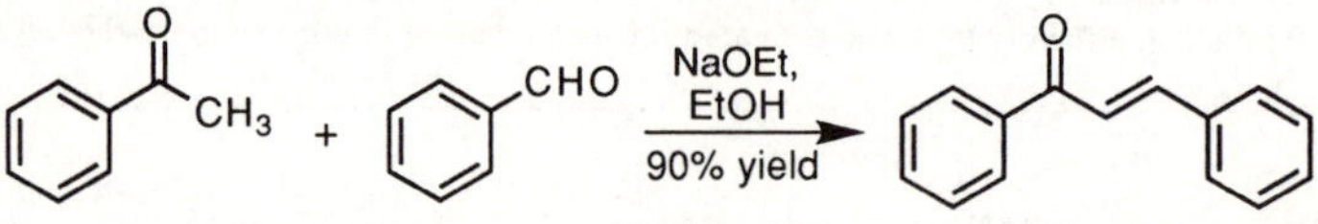

NaOEt, EtOH, 90% yield

29

A process closely related to the above is the synthesis of dibenzalacetone **30** from benzaldehyde and acetone (2:1 molar ratio) under base catalysis (the Claisen–Schmidt reaction) as described in Protocol 6.[34]

Protocol 6.
Synthesis of dibenzalacetone 30[34]

$$2PhCHO + CH_3COCH_3 \xrightarrow[H_2O\,/\,EtOH]{NaOH,} PhCH{=}CH{-}CO{-}CH{=}CHPh + H_2O$$

30

Caution! All procedures should be carried out in a well-ventilated hood, and disposable vinyl or latex gloves and chemical-resistant safety goggles should be worn.

Equipment

- Erlenmeyer flask (2 L)
- Beaker (500 mL)
- Mechanical stirrer
- Thermometer −10–110 °C
- Large cold-water bath (for Erlenmeyer flask)
- Large Büchner funnel

Materials

• Benzaldehyde[a] (FW 106.1) 101.2 mL, 1.0 mol	**combustible, irritant**
• Acetone (FW 58.1) 36.7 mL, 0.5 mol	**flammable, irritant**
• Sodium hydroxide (FW 40.0) 100.0 g, 2.5 mol	**corrosive, irritant**
• Ethanol *c.* 85 mL	**flammable**
• Distilled water *c.* 3 L	
• Ethyl acetate *c.* 275 mL	**flammable, irritant**
• Dibenzalacetone (product)	**unknown, assume toxic**

1. Clean all of the glassware carefully before starting the experiment.
2. Add a cooled solution (ice/water bath) of sodium hydroxide (100 g, 2.5 mol) dissolved in water (1 L) and ethanol (800 mL) to the Erlenmeyer flask surrounded by cold water and fitted with a mechanical stirrer.
3. Whilst keeping the solution at 20–25 °C and stirring vigorously, add one-half of a mixture of benzaldehyde (101.2 mL, 1.0 mol) and acetone (36.7 mL, 0.5 mol) slowly with stirring.
4. After about 2 min, yellow cloudiness is observed which soon becomes a flocculent precipitate.
5. After about 15 min add the rest of the mixed reagents, and rinse the beaker with a little ethanol which is added to the mixture.
6. Continue stirring for a further 30 min, then suction filter on a large Büchner funnel.
7. Wash the product thoroughly with water (4 × 500 mL) then separate the product and dry under vacuum at room temperature.
8. Recrystallise the crude product from hot ethyl acetate (using about 100 mL of solvent for each 40 g of material) to give dibenzalacetone (m.p. 110–111 °C, 88.0 g, 76% yield) as a white solid which displays the appropriate spectroscopic data.

[a]Distil benzaldehyde (b.p. 180–183 °C) before use.

The aldol condensation in an intramolecular sense followed by dehydration is often used to close five- and six-membered rings. The Robinson annulation[35] involves nucleophilic attack of an enolate on a vinyl ketone followed by aldol-type ring closure and subsequent dehydration. For example, treatment of the enolate of 2-methylcyclohexanone with methyl vinyl ketone followed by dehydration gives the cyclised product **31** in 63% overall yield.[36]

NaOEt, EtOH; aq. KOH

31

4.2 The Knoevanagel reaction

The condensation of aldehydes or ketones with compounds of the form Z–CH_2–Z′ or Z–CHR–Z′ is called the Knoevanagel reaction and is of wide scope and great synthetic value. Z and Z′ may be CHO, COR, COOR, CN, NO_2, SOR, SO_2R, or similar electron-withdrawing groups. In many cases dehydration usually follows the initial reaction giving alkenes with the appropriate electron-withdrawing Z and Z′ in the α-position. Often, the enolate precursor used is propanedioic acid (malonic acid). Reaction of malonic acid and benzaldehyde in the presence of base gives (*E*)-cinnamic acid **32** in 95% yield (Protocol 7).[37] In this reaction procedure, decarboxylation occurs to afford the mono-acid adduct, and this method thereby affords a useful procedure for the preparation of α,β-unsaturated carboxylic acids.

Protocol 7.
Synthesis of (*E*)-cinnamic acid 32

CHO + CO_2H CO_2H → pyridine, piperidine, reflux → CO_2H

32

Caution! All procedures should be carried out in a well-ventilated hood, and disposable vinyl or latex gloves and chemical-resistant safety goggles should be worn.

Equipment

- Oil bath
- One-necked, round-bottomed flask (100 mL)
- Water-jacketed reflux condenser
- Büchner flask (250 mL) and funnel

Materials

- Benzaldehyde[a] (FW 106.1) 3.0 mL, 29.5 mmol — **flammable, toxic**
- Malonic acid (FW 104.1) 3.10 g, 29.8 mmol — **irritant**
- Pyridine 5 mL — **flammable, irritant**
- Piperidine 0.1g — **corrosive, flammable**
- Light petroleum (b.p. 40–60 °C) 20 mL — **flammable**
- Hydrochloric acid (2 M) 70 mL — **corrosive**
- (*E*)-Cinnamic acid **32** (product) — **irritant**

1. Clean all of the glassware carefully before starting the experiment.

2. Dissolve malonic acid (3.10 g, 29.8 mmol) in pyridine (5 mL) in the round-bottomed flask equipped with condenser and warm gently to ensure dissolution.
3. Add benzaldehyde (3.0 mL, 29.5 mmol) dropwise to the flask followed by a catalytic quantity of piperidine (0.1 g).
4. Heat the reaction mixture to 100 °C. Reaction is indicated by the evolution of bubbles (CO_2).
5. Continue heating until the reaction becomes very slow (*c.* 30 min).
6. Allow the flask to cool then add hydrochloric acid 2 M (50 mL) and isolate the resultant solid by suction in a Büchner funnel.
7. Triturate the solid on the funnel with sequentially 2 M hydrochloric acid (20 mL), water (20 mL), and light petroleum (b.p. 40–60 °C) (20 mL).
8. Dry the crystals in an 80 °C oven to give (*E*)-cinnamic acid **32** (4.15 g, 28.0 mmol, 95% yield) as a white solid.

[a] Distil benzaldehyde (b.p. 180–183 °C) before use.

Another example of the Knoevanagel reaction is provided by the reaction of barbituric acid **33** and benzaldehyde to give benzalbarbituric acid **34** (Protocol 8).[38] In this instance there is no decarboxylation, and the product formed is therefore a trisubstituted alkene.

Protocol 8.
Synthesis of benzalbarbituric acid 34[38]

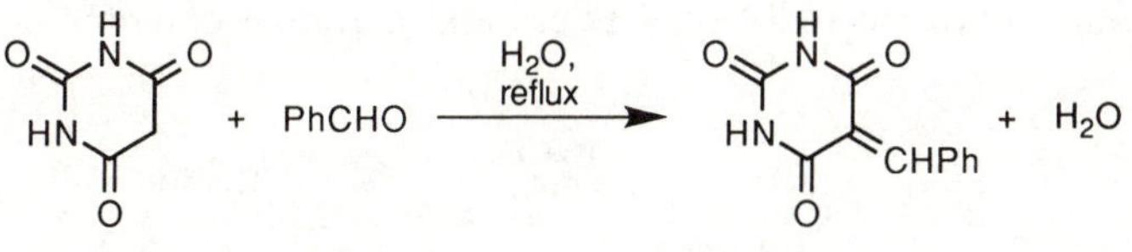

33 + PhCHO —(H_2O, reflux)→ **34** (CHPh) + H_2O

Caution! All procedures should be carried out in a well-ventilated hood, and disposable vinyl or latex gloves and chemical-resistant safety goggles should be worn.

Equipment

- Three-necked, round-bottomed flask (2 L)
- Water-jacketed reflux condenser
- Addition funnel (250 mL)
- Mechanical stirrer
- Oil bath
- Large Büchner funnel

Protocol 8. *Continued*

Materials

- Barbituric acid **33** (FW 128.1) 128 g, 1.0 mol — **irritant**
- Benzaldehyde[a] (FW 106.1) 110 mL, 1.08 mol — **flammable, toxic**
- Distilled water 1250 mL
- Hot water *c.* 500 mL
- Benzalbarbituric acid **34** (product) — **unknown, assume toxic**

1. Clean all of the glassware carefully before starting the experiment.
2. Add barbituric acid **33** (128 g, 1 mol) and distilled water (1250 mL) to the three-necked, round-bottomed flask equipped with a mechanical stirrer, reflux condenser, and addition funnel.
3. Heat the flask to 100 °C with stirring until the acid has dissolved then slowly add benzaldehyde (110 mL, 1.08 mol) via the addition funnel while heating and stirring are continued.
4. Heat the mixture at 100 °C for a further 1 h, then allow to cool.
5. Collect the solid in the Büchner funnel, wash with hot water (5 × 100 mL), and dry at 100 °C to give benzalbarbituric acid **34** (m.p. 254–256 °C, 205 g, 0.95 mol, 95% yield) as a very pale yellow solid which requires no further purification and displays the appropriate spectroscopic data.

[a]Distil benzaldehyde (b.p. 180–183 °C) before use.

4.3 The Stobbe condensation

The condensation of the anion derived from treatment of diethyl succinate with base and aldehydes or ketones followed by elimination to give an alkene is known as the Stobbe condensation,[39] and is a special case of the Claisen condensation between carboxylic esters and carbonyl compounds (Scheme 5.5).

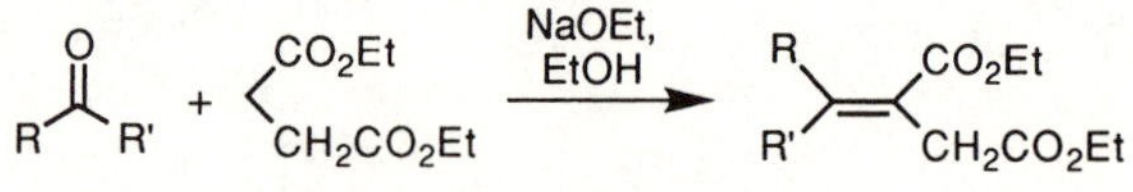

$$RC(O)R' + CH_2(CO_2Et)CH_2CO_2Et \xrightarrow{NaOEt,\ EtOH} RR'C{=}C(CO_2Et)CH_2CO_2Et$$

Scheme 5.5

4.4 The Perkin reaction

The Perkin reaction has frequently been used in the preparation of α,β-unsaturated carboxylic acids and derivatives and involves the condensation of aromatic aldehydes with acid anhydrides (Scheme 5.6).[40] Subsequent dehydration always occurs provided that the anhydride has two α-hydrogens. The base most frequently used in the reaction is the salt of the acid corresponding to the anhydride.

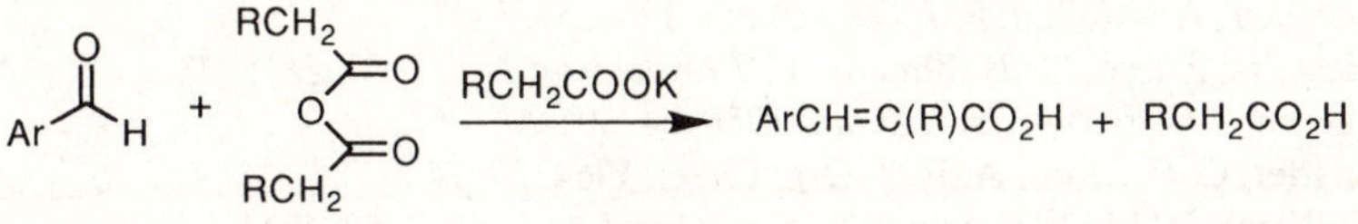

$$ArCHO + (RCH_2CO)_2O \xrightarrow{RCH_2COOK} ArCH{=}C(R)CO_2H + RCH_2CO_2H$$

Scheme 5.6

4.5 The Henry reaction

The use of nitroalkanes in aldol-type condensations to give β-hydroxy-nitro compounds which can then undergo dehydration to nitroalkenes is known as the Henry reaction (Scheme 5.7).[41]

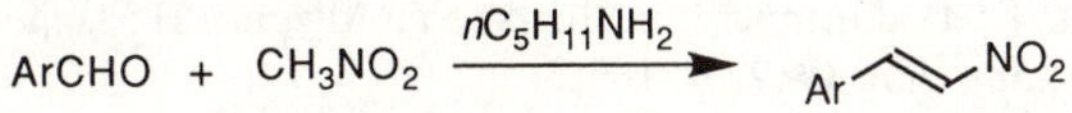

$$ArCHO + CH_3NO_2 \xrightarrow{nC_5H_{11}NH_2} ArCH{=}CHNO_2$$

Scheme 5.7

4.6 The Reformatsky reaction

The Reformatsky reaction involves the treatment of an aldehyde or ketone with zinc and a halide such as an α-halo ester to give a β-hydroxy ester. More recently, it has been found that in the presence of tri-*n*-butylphosphine, aldehydes such as **35** react with methyl bromoacetate and zinc to give (*E*)-α,β-unsaturated ester **36**.[42] A similar method for the formation of alkenes using potassium carbonate and catalytic amounts of dibutyl telluride regenerated by triphenyl phosphite has been reported.[43]

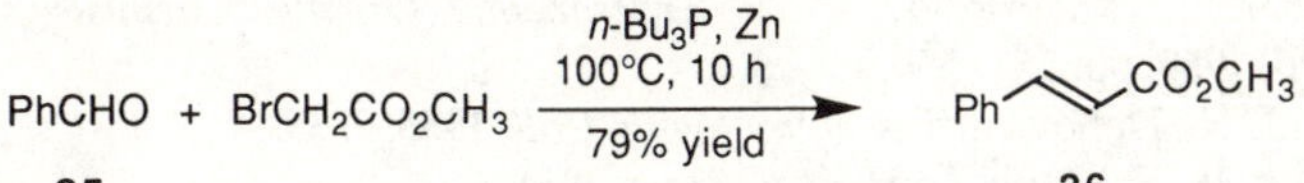

$$\underset{\mathbf{35}}{PhCHO} + BrCH_2CO_2CH_3 \xrightarrow[\text{79\% yield}]{n\text{-}Bu_3P,\ Zn,\ 100^{\circ}C,\ 10\ h} \underset{\mathbf{36}}{PhCH{=}CHCO_2CH_3}$$

References

1. Tu, C.-Y. J.; Lednicer, D. *J. Org. Chem.* **1987**, *52*, 5624.
2. Jeropoulos, S.; Smith, E. H. *J. Chem. Soc., Chem. Commun.* **1986,** 1621.
3. Dubey, S. K.; Kumar, S. *J. Org. Chem.* **1986**, *51*, 3407.
4. Tanida, H.; Irie, T. J. *Org. Chem.* **1987**, *52*, 5218.
5. Price, C. C.; Judge, J. M. *Org. Synth. Coll.* **1973**, *5*, 255.
6. Parrinello, G.; Stille, J. K. *J. Am. Chem. Soc.* **1987**, *109*, 7122.
7. Ichihara, A.; Miki, M.; Tazaki, H.; Sakamura, S. *Tetrahedron Lett.* **1987**, *28*, 1175.
8. Nishiguchi, T.; Machida, N.; Yamamoto, E. *Tetrahedron Lett.* **1987**, *28*, 4565.
9. Laali, K.; Gerzina, R. J.; Flajnik, C. M.; Geric, C. M.; Dombroski, A. M. *Helv. Chim. Acta* **1987**, *70*, 607.
10. Mehta, G.; Murthy, A. N.; Reddy, D. S.; Reddy, A. V. *J. Am. Chem. Soc.* **1986**, *108*, 3443.

11. Schwarz, A.; Madan, P. *J. Org. Chem.* **1986**, *51*, 5463.
12. Hess, T.; Zdero, C.; Bohlmann, F. *Tetrahedron Lett.* **1987**, *28*, 5643.
13. Salaun, J.; Fabel, A. *Org. Synth.* **1986**, *64*, 50.
14. Snyder, C. H.; Soto, A. R. *J. Org. Chem.* **1964**, *29*, 742.
15. Williams, R. M.; Maruyama, L. K. *J. Org. Chem.* **1987**, *52*, 4044.
16. Nace, H. R. *Org. React.* **1962**, *12*, 57.
17. Rastetter, W. H.; Nummy, L. J. *J. Org. Chem.* **1980**, *45*, 3149.
18. Kaiser, C.; Weinstock, *J. Org. Synth.* **1976**, *55*, 3.
19. Cope, A. C.; Ciganek, E. *Org. Synth. Coll.* **1963**, *4*, 612.
20. Crandall, J. K.; Apparu, M. *Org. React.* **1983**, *29*, 345.
21. Crandall, J. K.; Crawley, L. C. *Org. Synth.* **1973**, *53*, 17.
22. Reich, H. J.; Wollowite, S. *Org. React.* **1993**, *44*, 1.
23. Foster, S. J.; Rees, C. W. *J. Chem. Soc., Perkin Trans.* **1985**, 719.
24. Yoshimura, T.; Tsukurimichi, E.; Iizuka, Y.; Mizuno, H.; Isaji, H.; Shimasaki, C. *Bull. Chem. Soc. Jpn.* **1989**, *62*, 1891.
25. Jones, D. N.; Higgins, W. *J. Chem. Soc. C* **1970**, 81.
26. Shapiro, R. H. *Org. React.* **1976**, *23*, 405.
27. Shapiro, R. H.; Heath, M. J. *J. Am. Chem. Soc.* **1967**, *89*, 5734.
28. Adlington, R. M.; Barrett, A. G. M. *Acc. Chem. Res.* **1983**, *16*, 55.
29. Buckles, R. E.; Bader, J. M.; Thurmaier, R. J. *J. Org. Chem.* **1962**, *27*, 4523.
30. Grummitt, O.; Budewitz, E. P.; Chudd, C. C. *Org. Synth. Coll.* **1963**, *4*, 748.
31. Corey, E. J.; Hopkins, P. B. *Tetrahedron Lett.* **1982**, *23*, 1979.
32. Horton, D.; Tindall, Jr, C. G. *J. Org. Chem.* **1970**, *35*, 3558.
33. Heathcock, C. H. In *Comprehensive Organic Synthesis*; Trost, B. M.; Fleming, I., eds; Pergamon Press: New York, **1991**; Vol. 2, p. 150.
34. Conard, C. R.; Dolliver, M. A. *Org. Synth. Coll.* **1943**, *2*, 167.
35. Gawley, R. E. *Synthesis* **1976**, 777.
36. Marshall, J. A.; Fanta, W. I. *J. Org. Chem.* **1964**, *29*, 2501.
37. Harwood, L. M.; Moody, C. J. *Experimental Organic Chemistry*; Blackwell: Oxford, **1989**, pp. 557–559.
38. Speer, J. H.; Dabovich, T. C. *Org. Synth. Coll.* **1955**, *3*, 39.
39. Johnson, W. S.; Daub, G. H. *Org. React.* **1951**, *6*, 1.
40. Johnson, J. R. *Org. React.* **1942**, *1*, 210.
41. Baer, H. H.; Urbas, L. In *The Chemistry of the Nitro and Nitroso Groups*; Feuer, H., ed.; Wiley Interscience: New York, **1970**, Vol. 2, p. 150.
42. Shen, Y.; Xin, Y.; Zhao, J. *Tetrahedron Lett.* **1988**, *29*, 6119.
43. Huang, Y.-Z.; Shi, L.-L.; Li, S.-W.; Wen, X.-Q. *J. Chem. Soc. Perkin Trans. 1*, **1989**, 2397.

6

Reduction of alkynes

JOSHUA HOWARTH

1. Introduction

The conversion of alkynes into alkenes, by addition of two hydrogen atoms across the carbon–carbon triple bond, is invaluable in organic synthesis and has seen widespread use. There are several methods of performing this addition. The most popular method is partial hydrogenation of the triple bond by either heterogeneous or homogeneous catalysis which, in general, affords predominantly (*Z*)-alkenes. Another important procedure is the use of hydride reagents. In particular, the reduction of alkynes containing an α-hydroxyl group by aluminium hydrides, which produces (*E*)-alkenes, and the use of hydroboration followed by hydrolysis merit discussion (see also Chapter 7). Alkynes can also be reduced using dissolving metal reactions which have an essential role to play in the production of (*E*)-alkenes. The final category consists of a few miscellaneous methods, of which diimide reductions are the most important.

2. Reduction of alkynes with heterogeneous catalysts

The catalysts used for the semihydrogenation of the carbon–carbon triple bond are usually based on palladium or nickel on a support such as a polymer or calcium carbonate, although other metals, such as rhodium and iron, have also been used.[1,2]

Of the catalysts in this category the most famous and widely used is the Lindlar catalyst and variations of it.[3–5] In the catalyst's original form, palladium deposited on $CaCO_3$ was treated with $Pb(OAc)_2$ to lower its activity by affecting the morphology of the metal. Lindlar also suggested that quinoline should be used with his catalyst.[6] Kinetic studies have been carried out in order to determine the function of the quinoline,[7,8] and these suggest that the quinoline is adsorbed reversibly on the reactive sites, in competition with both the alkyne and product alkene. The critical conclusion is that the quinoline competes more effectively with the alkene. As the concentration of the alkyne diminishes, the quinoline occupies more of the surface than the alkene, but the amount of adsorbed alkene remains too low to affect the

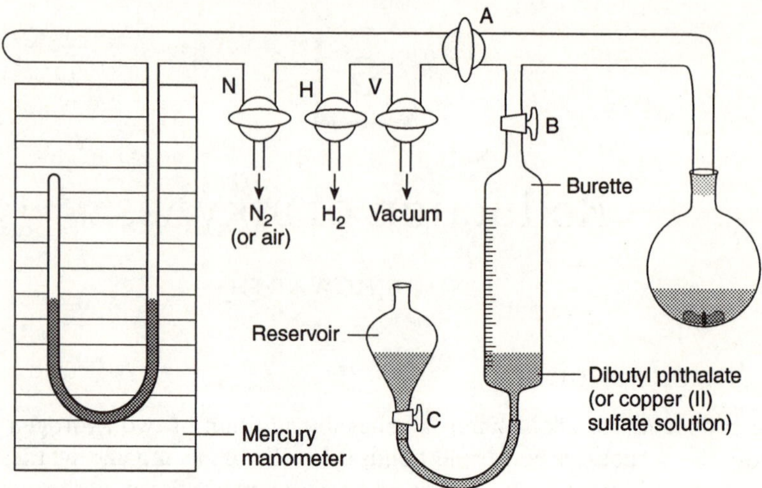

Fig. 6.1 An atmospheric hydrogenation apparatus.

composition of the product. Quinoline is unaffected by the usual mild experimental conditions and so its concentration remains constant throughout the duration of the reaction. The procedure for the use of Lindlar's catalyst is straightforward and is detailed in Protocol 2. An atmospheric pressure hydrogenation apparatus is essential and a simple version of one is shown in Fig. 6.1. The method of using the apparatus is given in Protocol 1.

Protocol 1.
The use of an atmospheric pressure hydrogenation apparatus

Caution! Hydrogen, especially in the presence of flammable solvents and active catalysts, presents a danger. Precautions should be taken against explosion and fire. For all procedures disposable vinyl or latex gloves and chemical-resistant safety goggles should be worn.

Equipment

- The equipment needed is shown in Fig. 6.1.

1. Attach the reaction flask equipped with magnetic stirrer to the apparatus.
2. Purge the system, i.e. evacuate and fill with hydrogen as follows.
   (a) Open tap A.
   (b) Close taps C, N, H, and V.
   (c) Turn on the water pump vacuum.
   (d) Regulate the hydrogen supply from the cylinder with a suitable regulator.

(e) Open tap V to evacuate the apparatus.

(f) Open tap C and allow the liquid to rise to the top of the burette, then close taps C and V. It may be necessary to close tap B if the level of solution rises too high.

(g) Fill the apparatus with hydrogen by opening tap H. Then open tap C and allow the liquid level to fall to the bottom of the burette. Close tap H.

(h) Repeat steps (a)–(g) twice more.

(i) Close tap A.

(j) Open taps V and N, then turn off the vacuum. Turn off the hydrogen supply.

3. Leave the reaction on for the desired length of time.[a]
4. At the end of the reaction remove the reaction flask and continue the work-up as instructed.

[a]It is often necessary to make a note of the volume of hydrogen used up as this indicates how far a reaction has gone to completion, or when the hydrogenation should be terminated.

Protocol 2.
Semihydrogenation of 3-phenyl-2-propyn-1-ol to give (*Z*)-3-phenyl-2-propen-1-ol using Lindlar's catalyst

Caution! Hydrogen, especially in the presence of flammable solvents and active catalysts, presents a danger. Precautions should be taken against explosion and fire. For all procedures disposable vinyl or latex gloves and chemical-resistant safety goggles should be worn.

$$Ph{-}C{\equiv}C{-}CH_2OH \xrightarrow[\text{quinoline}]{\text{Pd(5\% on CaCO}_3)} Ph{-}CH{=}CH{-}CH_2OH$$

Equipment

- One-necked, round-bottom flask (50 mL)
- Atmospheric pressure hydrogenation apparatus (see Fig. 6.1 for a suitable design)
- Magnetic stirrer.
- Magnetic stirrer bar

Materials

- 3-Phenyl-2-propyn-1-ol (FW 132.2) 3.54 g, 27 mmol
- Quinoline (FW 129.2) 1.0 mL, 1.09 g, 8.5 mmol — **irritant, harmful**
- Palladium 5% on calcium carbonate, 0.7 g
- Hexane — **highly flammable, harmful**
- A source of hydrogen

1. Place 3-phenyl-2-propyn-1-ol[a] (3.54 g, 0.027 mol) in the round-bottomed flask along with the magnetic stirrer bar.
2. Add quinoline (1.0 mL, 0.007 mol) and hexane (15 mL).

Protocol 2. *Continued*

3. Moisten 5% palladium on calcium carbonate (0.7 g) with hexane (5 mL) and add to the flask.
4. Attach the flask to the atmospheric pressure hydrogenation apparatus and stir the reaction, and then follow the instructions given in Protocol 1 for the introduction of hydrogen.
5. Follow the reaction by proton NMR[b] at half-hourly intervals.
6. When the reaction has gone to completion[b] remove the flask from the hydrogenation apparatus and filter the reaction mixture through a pad of pre-cleaned[c] Celite® to remove the catalyst. Retain the used palladium catalyst residue for recovery or disposal of the palladium.
7. Remove the hexane under reduced pressure and distil the residue, 92–94°C at 0.4 mmHg, to give (*Z*)-3-phenyl-2-propen-1-ol (3.5 g, 97%).

[a]This can be synthesised from phenylacetylene and paraformaldehyde,[9] but is widely available commercially.
[b]Watch for the appearance of the vinylic protons and the appearance of completely reduced material. The starting material will disappear just as the completely reduced material starts to appear.
[c]Use methanol to wash the Celite®. Run methanol (approximately 100 mL) through the Celite® under suction.

A simplified mechanism for the reduction of alkynes with hydrogen is given below, although studies have shown that the mechanism is more complicated than illustrated in Scheme 6.1.[5,10]

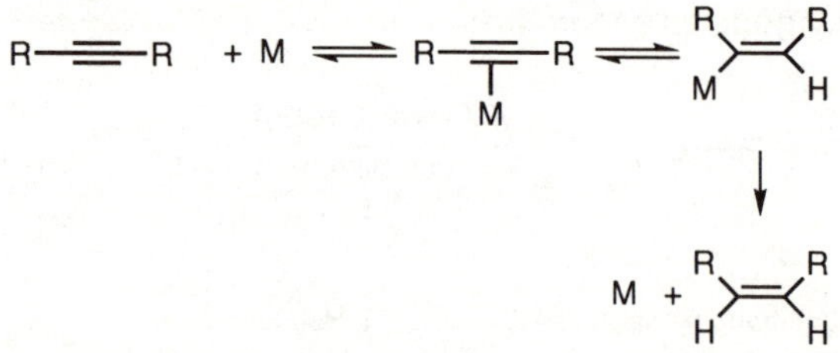

R—≡—R + M ⇌ R—≡—R (M) ⇌ R(M)C=C(R)H

↓

M + R(H)C=C(R)H

Scheme 6.1

Mononuclear complexes catalyse *syn*-addition of H_2 to alkynes to yield (*Z*)-isomers. The predominance of the (*Z*)-isomers as products of the heterogeneous palladium- or nickel-catalysed hydrogenation of disubstituted alkynes suggests that the main reaction path involves the alkyne bound to a single metal atom.[11]

The Lindlar catalyst has been used successfully in an enormous number of natural product syntheses. These range from β-carotene[12,13] to androstane[14] and encompass unsaturated fatty acids,[15] astaxanthin,[16] and methyl jasmonate.[17] There are several other heterogeneous catalysts based on palladium or nickel that differ enough from Lindlar's catalyst to merit discussion in their own right.

The naturally occurring clay, montmorillonite, can be used as a support for a bipyridinepalladium(II) acetate complex.[18] This catalyses the hydrogenation of alkynes more effectively than an analogous polymer-bound complex.[19] Hydrogenation of triple bonds, irrespective of their position, always precedes that of double bonds. The high (*Z*)-selectivity, normally greater than 95%, in the hydrogenation of triple bonds in variously substituted alkynes and in conjugated and unconjugated alkenynes demonstrates the versatility of this catalytic system. The (*Z*)-selectivity of palladium supported on tungsten has also been reported to be superior to the Lindlar catalyst.[20]

A palladium(II) salen complex has been used to mediate catalytic semihydrogenation.[21] However, there are only a few examples of its use and the structure of the catalyst has not been determined. In the only internal alkyne investigated, the (*Z*)-selectivity was quoted as 100%.

The reduction of $NiBr_2$·2DME with potassium–graphite affords highly dispersed nickel on the graphite surface.[22] When freshly prepared *in situ* this is an effective catalyst for semihydrogenation of alkynes to alkenes in the presence of ethylenediamine as a catalyst modifier. Both unconjugated and conjugated (*Z*)-alkenes can be obtained in most cases with selectivities in the range 96–99%.

The most widely used nickel catalyst for hydrogenation reactions is Raney nickel. It was first used to catalyse the addition of hydrogen to 1-heptyne, phenylacetylene, and 2-octyne giving the corresponding alkenes.[23] The selectivity of Raney nickel varies greatly with the substrate and mixtures of alkenes and alkanes are often produced.[24] It has been suggested that fresh Raney nickel is not as selective as aged reagent (non-pyrophoric) and that deactivation of the catalyst by copper acetate in methanol enhances the catalytic performance,[25] but the use of nickel catalysts in semihydrogenation has been almost confined to Raney nickel poisoned with piperidine and zinc acetate.[26]

Studies on the stereoselectivity of nickel catalysts are rather limited.[27] The available evidence suggests that the (*Z*)-isomer is usually produced with only about 80 to 90% selectivity, although there have been a few cases where this is 95% or higher.[28] It is perhaps because of this lack of selectivity that those natural product syntheses utilising Raney nickel for semihydrogenation have all been performed on isolated alkynes. Natural product syntheses using a nickel-mediated reduction include, civetone,[29] α-irone,[30] (*Z*)-6-undecanoic acid,[31] and an early synthesis of vitamin A ether.[32]

Partial reduction of alkynes to alkenes using an iron catalyst has been reported,[33] although only three substrates were tested. The primary reason for investigating the catalyst appears to be its inability to reduce alkenes to alkanes,[34] which removes the need for careful monitoring of reactions as is required in many catalytic procedures. Unfortunately, the requirement of high pressures and temperatures limits the use of iron as a convenient catalyst in everyday laboratory syntheses.

3. Reduction of alkynes with homogeneous catalysts

Compared with heterogeneous catalysis, which accounts for most of the investigation into the semihydrogenation of alkynes, the use of homogeneous catalysis is relatively neglected.

It is, however, possible to obtain high stereo- and chemoselectivity in the semihydrogenation of internal alkynes to (*Z*)-alkenes with complexes such as $[(\text{arene})\text{Cr(CO)}_3]$.[35] The example shown in Scheme 6.2 shows that 7-tetradecyne is reduced to the (*Z*)-alkene in quantitative yield.

$$C_6H_{13}{-}C{\equiv}C{-}C_6H_{13} \xrightarrow[120^\circ\text{C, 15 h, 100\%}]{H_2\text{, 70 atm. } [(\text{methyl benzoate})\text{Cr(CO)}_3]} (Z)\text{-}H_{13}C_6CH{=}CHC_6H_{13}$$

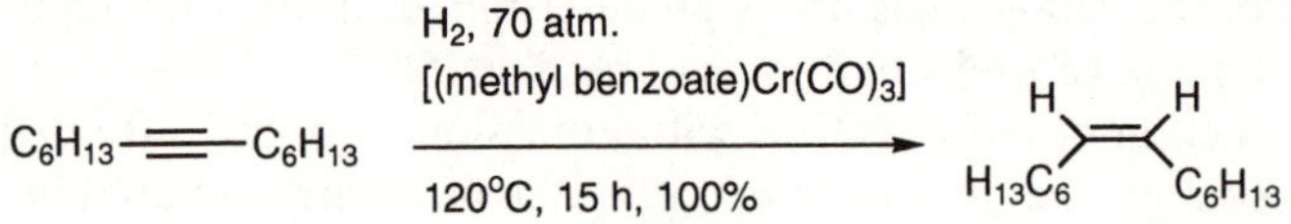

Scheme 6.2

One of the earliest and best-known semihydrogenation catalysts is Wilkinson's catalyst, chlorotris(triphenylphosphine) rhodium(I), which has been used to reduce 1-octynyl sulfoxide to (*Z*)-1-octenyl sulfoxide in quantitative yield.[36] In benzene, and with acidic alcohols such as 2,2,2-trifluoroethanol or phenol as co-solvents, the rate of hydrogenation of alkynes by Wilkinson's catalyst is accelerated, whereas that of alkenes remains unchanged. This allows selective hydrogenation of alkynes in the presence of alkenes.[37] Wilkinson's catalyst is commonly used for the hydrogenation of terminal alkynes, though other systems based on rhodium, iridium, ruthenium, or cobalt can perform this function.[38]

The iron cluster $[\{\text{CpFe}(\mu_3\text{-CO})\}_4]$ catalyses the selective hydrogenation of alkynes to alkenes at temperatures of 100–130 °C and 7 to 70 atmospheres of hydrogen.[39] Terminal alkynes are selectively hydrogenated to alkenes even in the presence of internal alkenes and alkynes (which are only slowly reduced to (*Z*)-alkenes), but nitro groups attached to aromatic rings and terminal alkenes are also reduced. Complexes of the type $[\text{Rh(diene)L}_n]\text{PF}_6$, where the diene is norbornadiene or 1,5-cyclooctadiene and L is a tertiary phosphine ($n = 2$ or 3), can be used for the reduction of internal alkynes to (*Z*)-alkenes. Selectivities are usually greater than 95%.[40,41]

Co-condensation of lanthanoid metal atoms with internal alkynes generates lanthanoid complexes of alkynes, which are catalysts for hydrogenation of potential synthetic value.[42] $[\{\text{Sm(1-hexyne)}\}_n]$ or $[\{\text{Er(3-hexyne)}\}_n]$ catalyse hydrogenation of hex-3-yne to (*Z*)-hex-3-ene with a (*Z*)-selectivity of 97%, at room temperature and atmospheric pressure.

3.1 Reduction with transition metal hydride reagents

Although the major product in a reduction of an alkyne is generally the (*Z*)-isomer, there are several catalysts which have been developed that give the (*E*)-isomer as the major or sole product. These catalysts have so far been based on rhodium. The complex $[RhH_2(OC(=O)OH)(PPr^i_3)_2]$ gives dimethyl fumarate and (*E*)-stilbene on reduction of their respective alkyne precursors as shown in Scheme 6.3.[43] This complex also catalyses the isomerisation of (*Z*)-stilbene to (*E*)-stilbene although the isomerisation is eight times slower than hydrogenation.

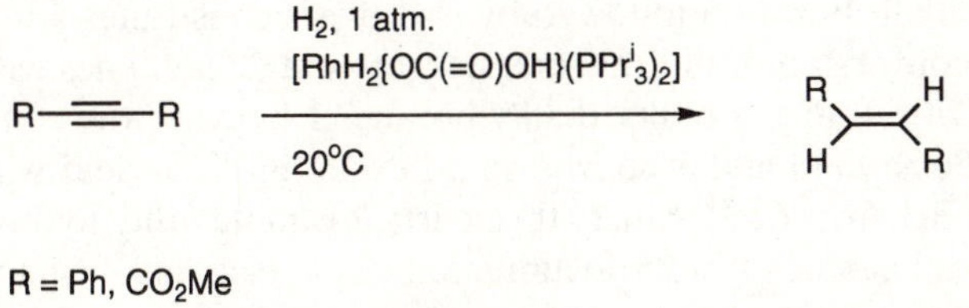

Scheme 6.3

The dinuclear rhodium hydride complex $[\{(\mu\text{-H})Rh\{P(OPr^i)_3\}_2\}_2]$ is a catalyst for the stereoselective hydrogenation of dialkylalkynes and diarylalkynes, and it has been shown that the catalyst converts the alkynes into (*E*)-isomers as initial products.[44] The alkyne addition product has been isolated; its structure shows the vinyl group σ-bonded to one rhodium atom and π-bonded to the other, as illustrated in Scheme 6.4. This structure resembles one in an earlier proposal to explain the formation of (*E*)-isomers and alkanes.[10,45] Unfortunately the lifetime of the dinuclear rhodium hydride catalyst is short under the reaction conditions.[46]

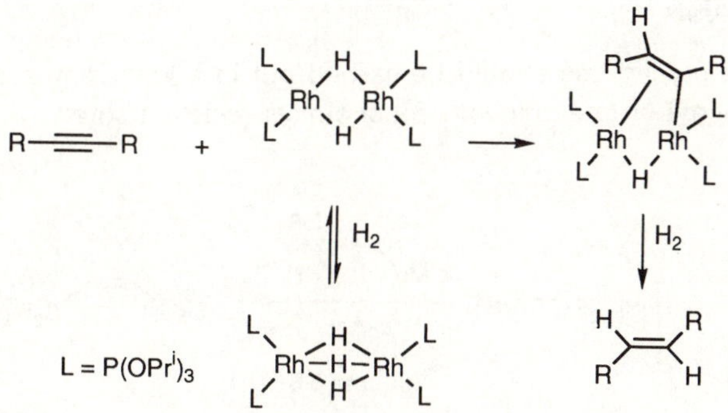

Scheme 6.4

With $[RhCl_3(Py)_3]$, DMF, and $NaBH_4$, diphenylacetylene is converted into (*E*)-stilbene as the only product.[47] However, with other alkynes, such as 2-butyne-1,4-diol, the major product is the (*Z*)-alkene.

Selective reductions of several substituted alkynes using cationic rhodium dihydride and monohydride complexes have been successful.[1] In a direct comparison, a system based on a cationic rhodium precursor was shown to be far superior to the Lindlar-type heterogeneous catalyst. Indeed, in some cases the (*Z*)-selectivity shown by these cationic complexes is 99%. However, there is poor selectivity in the simple semihydrogenation of 2-butyne-1,4-diol.[1]

4. Reduction of alkynes using boron reagents

The use of hydroboration followed by hydrolysis constitutes a method which allows the conversion of internal alkynes into the (*Z*)-alkenes with high yield and selectivity. Using a bulky dialkylborane, 1-hexyne was converted to 1-hexene in 90% yield and 3-hexyne to 3-hexene in 82% yield with 99% (*Z*)-selectivity, Scheme 6.5.[48] Since this early example, the hydroboration of alkynes has become a well-documented area (see also Chapter 7).[49] In general, protonolysis of the resulting alkenyl dialkylboranes is carried out utilising an excess of acetic acid,[50] and a simple experimental procedure for this is listed in Protocol 3. This procedure can be incompatible with various acid-sensitive groups, and so a method for hydrolysis under basic conditions has been developed.[51] This involves the treatment of an alkenyltrialkylborate, formed by addition of *n*-BuLi, with aqueous sodium hydroxide and results in the corresponding alkene with excellent yields and purity. A general example of its application is given in Scheme 6.6.

Protocol 3.
Conversion of 3-hexyne into 3-hexene by hydroboration followed by acid hydrolysis

Caution! All procedures should be carried out in a well-ventilated hood, and disposable vinyl or latex gloves and chemical-resistant safety goggles should be worn.

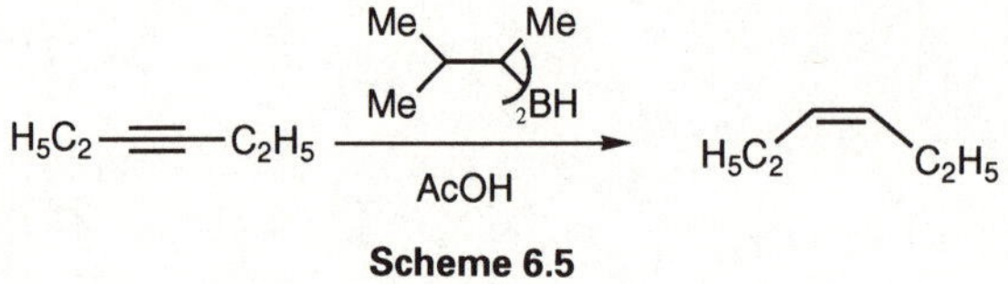

$H_5C_2-C\equiv C-C_2H_5 \xrightarrow[\text{AcOH}]{(\text{Me}_2\text{CHCH(Me)})_2\text{BH}} H_5C_2-CH=CH-C_2H_5$

Scheme 6.5

Equipment

- One-necked, round-bottomed flask (100 ml)
- Glass syringes with needle-lock Luers (5 ml and 10 ml)
- Four six-inch, medium-gauge needles
- Nitrogen bubbler
- Septum
- Magnetic stirrer
- Magnetic stirrer bar
- Ice bath

Materials

- 3-Hexyne (FW 82.15) 3.77 mL, 33 mmol — **highly flammable**
- Sodium borohydride (FW 37.8) 1.0 M in diglyme/9.22 mL, 9.22 mmol — **flammable, moisture, sensitive**
- Boron trifluoride etherate (FW 141.9) 1.49 mL, 12 mmol — **highly flammable, corrosive moisture sensitive**
- Dry ethylene glycol — **harmful, moisture sensitive**
- Glacial acetic acid (FW 60.1) — **corrosive, flammable**
- Source of dry nitrogen

1. Clean all glassware and syringes and dry for at least 4 h in a 120 °C electric oven. Whilst still hot place a septum in the top and allow nitrogen, which should pass through a nitrogen bubbler, into the system via a needle. Once the flask is cool place another needle in the septum and allow nitrogen to flush the flask for 5 min. Remove the extra needle.
2. Add the sodium borohydride (1.0 M in diglyme) (9.22 mL) to the flask, via the septum, using a syringe (10 mL).
3. Whilst stirring the solution, add 3-hexyne (3.77 mL, 0.033 mol), dropwise via a syringe.
4. Immerse the flask in ice.
5. Using a syringe (5 mL) add borontrifluoride etherate (1.49 mL, 0.012 mol) over 1 h.
6. After a further 30 min at room temperature add ethylene glycol (2 mL) to destroy the residual hydride.
7. Using a syringe (10 mL) add glacial acetic acid (6.7 mL), maintaining the nitrogen atmosphere, and leave stirring overnight.
8. Pour the solution into ice–water (30 mL) and extract with ether (4 × 30 mL). Combine the organic layers and dry ($MgSO_4$).
9. After removal of the solids by filtration and removal of solvent *in vacuo*, distil the residue (b.p. 67 °C at atmospheric pressure). The yield of (*Z*)-3-hexene should be around 70%.

Brown and Molander[52] have investigated a method for the protonolysis of alkenyldialkylboranes using an equivalent of methanol and a catalytic amount (1–5 mol%) of either sodium methanoate or acetic acid in THF, as shown in Scheme 6.7. The hydrolysis is close to being neutral and offers flexibility. It can be used for both the protonolysis of terminal and internal alkenyldialkylboranes. Yields and (*Z*)-selectivity are high. The method is also applicable to more functionalised alkynes, and is shown in Scheme 6.8, where hydroboration has been utilised in the conversion of alkynylsilanes to (*Z*)-vinylsilanes. This non-oxidative method, involving 9-borabicyclo[3.3.1] nonane, produces the (*Z*)-vinylsilane in good yields, 60–74%, high purity, 98%, and with excellent (*Z*)-selectivities, >99%.

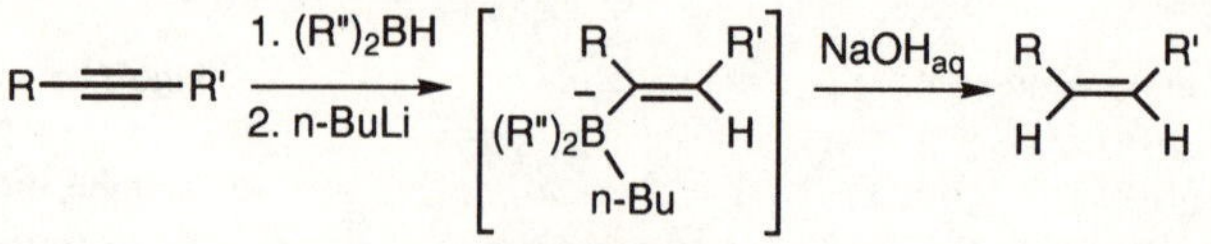

Scheme 6.6

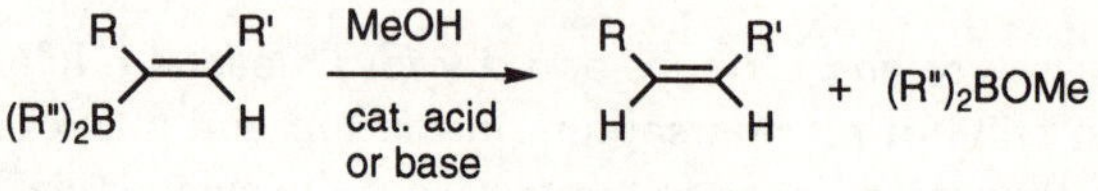

Scheme 6.7

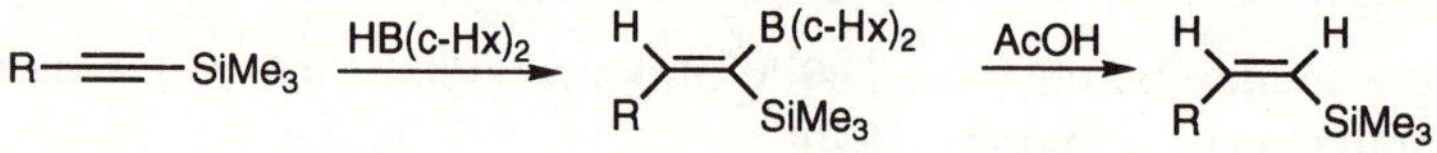

Scheme 6.8

Complex conjugated diynes adjacent to a sulfonyl group have been reduced using sodium borohydride under mild alkaline conditions.[53] The reduction required the use of $NaBH_4$ in a THF/methanol solution with the presence of either triethylamine or Magox (MgO). The method gave (*Z*)-selectivity of 85% and established a route to selectively placing deuterium in the product, Scheme 6.9.

The iron cluster $[Fe_4S_4(SPh)_4]^{2-}$ was found to catalyse the reduction of diphenylacetylene to (*Z*)-stilbene in the presence of $NaBH_4$.[54] The *Z*-selectivity is 95%, but the yields, isomer selectivity, and rates of reaction are heavily dependent on the sulfur ligands and co-solvents, such as methanol.

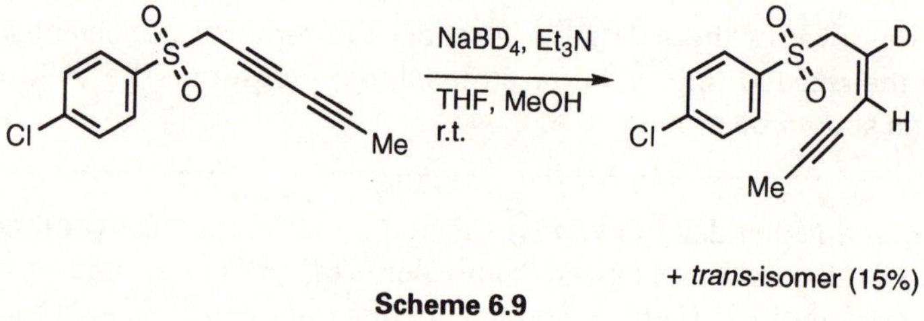

+ *trans*-isomer (15%)

Scheme 6.9

5. Dissolving metal reductions of alkynes

Although there are many methods for the reduction of alkynes to the (*Z*)-isomer, the number of ways to reduce an alkyne to the (*E*)-isomer are limited. Apart from the restricted use of aluminium hydride reagents with alkynes having an α-hydroxyl group (see Chapter 7), the main approach to producing (*E*)-alkenes involves dissolving metal reductions ((*Z*)-alkenes can also be produced using this method[55]). Several metals have been examined in dissolving metal reductions, as well as a variety of solvents and co-reactants. Terminal alkynes cannot be reduced by the standard sodium/liquid ammonia procedure as they are converted to acetylide ions under these conditions. They can, however, be reduced by addition of ammonium sulfate to the reaction, which acts as a proton source.[55] Internal alkynes can be reduced to the (*E*)-isomer using the liquid ammonia/low-molecular-weight amine–alkali metal system.

Generally the reduction of alkynes to (*E*)-alkenes using sodium in liquid ammonia has been confined to investigating homologous series of alkynes.[49,56] In these early reports very little attention has been paid to the exact constitution of the isomer ratios. One of the first synthetic uses was in the production of nine-membered carbocycles containing an (*E*)-alkene.[57] Although this method of reduction is not commonly used in the synthesis of natural products, there are cases where it has been used highly successfully, as in the synthesis of insect sex attractants. Alkenol acetate sex attractants require the isomeric purity of the (*E*)-double bond to be extremely high for conclusions with respect to field testing to be meaningful, as the (*Z*)-isomer might have a masking effect. High (*E*)-selectivity, 90%, can be obtained if careful attention is paid to the proportions of reactants used.[58] Other researchers[59] working with these attractants found at least 80% (*E*)-selectivity in several systems. In three examples in which an equivalent of *t*-butanol has been added to the sodium/liquid ammonia reduction mixture, (*E*)-selectivities of 100% have been claimed, Scheme 6.10.[60] Protocol 4 provides a representative procedure for the sodium/liquid ammonia reduction of an alkyne.

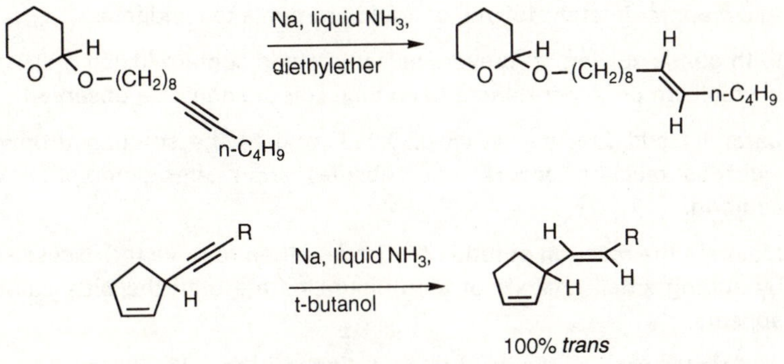

Na, liquid NH_3, diethylether

$O-(CH_2)_8$, n-C_4H_9

Na, liquid NH_3, t-butanol

100% *trans*

Scheme 6.10

Protocol 4.
Reduction of oct-2-yne using sodium in liquid ammonia to give (*E*)-oct-2-ene

Caution! All procedures should be carried out in a well-ventilated hood, and disposable vinyl or latex gloves and chemical-resistant safety goggles should be worn.

$$C_5H_{11}-\!\!\equiv\!\!-CH_3 \xrightarrow{Na/NH_3} C_5H_{11}CH{=}CHCH_3$$

Equipment

- Three-necked, round-bottomed flask (250 mL)
- Pressure-equalising addition funnel (25 mL)
- Dry ice condenser
- Inlet tube and adapter
- Calcium oxide drying tube
- Magnetic stirrer
- Magnetic stirrer bar
- Dry ice Dewar bowl

Materials

- Sodium (FW 23.0) 350 mg, 15 mmol — **corrosive, flammable solid, dangerous when wet**
- A source of dry ammonia (FW 18.0) — **toxic**
- A source of dry nitrogen gas
- 2-Octyne

1. Clean all glassware and dry for at least 4 h in a 120°C electric oven. Assemble the apparatus whilst still hot, as shown in Fig. 6.2. Flush the system with nitrogen.
2. Place the Dewar bowl around the flask and cool the flask using a dry ice/acetone mixture. Allow dry nitrogen into the system whilst carrying out this operation. Place crushed dry ice in the condenser well.
3. Introduce dry ammonia gas to the flask via the inlet tube and adapter.[a]
4. Allow approximately 150 mL of liquid ammonia to condense.
5. With gentle stirring, to prevent splashing, add sodium (0.350 g, 0.015 mol) in small lumps. A permanent deep-blue colour should be observed.
6. Carefully add 2-octyne (1.45 g, 0.013 mol) to the solution dropwise. A vigorous reaction occurs, but subsides soon after completion of the addition.
7. Maintain the reaction at reflux for 2 h and then remove the excess sodium by adding small crystals of ammonium nitrate until the blue colour disappears.
8. Add water very cautiously, with cooling, as there is a vigorous reaction with the sodamide by-product.

9. Extract the aqueous layer with dichloromethane (3 × 100 mL) and dry the combined organic layers (Na_2SO_4).
10. Remove the solids by filtration and remove the solvent *in vacuo*. Distil the residue to give (*E*)-2-octene b.p. 129°C.

[a]A suitable reaction set-up is shown in Fig. 6.2.

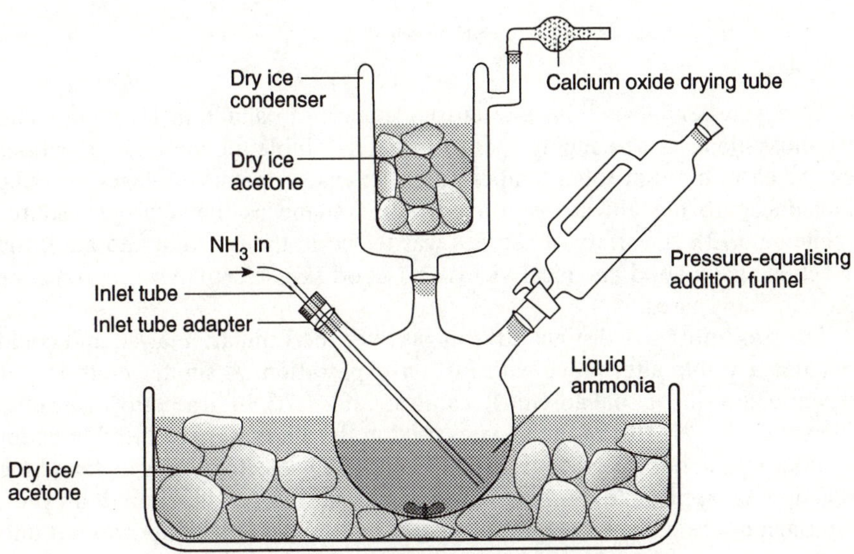

Fig. 6.2 Apparatus suitable for liquid ammonia reactions.

Variations of the reaction, such as ytterbium in liquid ammonia,[61] calcium with mixed amine solutions,[62] and lithium in methylamine,[63] have been utilised but offer no general advantages. There are dissolving metal reductions that do not involve the use of amines. Usually, these involve the dissolution of the alkali metal in an ethereal solvent, which is most commonly THF.[64] In the presence of a catalytic amount of a crown ether, a sodium–potassium alloy in THF successfully reduces alkynes to alkenes with (*E*)-selectivity of 90%.[65] However, in some cases alkanes are formed.

6. Alternative methods for the reduction of alkynes

There are several methods for reducing alkynes to alkenes that do not fall into any of the earlier categories, and it is in this section that these procedures are discussed.

The reduction of alkynes can be effected with the use of a palladium catalyst in the presence of trialkylammonium formates (Scheme 6.11).[66] The transfer hydrogenation process can potentially provide advantages over catalytic hydrogenation; it can be carried out using standard chemical apparatus and an exact amount of reducing agent can be measured out. However, terminal alkynes give considerable quantities of polymer side-products along with the alkene, and alkanes are produced alongside the required (*Z*)-alkene.

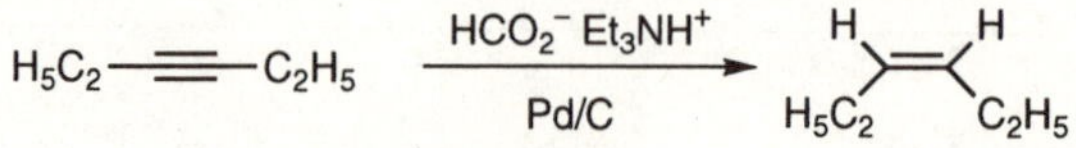

$$H_5C_2{-}C{\equiv}C{-}C_2H_5 \xrightarrow[\text{Pd/C}]{HCO_2^-\ Et_3NH^+} (Z)\text{-}H_5C_2CH{=}CHC_2H_5$$

Scheme 6.11

Tour and Pendalwar[67] have discussed the use of palladium(II) acetate and triethoxysilane as a reducing agent for water-soluble alkynes. The reactions can be carried out at room temperature. An exact amount of reagent may be introduced to the alkyne by using triethoxysilane as the hydrogen source (Scheme 6.12). A variety of alkynes was tested in the reaction and all of the reactions gave good chemical yields and good (*Z*)-selectivities, approaching 95% in many cases.

The procedure has distinct advantages over the Lindlar catalyst, and could become a viable alternative with further exploration. A similar method was developed using a palladium(0) catalyst and 1,1,3,3-tetramethyldisiloxane (Scheme 6.13).[68] Most of the alkynes used in the study were reduced in under an hour and in one case within a minute. The (*Z*)-selectivity was high and the reaction is applicable both to diynes and variously substituted alkynes, although not selectively. It is unfortunate that the (*Z*)-selectivity was not universal and several problems must be removed before the method is of real synthetic utility.

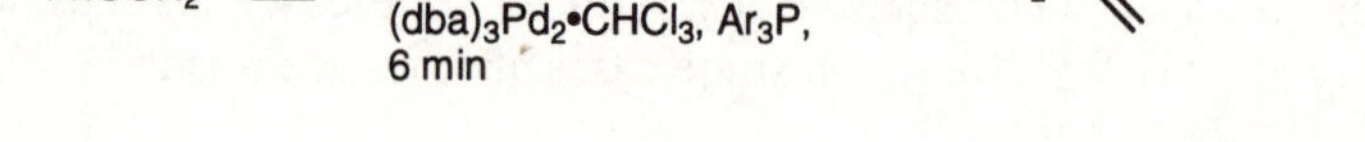

$$n\text{-}C_4H_9{-}C{\equiv}C{-}(CH_2)_5CO_2H \xrightarrow[\text{triethoxysilane}, H_2O]{5\text{ mol\% }Pd(OAc)_2,} (Z)\text{-}n\text{-}C_4H_9CH{=}CH(CH_2)_5CO_2H$$

Scheme 6.12

$$PhOCH_2{-}C{\equiv}CH \xrightarrow[(dba)_3Pd_2{\bullet}CHCl_3,\ Ar_3P,\ 6\text{ min}]{(Me_2SiH)_2O,\ AcOH,\ PhH,} PhOCH_2{-}CH{=}CH_2$$

$$PhCH{=}CHC(O)OCH_2C{\equiv}C{-}TMS \xrightarrow[(dba)_3Pd_2{\bullet}CHCl_3,\ Ar_3P]{(Me_2SiH)_2O,\ AcOH,\ PhH,} PhCH{=}CHC(O)OCH_2CH{=}CH{-}TMS$$

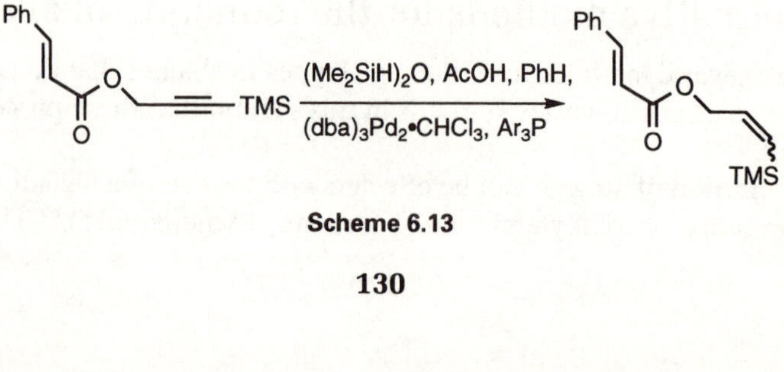

Scheme 6.13

Other transition metal-based reductions involve low-valent niobium and tantalum with zinc and methanol/water.[69] The conversion of alkynes into alkenes is achieved with good (*Z*)-selectivity, in the range of 90 to 100%. Another low-valent transition metal reduction is the homogeneous reduction of alkynes using chromous sulfate in water or water/DMF at room temperature to give (*E*)-alkenes in high yields and it has obvious merit.[70] The mechanism that is most probable is shown in Scheme 6.14. This is a simple and clean way of producing (*E*)-alkenes. The facile procedure is detailed in Protocol 5.

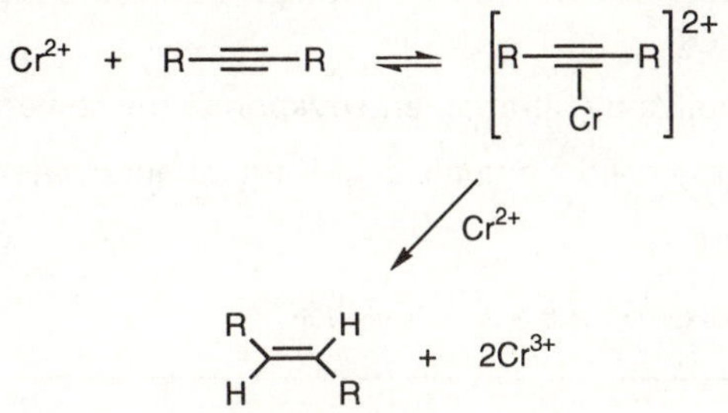

$$Cr^{2+} + R{-}C{\equiv}C{-}R \rightleftharpoons [R{-}C{\equiv}C{-}R \cdot Cr]^{2+} \xrightarrow{Cr^{2+}} (E)\text{-}RHC{=}CHR + 2Cr^{3+}$$

Scheme 6.14

Protocol 5.
Preparation of (*E*)-alkenes using chromium(II) sulfate

Caution! All procedures should be carried out in a well-ventilated hood, and disposable vinyl or latex gloves and chemical-resistant safety goggles should be worn.

$$Ph{-}C{\equiv}C{-}CO_2H \xrightarrow[DMF/H_2O]{CrSO_4} (E)\text{-}PhCH{=}CHCO_2H$$

Equipment

- Two-necked, round-bottomed flask (500 mL)
- Pressure-equalising addition funnel
- Magnetic stirrer
- Magnetic stirrer bar
- Septum

Materials

- Ammonium sulfate
- Phenylpropiolic acid (FW 146.1) 2.50 g, 17 mmol — **irritant**
- DMF 100 mL — **harmful, flammable**
- Chromium(II) sulfate hydrate[a] (FW 238.1 for $CrSO_4 \cdot 5H_2O$) 18.43 g, 47 mmol — **harmful, corrosive**

1. Clean the glassware. Attach a septum to one neck of the flask, and blow nitrogen through the system via a needle.
2. Place phenylpropiolic acid (2.50 g, 17 mmol) in the flask.

Protocol 5. *Continued*

3. Add DMF (100 mL) and water (100 mL) and stir.
4. Dissolve the chromium(II) sulfate (18.43 g, 47 mmol) in water (100 mL) and add this solution via the stoppered addition funnel.
5. Leave the reaction for 3 h and then saturate the solution with ammonium sulfate. (Add 20 g.)
6. Extract the solution with ether (4 × 100 mL). Combine the organic layers and dry ($MgSO_4$).
7. Remove the solids by filtration and evaporate the solvent *in vacuo*.
8. Recrystallise the solid cinnamic acid from ethanol/water until constant m.p. 133°–134°C.

[a]For the preparation of $CrSO_4{\cdot}5H_2O$, see reference 70b.

As previously stated, zinc can be used successfully for the partial reduction of alkynes. It has also been used to great effect without the presence of molecular hydrogen. The use of zinc powder, activated by treatment with 1,2-dibromoethane in absolute ethanol, is extremely promising as a future common method for partial reduction.[71,72] In most cases a large range of substrates give yields of the (*Z*)-isomer around 95% and usually in reaction times of under 3 h.

Diimide reductions are another good way of reducing symmetrical alkynes to the corresponding alkene. Alkynes are in general easily reduced, but cannot be reduced in the presence of symmetrical double bonds, except in special cases.[73] The high reactivity of non-polar triple bonds allows their reduction by diimide to be carried out under very mild conditions. This allows alkynes to be reduced in the presence of other functionalities, which, under many of the previously described reduction methods, would also react. For example, allylic and benzylic functions do not undergo hydrogenolysis with diimide reductions, and heteroatom bonds such as N–N and O–O are not reductively cleaved, as is often the case in catalytic hydrogenations. Moreover, catalytic hydrogenations are often complicated by lack of position selectivity.

Alkynes undergo reduction to produce (*Z*)-alkenes, which in turn can undergo further reduction to alkanes. With non-terminal dialkylalkynes, the reactivity of the triple bond is sufficiently comparable to that of alkenes produced to make partial reduction unfeasible. However, certain heteroatom-substituted terminal alkynes can undergo partial reduction. The comparatively lower reactivity of such alkenes allows the isolation of the (*Z*)-alkenes in good yields. This is shown in Protocol 6 with a 1-iodoalkyne as an example.[74,75]

Protocol 6.
The use of diimide in alkyne reductions

Caution! All procedures should be carried out in a well-ventilated hood, and disposable vinyl or latex gloves and chemical-resistant safety goggles should be worn.

$$\text{R}{-}{\equiv}{-}\text{I} \xrightarrow[\text{pyridine/HOAc}]{\text{KO}_2\text{CN=NCO}_2\text{K}} \text{R}{-}\text{CH=CH}{-}\text{I}$$

Equipment

- Two-necked, round-bottomed flask (500 mL)
- Pressure-equalising addition funnel
- Magnetic stirrer
- Magnetic stirrer bar
- Septum

Materials

• Iodoacetylene[a] (0.06 mol)	**assume toxic**
• Pyridine 25 mL	**highly flammable, harmful**
• Dipotassium azodicarboxylate[b] 30 g, 0.155 mol	**assume toxic**
• Methanol 75 mL	**highly flammable, toxic**
• Glacial acetic acid 17.8 mL	**corrosive, flammable**
• Hydrochloric acid 5% in water 20 mL	**corrosive**

1. Clean the glassware and dry for at least 4 h in a 120 °C electric oven.
2. Attach a septum to one neck of the flask, and blow nitrogen through the system via a needle. Attach the addition funnel to the other neck.
3. Place the iodoacetylene[a] in the flask and dissolve in methanol (75 mL).
4. Add pyridine (25 mL) and dipotassium azodicarboxylate (12.0 g, 0.062 mol).
5. Add glacial acetic acid (7.5 mL) via the addition funnel to the mixture over 2 h with stirring at room temperature.
6. After 8 h add another portion of dipotassium azodicarboxylate (18 g, 0.093 mol), and another portion of glacial acetic acid (10.3 mL) in the same manner.
7. When no starting material can be detected (by TLC) destroy the remaining diimide precursor by carefully adding 5% hydrochloric acid (100 mL) with vigorous stirring.
8. Separate the organic layer and extract the aqueous layer diethyl ether (2 × 100 mL). Combine the organic layers and wash with 5% hydrochloric acid followed by 5% sodium bicarbonate and then dry ($MgSO_4$).
9. Remove the solids by filtration and the solvent *in vacuo* and then distil the residue.

[a]See reference 75 for a general procedure to produce iodoalkynes.
[b]See reference 76b for the synthesis of dipotassium azodicarboxylate.

Diimide may also be generated by the oxidation of hydrazine with O_2/cat Cu(II) or H_2O_2/cat Cu(II). In general, it is believed that there is a concerted symmetrical transfer of hydrogen from the (*Z*)-diimide to the alkyne, via a cyclic transition state as shown in Scheme 6.15.[76] The reaction displays autocatalytic behaviour and is catalysed by low concentrations of ethanoic acid, which is thought to aid the equilibration between the (*Z*)- and (*E*)-diimide.

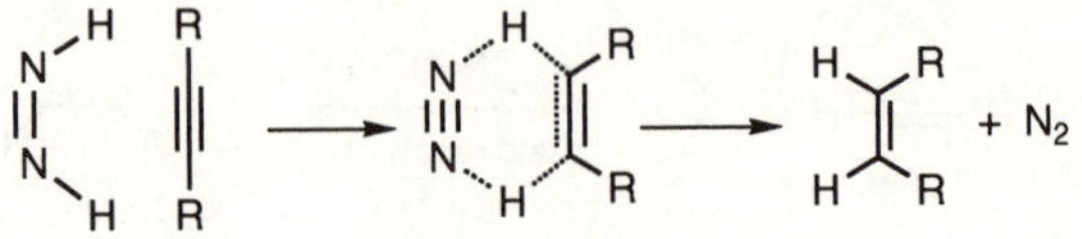

Scheme 6.15

Electrochemical methods of reducing alkynes to alkenes have been investigated. A series of dialkylalkynes was reduced using an electric current through a methylamine solution of lithium chloride and the alkyne.[77] The major product was the (*E*)-isomer and is produced in the solution not at the electrode. It is thought that the active reducing agent is lithium produced *in situ* and dissolved in the amine solvent. The electrochemical method of reduction has the advantage over the standard alkali metal/amine system in that direct handling of the metals can be avoided. Dialkylalkynes can also be electrochemically reduced to (*Z*)-alkenes at a spongy nickel cathode in a methanol solution of sulfuric acid or potassium hydroxide.[78]

References

1. Schrock, R. R.; Osborn, J. A. *J. Am. Chem. Soc.* **1976**, *98*, 2143–2147.
2. Thompson, Jr, A. F.; Wyatt, S. B. *J. Am. Chem. Soc.* **1940**, *62*, 2555–2556.
3. Rylander, P. N. *Hydrogenation Methods*; Academic Press: London, **1985**.
4. Nishimura, S.; Takagi, U. *Catalytic Hydrogenation Application to Organic Synthesis*; Tokyo Kagaku Dojin: Tokyo, **1987**.
5. Henrick, C. A. *Tetrahedron* **1977**, *33*, 1845–1889.
6. Dear, R. E. A.; Pattison, F. L. M. *J. Am. Chem. Soc.* **1963**, *85*, 622–626.
7. Kieboom, A. P. G.; van Rantwijik, F. *Hydrogenation and Hydrogenolysis in Synthetic Organic Chemistry*; Delft University Press; Delft, **1977**.
8. Steenhoek, A.; van Wijngaarden, B. H.; Pabon, H. J. J. *Recl. Trav. Chim. Pays-Bas* **1971**, *90*, 961–973.
9. Denis, J. N.; Greene, A. E.; Serra, A. A.; Luche, M. J. *J. Org. Chem.* **1986**, *51*, 46–50.
10. Bond, G. C.; Wells, P. B. *Adv. Catal.* **1964**, *15*, 91–95.
11. Collman, J. P.; Hegedus, L. S.; Norton, J. R.; Finke, R. G. *Principles and Applications of Organotransition Metal Chemistry*; University Science Books: Mill Valley, CA, **1987**.
12. Karrer, P.; Eugster, C. H. *Helv. Chem. Acta* **1950**, *33*, 1172–1174.
13. Milas, N. A.; Davies, P.; Belic, I.; Fles, D. A. *J. Am. Chem. Soc.* **1950**, *72*, 4844.

14. Djerassi, C.; Yashin, R.; Rosencranz, G. *J. Am. Chem. Soc.* **1950**, *72*, 5750–5751.
15. Taylor, W. R.; Strong, F. M. *J. Am. Chem. Soc.* **1950**, *72*, 4263–4265.
16. Surmatis, J. D.; Thommen, R.; *J. Am. Chem. Soc.* **1966**, *32*, 180–184.
17. Kosugi, H.; Kitaoka, M.; Tagami, K.; Uda, H. *Chem. Lett.* **1985**, 805–808.
18. Choudary, B. M.; Vasantha, G.; Sharma, M.; Bharathi, P. *Angew. Chem., Int. Ed. Engl.* **1989**, *28*, 465–466.
19. (a) Card, R. J.; Leisner, C. E.; Neckers, D. C. *J. Org. Chem.* **1979**, *44*, 1095–1098. (b) Card, R. J.; Neckers, D. C. *Inorg. Chem.* **1978**, *17*, 2345–2349.
20. Ulan, J. G.; Maier, W. F. *J. Org. Chem.* **1987**, *52*, 3132–3142.
21. Kerr, J. M.; Suckling, C. J. *Tetrahedron Lett.* **1988**, *29*, 5545–5548.
22. Savoia, D.; Tagliavani, E.; Trombini, C.; Umani-Ronchi, A. *J. Org. Chem.* **1981**, *46*, 5340–5343.
23. Dupont, G. *Bull. Soc. Chim. Fr.* **1936**, (*5*) *3*, 1030–1035.
24. Crombie, L. *J. Chem. Soc.* **1955**, 3510–3512.
25. Elsner, B. B.; Paul, P. F. *J. Chem. Soc.* **1953**, 3156–3160.
26. Oroshnik, W.; Karmas, G.; Mebane, A. D. *J. Am. Chem. Soc.* **1952**, *74*, 3807–3813.
27. Marvell, E. N.; Li, T. *Synthesis* **1973**, 457–468.
28. Hoff, M. C.; Greelee, K. W.; Boord, C. E. *J. Am. Chem. Soc.* **1951**, *73*, 3329–3336.
29. Stoll, M.; Hulstkamp, J.; Rouve, A. *Helv. Chim. Acta* **1948**, *31*, 543–553.
30. Grutter, H.; Helg, R.; Schinz, H. *Helv. Chim. Acta* **1952**, *35*, 771–775.
31. Ahmad, A.; Strong, F. M. *J. Am. Chem. Soc.* **1948**, *70*, 1699–1700.
32. Oroshnik, W.; Mebane, A. D. *J. Am. Chem. Soc.*, **1954**, *76*, 5719–5736.
33. Thompson, Jr, A. F.; Wyatt, S. B. *J. Am. Chem. Soc.* **1940**, *62*, 2555–2556.
34. Hubert, A. J. *J. Chem. Soc.* **1965**, 6669–6674.
35. Sodeoka, M.; Shibasaki, M. *J. Org. Chem.* **1985**, *50*, 1147–1149, 3246.
36. Trost, B. M.; Braslau, R. *Tetrahedron Lett.* **1989**, *30*, 4657–4660.
37. Candlin, J. P.; Oldham, A. R. *Faraday Discuss. Chem. Soc.* **1968**, *46*, 60.
38. Jardine, I.; McQuillin, F. J. *Tetrahedron Lett.* **1966**, 4871–4875.
39. Pregaglia, G. F.; Andreetta, A.; Ferrari, G. F.; Montrasi, G.; Ugo, R. *J. Organomet. Chem.* **1971**, *33*, 73.
40. Schrock, R. R.; Osborn, J. A. *J. Am. Chem. Soc.* **1971**, *93*, 3089–3091.
41. Schrock, R. R.; Osborn, J. A. *J. Am. Chem. Soc.* **1976**, *98*, 2143–2147.
42. Evans, W. J.; Engerer, S. C.; Coleson, K. M. *J. Am. Chem. Soc.* **1981**, *103*, 6672–6677.
43. Yoshida, T.; Youngs, W. J.; Sakaeda, T.; Ueda, T.; Ibers, J. A. *J. Am. Chem. Soc.* **1983**, *105*, 6273–6278.
44. Burch, R. R.; Shusterman, A. J.; Muetterties, E. L.; Teller, R. G.; Williams, J. M. *J. Am. Chem. Soc.* **1983**, *105*, 3546–3556.
45. Linstead, R. P.; von Doering, E.; Davis, S. B.; Levine, P.; Whetstone, R. B. *J. Am. Chem. Soc.* **1942**, *64*, 1985–1991.
46. Burch, R. R.; Shusterman, A. J.; Muetterties, E. L.; Teller, R. G.; Williams, J. M. *J. Am. Chem. Soc.* **1982**, *104*, 4257–4258.
47. Abley, P.; McQuillin, F. J. *J. Chem. Soc., Chem. Commun.* **1969**, 1503.
48. Brown, H. C.; Zweifel, G. *J. Am. Chem. Soc.* **1959**, *81*, 1512.
49. Brown, H. C.; Campbell, Jr, J. B. *Aldrichim. Acta.* **1981**, *14*, 3–11.
50. Zweifel, G.; Clark, G. M.; Polston, N. L. *J. Am. Chem. Soc.* **1971**, *93*, 3395–3399.
51. Negishi, E.; Chiu, K. W. *J. Org. Chem.* **1976**, *41*, 3484–3486.

52. Brown, H. C.; Molander, G. A. *J. Org. Chem.* **1986**, *51*, 4512–4514.
53. Thyagarajan, B. S.; Chander, R.; Santillan, A. *Synth. Commun.* **1990**, *20 (22)*, 3477–3488.
54. Itoh, T.; Negano, T.; Hirobe, M. *Tetrahedron Lett.* **1980**, *21*, 1343–1346.
55. Henne, A. L.; Greenlee, K. W. *J. Am. Chem. Soc.* **1943**, *65*, 2020–2023.
56. (a) Campbell, K. N.; Eby, L. T. *J. Am. Chem. Soc.* **1941**, *63*, 216–219, 2683–2685. (b) Raphael, R. A.; Dobson, N. A. *J. Chem. Soc.* **1955**, 3558–3560. (c) Newman, S. M.; Waltcher, I.; Ginsberg, H. F. *J. Org. Chem.* **1952**, 962–970.
57. Blomquist, A. T.; Liu, L. H.; Bonner, J. C. *J. Am. Chem. Soc.* **1952**, *74*, 3643–3647.
58. Warthen, Jr, J. D.; Jacobson, M. *Synthesis* **1973**, 616–617.
59. Schwarz, M.; Waters, R. M. *Synthesis* **1972**, 567–568.
60. Boland, W.; Hansen, V.; Jaenicke, L. *Synthesis* **1979**, 114–116.
61. White, J. D.; Gerald, L. L. *J. Org. Chem.* **1978**, *43*, 4555–4556.
62. Benkeser, R. A.; Belmonte, F. G. *J. Org. Chem.* **1984**, *49*, 1662–1664.
63. Fried, J.; Heim, S.; Etheredge, S. J.; Sunder-Plassmann, P.; Santhanakrishnan, T. S.; Himizu, J.; Lin, C. H. *J. Chem. Soc., Chem. Commun.* **1968**, 634–635.
64. Levin, G.; Jagur-Grodzinski, J.; Szwarc, M. *J. Org. Chem.* **1970**, *35*, 1702.
65. Mathre, D. J.; Guida, W. C. *Tetrahedron Lett.* **1980**, *21*, 4773–4776.
66. Cortese, N. A.; Heck, R. F. *J. Org. Chem.* **1978**, *43*, 3985–3987.
67. Tour, J. M.; Pendalwar, S. L. *Tetrahedron Lett.* **1990**, *33*, 4719–4722.
68. Trost, B. M.; Breslau, R. *Tetrahedron Lett.* **1989**, *30*, 4657–4660.
69. Kataoka, Y.; Takai, K.; Oshima, K.; Utimoto, K. *Tetrahedron Lett.* **1990**, *31*, 365–368.
70. (a) Castro, C. E.; Stephens, R. D. *J. Am. Chem. Soc.* **1964**, *86*, 4358–4363. (b) Castro, C. E.; Kray, W. C. *J. Am. Chem. Soc.* **1963**, *85*, 2768.
71. Aerssens, M. H. P. J.; van der Heiden, R.; Heus, M.; Bransdsma, L. *Synth. Commun.* **1990**, *20 (22)*, 3421–3425.
72. Sondengam, B. L.; Charles, G.; Akam, T. M. *Tetrahedron Lett.* **1980**, *21*, 1069–1070.
73. Pasto, D. J.; Taylor, R. T. *Org. React.* **1991**, **40**, 91–155.
74. Kluge, A. F.; Untch, K. G.; Ried, J. H. *J. Am. Chem. Soc.* **1972**, *94*, 9256–9258.
75. Luthy, C.; Konstantin, P.; Untch, J. G. *J. Am. Chem. Soc.* **1978**, *100*, 6211–6217.
76. (a) van Tamelen, E. E.; Dewey, R. S.; Timmons, R. J. *J. Am. Chem. Soc.* **1961**, *83*, 3725–3726. (b) van Tamelen, E. E.; Dewey, R. S.; Lease, M. F.; Pirkle, W. H. *J. Am. Chem. Soc.* **1961**, *83*, 4302. (c) Corey, E. J.; Pasto, D. J.; Mock, W. L. *J. Am. Chem. Soc.* **1961**, *83*, 2957–2958.
77. Benkeser, R. A.; Tincher, C. A. *J. Org. Chem.* **1968**, *33*, 2727–2730.
78. Campbell K. N.; Young, E. E. *J. Am. Chem. Soc.* **1943**, *65*, 965–967.

7

Reaction of alkynes with organometallic reagents

EI-ICHI NEGISHI and DANIELE CHOUEIRY

1. Introduction

Hydrometallation and carbometallation may be defined as additions of hydrogen–metal and carbon–metal bonds, respectively, to carbon–carbon multiple bonds. Of interest here are those involving alkynes (Scheme 7.1). Such addition reactions can be *syn-*, *anti-*, and non-stereoselective.

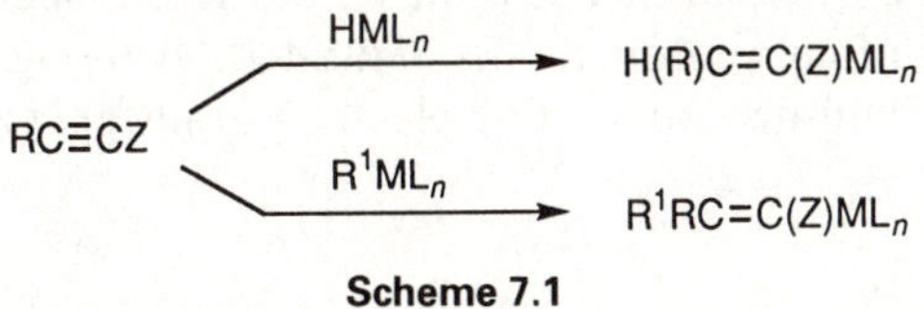

Scheme 7.1

In cases where a low-lying metal empty orbital is readily available, *syn*-stereoselective addition may be observed. Such reactions are most likely to be concerted processes involving the HOMO–LUMO interaction schemes shown in Scheme 7.2. On this basis, a wide variety of main group and transition metal complexes should be, in principle, capable of undergoing both hydrometallation and carbometallation. This has indeed been shown to be the case. In reality, however, various other factors, such as steric hindrance, aggregation of metal hydrides and organometals, deactivation of metal complexes through solvation and other intra- and intermolecular interactions, and instability of the products leading to further transformations, impose restrictions rendering many of these reactions of limited practical utility. At present, hydroboration,[1,2] hydroalumination,[3,4] hydrozirconation,[5] and transition metal-catalyzed hydrosilation[6] represent practically useful *syn*-hydrometallation reactions capable of producing alkenylmetals as discrete products, while carbocupration[7,8] and zirconium-catalyzed carboalumination represent related *syn*-carbometallation[7–10] reactions. Allylmetallation and related reactions that can proceed via six-centered transition states rather than four-centered transition states implied in Scheme 7.2 represent significant exceptions to the generalization presented above. Thus, practically useful *syn*-allylmetallation reactions have been observed even with those metals which do not readily participate in more usual carbometallation reactions, such as boron[7,8,11] and zinc.[7,8]

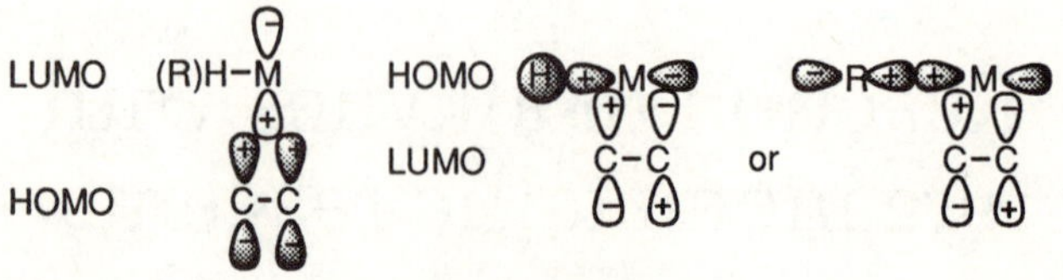

LUMO (R)H–M HOMO C–C HOMO M C–C LUMO or

Scheme 7.2

In addition to the *syn*-addition reactions mentioned above, there are some synthetically useful *anti*-addition reactions of metal hydrides and organometals to alkynes. Of particular synthetic importance are those involving propargylic alcohols and related derivatives. More specifically, *anti*-hydroalumination[12,13] and copper-catalyzed *anti*-carbomagnesiation[14,15] have been developed into highly regio- and stereoselective routes to tri- and tetrasubstituted allylic alcohols which are not only synthetically valuable but also difficult to prepare by other means (eqn 1). In these reactions, the reagents, e.g. $LiAlH_4$, are often coordinatively saturated. Although little is known about their mechanisms, some one-electron transfer processes may be operating.

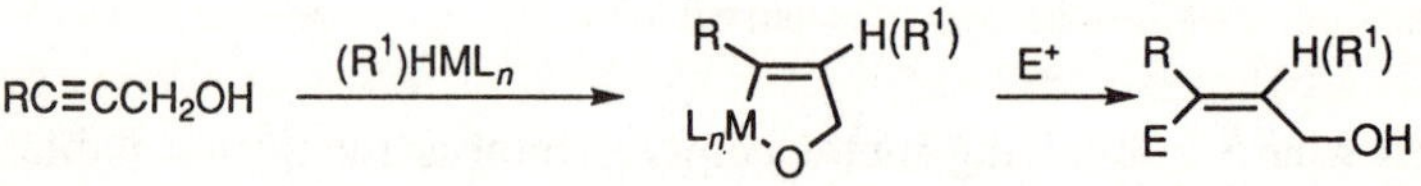

$$\mathrm{RC{\equiv}CCH_2OH} \xrightarrow{(\mathrm{R^1})\mathrm{HML}_n} \text{[cyclic } \mathrm{L}_n\mathrm{M{-}O} \text{ intermediate; R, H(R}^1)] \xrightarrow{\mathrm{E^+}} \text{R(E)C=C(H(R}^1))\mathrm{CH_2OH} \quad (1)$$

There are also some non-stereoselective hydrometallation reactions which are of considerable synthetic utility. Hydrostannation[16] of alkynes is a representative example. Although the ability of stannanes to undergo concerted *syn*-addition to alkynes cannot be totally ruled out, hydrostannation appears to be in most cases a radical or ionic process.

Finally, it should be pointed out here that the scope of carbometallation is much broader than that indicated in Scheme 7.1. Thus, it may be cyclic (eqn 2). The metal-containing reagents may also be metal–carbene complexes[17] (eqn 3) and metallacycles such as metallacyclopropanes and metallacyclopropenes[18,19] (Scheme 7.3). These reactions are, however, not discussed further in this chapter. In the following sections, several hydrometallation and carbometallation reactions of established synthetic utility are discussed with emphasis on some representative experimental procedures.

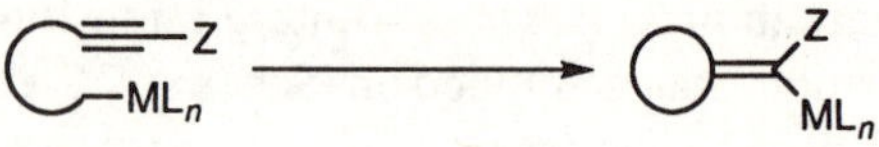

$$\text{[ring]}{\equiv}\mathrm{Z},\ {-}\mathrm{ML}_n \longrightarrow \text{[ring]}{=}\mathrm{C(Z)(ML}_n) \quad (2)$$

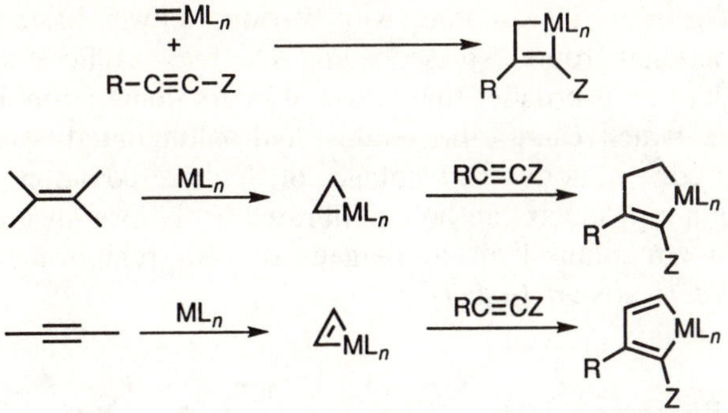

$$=ML_n + R{-}C{\equiv}C{-}Z \longrightarrow \text{(metallacyclobutene with } ML_n\text{, R, Z)} \quad (3)$$

Scheme 7.3

2. *syn*-Stereoselective hydrometallation reactions

2.1 Overview

A dozen or so main group metals and essentially all transition metals are known to participate in *syn*-hydrometallation. However, hydroboration,[1,2] hydroalumination,[3,4] hydrozirconation,[5] and hydrosilation[6] catalyzed by transition metal complexes, most commonly platinum compounds, are about the only currently well-developed and widely used *syn*-hydrometallation reactions of alkynes. Extensive reviews are available on each of these reactions, and they should be consulted for further details. The regiochemistry of *syn*-hydrometallation of terminal alkynes generally is as shown in eqn 4. Furthermore, the regioselectivity approaches 100% in most cases. On the contrary, the regioselectivity of the corresponding reactions of internal alkynes varies. Unless the two substituents in alkynes are widely dissimilar, regioisomeric mixtures usually result (eqn 5). As shown in eqs 4 and 5, *syn*-hydrometallation can produce (*E*)-disubstituted and trisubstituted alkenes.

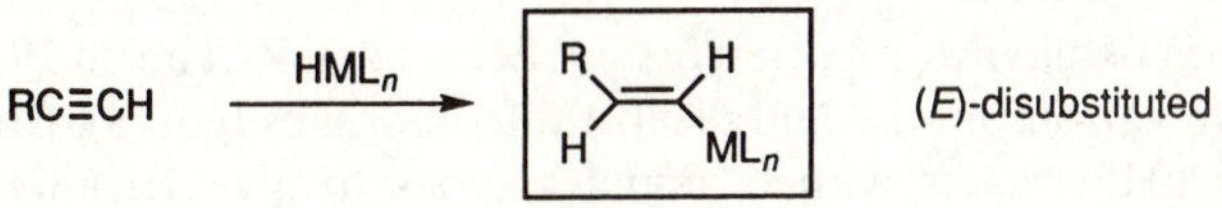

$$RC{\equiv}CH \xrightarrow{HML_n} \text{(R)(H)C=C(H)(ML}_n) \quad (E)\text{-disubstituted} \quad (4)$$

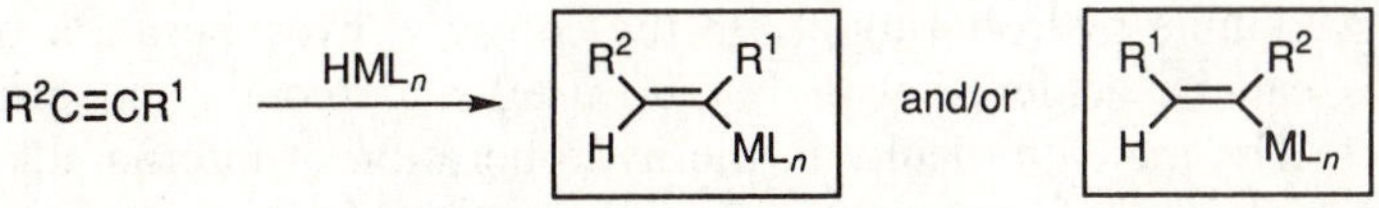

$$R^2C{\equiv}CR^1 \xrightarrow{HML_n} \text{(R}^2\text{)(H)C=C(R}^1\text{)(ML}_n) \text{ and/or } \text{(R}^1\text{)(H)C=C(R}^2\text{)(ML}_n) \quad (5)$$

Of those *syn*-hydrometallation reactions involving boron, aluminium, and zirconium, hydroboration is the most widely applicable and the most tolerant of various functional groups, while the scope of hydroalumination is the narrowest. For example, hydroalumination of internal alkynes is often accompanied by carboalumination leading to the formation of dimeric products[3,4] (eqn 6). The relatively high Lewis acidity and hydriding ability of alanes can lead to undesirable interactions with various Lewis basic and readily reducible functional groups. The scope and reactivity profile of hydrozirconation are similar to but broader than those of hydroalumination. For example, ethoxyethyne, which cannot be readily hydroaluminated, can be cleanly hydrozirconated.[20] Another advantage of hydrozirconation is that the alkenylzirconium products can be equilibrated in the presence of a catalytic amount of a zirconium hydride reagent to give regioisomerically purer alkenylmetal derivatives[21] (eqn 7).

$$\mathrm{RC{\equiv}CR} \xrightarrow{\mathrm{HAlR^1_2}} \mathrm{R(H)C{=}C(R)AlR^1_2} \xrightarrow{\mathrm{RC{\equiv}CR}} \mathrm{R(H)C{=}C(R){-}C(R){=}C(R)AlR^1_2} \qquad (6)$$

$$\mathrm{R^2C{\equiv}CR^1} \xrightarrow{\mathrm{HZrCp_2Cl}} \mathrm{R^2(H)C{=}C(R^1)ZrCp_2Cl} + \mathrm{R^1(H)C{=}C(R^2)ZrCp_2Cl} \quad (\rightleftharpoons \text{ cat. } \mathrm{HZrCp_2Cl}) \qquad (7)$$

The overall synthetic utility of *syn*-hydrometallation reactions cannot be determined by their scope alone. It also significantly depends on the usefulness of the alkenylmetal products. In the latter sense, alkenylalanes tend to be generally most reactive and versatile as carbon nucleophiles. The relative nucleophilicity order is roughly: aluminium > zirconium > boron. Alkenylboranes, however, are capable of undergoing a wide variety of 1,2-migration reactions,[1,2] which make them a uniquely valuable class of reagents for preparing alkenes. In this sense, the synthetic utility of alkenylsilanes[22] as intermediates for the preparation of metal-free alkenes is currently more limited than that of the other alkenylmetals.

2.2 Hydroboration

Hydroboration of alkynes, reported first by Brown and Zweifel in 1959,[23] can be achieved with various mono- and disubstituted boranes (eqn 8). Borane itself, e.g. BH_3 in THF, reacts with terminal alkynes to give 1,1,-diboraalkanes. Even some relatively unhindered dialkylboranes, e.g. 9-BBN, tend to undergo double hydroboration. On the contrary, hydroboration of internal alkynes can be achieved with borane itself and mono- and disubstituted boranes. The major difficulty in the hydroboration of internal alkynes is to attain a high level of regioselectivity. However, this problem can be overcome through the use of highly hindered disubstituted boranes, e.g. dimesitylborane (Mes_2BH) as indicated by the results shown in eqn 9.[24]

Borane in THF, $BH_3{\cdot}SMe_2$, 9-BBN, $Cl_2BH{\cdot}SMe_2$, $Br_2BH{\cdot}SMe_2$, catecholborane, and dimesitylborane (Mes_2BH) are commercially available, while thexylborane ($Me_2CHCMe_2BH_2$), disiamylborane ($(Me_2CHCHMe)_2BH \equiv Sia_2BH$), dicyclohexylborane, and dipinanylborane can be readily generated

$$RC\equiv CH \xrightarrow{H_2BR^1} (E)\text{-}RCH{=}CHBHR^1 \qquad RC\equiv CH \xrightarrow{HBR^1_2} (E)\text{-}RCH{=}CHBR^1_2 \tag{8}$$

$$RC\equiv CMe \xrightarrow{HB<} \underset{(I)}{RCH{=}C(Me)B<} + \underset{(II)}{RC(B<){=}CHMe} \tag{9}$$

Reagent	Ratio (I/II) R = n-Pr	Ratio (I/II) R = Ph
BH_3•THF	40:60	74:26
Sia_2BH	39:61	19:81
9-BBN	22:78	65:35
Br_2BH•SMe_2	25:75	64:36
Mes_2BH	10:90	2:98

in situ from $BH_3{\cdot}THF$ or $BH_3{\cdot}SMe_2$ and the corresponding alkenes.[1,2] Detailed experimental procedures for the reactions of (i) 1-octyne with disiamylborane, (ii) cyclohexylethyne with catecholborane, (iii) 3-hexyne with $ClBH_2{\cdot}OEt_2$, (iv) diphenylacetylene with disiamylborane, (v) 1-hexyne with dicyclohexylborane, (vi) 1-bromo-1-hexyne with thexyl(cyclohexyl)borane, and (vii) 1-chloro-1-hexyne with thexylborane have been described.[1] The alkenylborane products generated in these reactions have been converted to a variety of organic and organometallic compounds[1,2] including (*E*)- and (*Z*)-alkenyl halides containing iodine, bromine, and chlorine, (*E*)- and (*Z*)-disubstituted alkenes, (*E*,*E*)-, (*E*,*Z*)-, and (*Z*,*Z*)-conjugated dienes, (*E*)-1,2,3-butatrienes, and (*E*)-alkenylmercuries via protonolysis, halogenolysis, 1,2-migration reactions, palladium-catalyzed cross coupling, and transmetallation.

A procedure for the conversion of an alkyne into an alkene via hydroboration is given in Chapter 6 (Protocol 3). Palladium-catalyzed conversion of a vinylborane into a diene is described in Chapter 8 (Protocol 5).

2.3 Hydroalumination

Wilke and Müller[25] reported in 1956 that terminal alkynes reacted with dialkylalanes to give the corresponding (*E*)-alkenylalanes, although the reaction of acetylene itself was not clean. In most cases, a few to several per cent each of alkynylalanes and 1,1-dialuminoalkanes are formed as by-products, which limits the yield of the desired products to the 90–95% range (eqn 10). This reaction has been extensively reviewed.[3,4] Diisobutylaluminium hydride (DIBAL) is by far the most commonly used reagent, although Et_2AlH and other dialkylaluminium hydrides have also been used. The reaction is normally carried out either neat or in non-polar solvents, e.g. hexanes. Selection of reaction medium is very important. Thus, the reaction shown in eqn 10 will give the alkynylalane as the major product, if it is run in THF. Ether appears to be a preferred solvent for *syn*-hydroalumination of alkynylsilanes. In hydrocarbons this reaction gives predominantly the *anti*-addition products (eqn 11). Some representative results showing various reactivity patterns of hydroalumination are summarized in Table 7.1. Detailed procedures for conversion via hydroalumination with DIBAL of (i) *t*-butyl(phenyl)acetylene to (*Z*)-β-*t*-butylstyrene, (ii) 1-octyne to (*E*)-1-bromo-1-octene, (iii) 1-octyne to (*E*)-1-iodo-1-octene, and (iv) 1-(trimethylsilyl)-1-hexyne to (*E*)-1-bromo-1-(trimethylsilyl)-1-hexene have been described.[4]

$$n\text{-BuC}\equiv\text{CH} \xrightarrow[50^\circ\text{C, 4 h}]{i\text{-Bu}_2\text{AlH}} (E)\text{-}n\text{-BuCH=CHAl}i\text{-Bu}_2\ (90\%) + n\text{-BuC}\equiv\text{CAl}i\text{-Bu}_2\ (6\%) + n\text{-BuCH}_2\text{CH(Al}i\text{-Bu}_2)_2\ (4\%) \quad (10)$$

$$\text{RC}\equiv\text{CSiMe}_3 \xrightarrow[\text{Et}_2\text{O}]{i\text{-Bu}_2\text{AlH}} \text{R(H)C=C(SiMe}_3)(\text{Al}i\text{-Bu}_2)\ (\text{R, AlBu}_2\ cis) ; \quad \text{RC}\equiv\text{CSiMe}_3 \xrightarrow[\text{hydrocarbon}]{i\text{-Bu}_2\text{AlH}} \text{R(H)C=C(Al}i\text{-Bu}_2)(\text{SiMe}_3)\ (\text{R, SiMe}_3\ cis) \quad (11)$$

Table 7.1 Hydroalumination of alkynes with DIBAL[3,4,25]

Alkyne	**Temp (°C)**	**(*E*)-1-Alkenyl- (%)**	**(*Z*)-1-Alkenyl- (%)**	**Alkynyl- (%)**
n-BuC≡CH	50	90	0	6
PhC≡CH	20	71	0	29
PhC≡CBr	100	0	0	100
PhC≡CSiMe$_3$	60	96	4	0
PhC≡CSnMe$_3$	20	0	0	100

2.4 Hydrozirconation

The reaction of alkynes with chlorohydridozirconocene ($HZrCp_2Cl$) to give alkenylzirconocene derivatives (eqn 12) was discovered by Wailes *et al.*[26] and developed mainly by Schwartz and co-workers.[5,21] This reaction too is nearly 100% *syn*-stereoselective. Terminal alkynes react regioselectively, whereas internal alkynes tend to give regioisomeric mixtures which can be equilibrated generally to improve the regioisomeric ratios in the presence of a catalytic amount of $HZrCp_2Cl$, but the overall yields generally decrease during this isomerization process. The reaction can be carried out in a variety of solvents that are inert to $HZrCp_2Cl$, and it is relatively insensitive to solvents. The reaction of 1-octyne with $HZrCp_2Cl$ has been described in detail.[27]

$$\mathrm{HZrCp_2Cl} \xrightarrow{\mathrm{RC{\equiv}CH}} (E)\text{-}\mathrm{RCH{=}CH{-}ZrCp_2Cl}$$
$$\mathrm{HZrCp_2Cl} \xrightarrow{\mathrm{R^2C{\equiv}CR^1}} \mathrm{R^2(H)C{=}C(R^1)ZrCp_2Cl} + \mathrm{R^1(H)C{=}C(R^2)ZrCp_2Cl} \quad (12)$$

The main inconvenience associated with this reaction lies in the preparation of $HZrCp_2Cl$ which is practically insoluble in most organic solvents. Although it is commercially available, it is often prepared shortly before its use because it is sensitive to air, moisture, and light and therefore somewhat difficult to store. The most commonly used method involves reduction of commercially available Cl_2ZrCp_2, relatively stable to air, moisture, heat, and light, with $NaAlH_2(OCH_2CH_2OCH_3)_2$ or $LiAlH_4$ under nitrogen or argon followed by filtration and washing with organic solvents, e.g. THF.[28] To alleviate difficulties associated with H_2ZrCp_2 formed as a by-product, the treatment of the initial product with dichloromethane has been recommended and described in detail.[29] Also noteworthy is the development of procedures involving *in situ* generation of $HZrCp_2Cl$ or its equivalents by treatment of Cl_2ZrCp_2 with *t*-BuMgCl,[30,31] $LiAlH_4$, and other related reagents.[32] Although many satisfactory results have been obtained using the *in situ* generation procedures, their synthetic equivalency to those involving pure, isolated $HZrCp_2Cl$ must be carefully established in each individual case. The scope of hydrozirconation of alkynes is significantly broader than that of hydroalumination. Thus, for example, some alkynes which cannot be cleanly hydroaluminated, such as acetylene itself, alkoxyethynes, and alkynylstannanes, can be cleanly hydrozirconated.

Protocol 1.
Synthesis of (*E*)-4-iodo-4-octene. Reaction of 4-octyne with *t*-BuMgCl/Cl_2ZrCp_2[30]

Caution! All procedures should be carried out in a well-ventilated hood, and disposable vinyl or latex gloves and chemical-resistant safety goggles should be worn.

$$\text{4-octyne} \xrightarrow[\text{2. } I_2]{\text{1. } t\text{-BuMgCl / }Cl_2ZrCp_2} (E)\text{-4-iodo-4-octene}$$

Equipment

- Magnetic stirrer
- One-necked, round-bottomed flask (250 mL) with a side-arm
- Tubing adapter
- Septum (for the side-arm)
- Magnetic stirring bar
- Glass syringes with needle-lock Luers (volume appropriate for quantities of solutions to be transferred)
- Six-inch medium gauge needles

Protocol 1. *Continued*

- Nitrogen or argon line
- Condenser
- Oil bath
- Water-jacketed, short-path distillation apparatus

Materials

- Dry benzene[a,b] 40 mL — **flammable, cancer suspect agent**
- Dichlorobis(η^5-cyclopentadienyl)zirconium[c,d] (FW 292.32) 5.84 g, 20 mmol — **irritant, moisture sensitive**
- *t*-Butylmagnesium chloride[c] 2.0 M in ether, 10 mL, 20 mmol — **flammable, air and moisture sensitive**
- 4-Octyne[c] (FW 110.20) 2.20 g, 2.93 mL, 20 mmol — **flammable, irritant**
- Iodine (FW 253.81) 7.61 g, 30 mmol — **corrosive, toxic**
- Dry THF[e] 30 mL — **flammable, irritant**
- Hydrochloric acid 3.0 M — **corrosive**
- Pentane for extraction — **flammable, irritant**

1. Clean glassware, syringes, needles, and stirring bar and dry for at least 4 h in a 120 °C electric oven before use.
2. Assemble the flask equipped with the stirring bar, condenser, adapter, and septum while the glassware is still hot, and flush the system with nitrogen or argon (see Chapter 1, Figs 1.1–1.5).
3. Place the system under a static pressure of nitrogen or argon, detach the condenser and charge the flask with dichlorobis(η^5-cyclopentadienyl)zirconium (5.84 g, 20 mmol), then gently flush nitrogen or argon through the system to replace any air that might have been introduced.
4. Add benzene (**caution**!) using a syringe through the septum in the side-arm.
5. Cover the system with aluminium foil[e] and add dropwise the solution of *t*-butylmagnesium chloride in ether (10 mL, 20 mmol) at 25 °C (**caution**! see Chapter 1, Protocol 1).
6. Heat the reaction mixture at 50 °C for 1 h.
7. Add 4-octyne (2.93 mL, 20 mmol) to the resulting clear yellow solution and stir for 6 h at 50 °C.
8. Cool the reaction mixture to −78 °C.[f]
9. Add slowly a solution of iodine (7.61 g, 30 mmol) in THF (30 mL) and let the system warm to 25 °C.
10. Add hydrochloric acid (3.0 M) to quench the reaction mixture.
11. Add pentane, transfer the mixture to a separatory funnel, and separate the organic layer from the aqueous layer.
12. Extract the aqueous layer with pentane. Combine the organic layers, wash with aqueous $NaHCO_3$, $Na_2S_2O_3$, and brine, then dry ($MgSO_4$), and filter.

13. Concentrate the filtrate under reduced pressure using a rotary evaporator (25°C/20 mmHg).

14. Distil the crude product to obtain (*E*)-4-iodo-4-octene (b.p. 60–65°C/0.1 mmHg; 3.80 g, 80% yield; displays appropriate spectroscopic and analytical data).

[a]Distilled from lithium aluminium hydride under an inert atmosphere.
[b]May be replaced by toluene.
[c]Commercially available.
[d]This compound may be handled in air. However, it appears desirable to minimize its exposure to air, moisture, and light.
[e]Alkynyl- and alkenylzirconium derivatives as well as chlorohydridozirconium are sensitive to air, moisture, and light.
[f]The shrinkage in volume must be compensated for with nitrogen or argon.

3. *anti*-Stereoselective hydrometallation. Hydroalumination of propargyl alcohols

$$RC{\equiv}CR \xrightarrow{H\bar{A}lX_3} (R)(H)C{=}C(\bar{A}lX_3)(R) \tag{13}$$

Internal alkynes react with $LiAlH_4$[33] and other tetracoordinate aluminium hydrides[34] to give the corresponding *anti*-addition products (eqn 13). These reactions are generally much more sluggish than the corresponding *syn*-hydroalumination reactions, limiting their synthetic utility. However, those alkynes that contain electron-withdrawing substituents, such as alkynyl, CN, and Cl, readily react with $LiAlH_4$ even at or below room temperature.[4] Particularly attractive are the reactions of propargylic alcohols with $LiAlH_4$ and related compounds.[35] In cases where either a combination of $LiAlH_4$ and NaOMe[12] or $Na_2AlH_2(OCH_2CH_2OCH_3)_2$ (Red-Al)[36,37] is used, γ-aluminoallyl alcohols are obtained in a regio- and stereoselective manner. On the contrary, the use of either a combination of $LiAlH_4$ and $AlCl_3$ or preferably *n*-BuLi for lithiation of alcohols followed by *i*-Bu_2AlH[13] leads to the formation of β-aluminoallyl alcohols (eqn 14). These provide valuable synthons for essentially 100% regio- and stereoselective syntheses of trisubstituted alkenes of terpenoid origin.

$$RC{\equiv}CCH_2OH \xrightarrow{LiAlH_4,\ NaOMe\ \text{or Red-Al}} \text{(cyclic alkenyl-Al–O species, R and H on the alkene)} \xrightarrow{I_2} (R)(I)C{=}C(H)CH_2OH$$

$$RC{\equiv}CCH_2OH \xrightarrow{LiAlH_4,\ AlCl_3\ \text{or } n\text{-BuLi then DIBAL}} \xrightarrow{I_2} (R)(H)C{=}C(I)CH_2OH \tag{14}$$

Protocol 2.
Synthesis of (*Z*)-3-iodo-2-buten-1-ol.[38] Reaction of 2-butyn-1-ol with Red-Al[39]

Caution! All procedures should be carried out in a well-ventilated hood, and disposable vinyl or latex gloves and chemical-resistant safety goggles should be worn.

$$H_3C{-}C{\equiv}C{-}CH_2OH \xrightarrow[\text{2. } I_2]{\text{1. Red-Al}} (Z)\text{-}H_3C(I)C{=}CH{-}CH_2OH$$

Equipment

- Magnetic stirrer
- One-necked, round-bottomed flask (1 L) with a side-arm
- Tubing adapter
- Septum
- Magnetic stirring bar
- Glass syringes with needle-lock Luers (volume appropriate for quantities of solutions to be transferred)
- Six-inch medium-gauge needles
- Nitrogen or argon line

Materials

- Dry THF 300 mL — **flammable, irritant**
- 2-Butyn-1-ol[a] (FW 70.09) 7.01 g, 7.48 mL, 100 mmol — **irritant**
- Sodium bis(2-methoxyethoxy)dihydridoaluminate (Red-Al)[a] 3.4 M in toluene, 38.24 mL, 130 mmol — **flammable, moisture-sensitive**
- Iodine (FW 253.81) 50.76 g, 200 mmol — **corrosive, toxic**
- Ether for extraction — **flammable, irritant**
- Technical hexane — **flammable, irritant**

1. Clean glassware, syringes, needles, and stirring bar and dry for at least 4 h in a 120 °C electric oven before use.
2. Assemble the round-bottomed flask equipped with the magnetic stirring bar, adapter, and septum while hot and flush the system with nitrogen or argon.
3. Charge the flask with 2-butyn-1-ol (7.48 mL, 100 mmol) using a syringe through the septum in the side-arm.
4. Add THF (200 mL).
5. Cool the mixture to 0°C[b] and add dropwise the Red-Al (38.23 mL, 130 mmol).
6. Stir for 2 h.
7. Add a solution of iodine (50.76 g, 200 mmol) in THF (100 mL) and stir the mixture for 2 h.
8. Add to the reaction mixture aqueous NH_4OH and aqueous $Na_2S_2O_3$.
9. Add ether and filter the mixture through a Celite® pad.

10. Transfer the mixture to a separatory funnel, and separate the organic layer from the aqueous layer.
11. Extract the aqueous layer with ether. Combine the organic layers, wash with brine, then dry ($MgSO_4$), and filter.
12. Concentrate the filtrate under reduced pressure using a rotary evaporator (25°C/20 mmHg).
13. Filter through a silica gel pad (*n*-hexane/ether 4:1 v/v) and concentrate to obtain (*Z*)-3-iodo-2-buten-1-ol (17.03 g, 86% yield; displays appropriate spectroscopic and analytical data).

[a]Commercially available.
[b]The shrinkage in volume must be compensated for with nitrogen or argon.

4. Stereorandom hydrometallation. Hydrostannation

There are some hydrometallation reactions in which the stereochemistry is strongly substrate-dependent and tends to be non-selective. In some special cases, however, they can be highly stereoselective, although either *syn*- or *anti*-selectivity may be observed. Currently, the most representative example of this class of reactions is hydrostannation.[16] The hydrostannation reaction of alkynes is normally carried out at 50–100°C with or without the use of radical initiators. This reaction most probably proceeds via radical and/or polar mechanisms. Although a mixture of regio- and stereoisomers is usually obtained, some special cases, such as hydrostannation of alkynes containing CH_2OTHP,[40] $SnBu_3$,[41] and CN[16] (eqns 15–17), as well as some cyclic hydrostannation reactions such as that shown in eqn 18,[42] are almost totally regio- and stereoselective.

$$HC{\equiv}CCH_2OTHP \xrightarrow[80°C]{HSnBu_3} (E)\text{-}Bu_3Sn(H)C{=}C(H)CH_2OTHP \qquad (15)$$

$$HC{\equiv}CSnBu_3 \xrightarrow{HSnBu_3} (E)\text{-}Bu_3Sn(H)C{=}C(H)SnBu_3 \qquad (16)$$

$$HC{\equiv}CCN \xrightarrow[20°C]{HSnEt_3} H_2C{=}C(CN)SnBu_3 \qquad (17)$$

$$\text{1,4-pentadiyne} \xrightarrow{H_2SnBu_2} \text{1,1-dibutylstannacyclohexa-2,5-diene} \qquad (18)$$

5. *syn*-Stereoselective carbometallation

5.1 Carbocupration

$$\mathrm{RC{\equiv}CZ} \xrightarrow[\text{Z = H, C, or heteroatom group}]{\mathrm{R^1CuL}_n} \mathrm{(R)(R^1)C{=}C(Z)(CuL}_n) \qquad (19)$$

Various types of organocoppers react with alkynes to give *syn*-addition products[7,8] (eqn 19). Both regio- and stereoselectivities can be essentially 100%. Alkenylcoppers are versatile synthetic intermediates. In addition to protonolysis, deuterolysis, and halogenolysis, the copper-containing group can be substituted for those containing carbon, silicon, tin, phosphorus, sulfur, and so on. Substitution with a carbon group can be achieved by direct cross coupling, palladium catalyzed cross-coupling, and additions to aldehydes, ketones, CO_2, CS_2, ArNCO, epoxides, and enones.[7,8] Those organocoppers that have been successfully used for this purpose include $RCu{\cdot}MgX_2$, R_2CuMgX, R_2CuLi, RCu(Y)Li, where Y = OBu-*t*, SPh, CN, etc. The choice of reagents depends on the alkyne structure, and the reaction conditions must be carefully chosen for a given combination of reactants. A few representative examples are shown in eqns 20[43] and 21.[44] The reaction shown in eqn 21 represents the currently most general route to β,β′-disubstituted alkenylmetals via carbometallation. There are, however, some significant limitations. Most critically, methylcupration is too sluggish to be practically useful. Neither allylcupration nor benzylcupration has been reported. The regioselectivity of carbocupration of alkynes containing proximal heteroatom substituents, e.g OR, SR, NR_2, and PR_2, is unpredictable and low in some cases.[45] This reaction has been extensively reviewed,[7,8] and some detailed procedures corresponding to eqns 20[46] and 21[47] have been described.

$$2\ \mathrm{HC{\equiv}CH} \xrightarrow[-50\ \text{to}\ -20^\circ\mathrm{C}]{\mathrm{R_2CuLi}} (\mathrm{R{-}CH{=}CH{-}})_2\mathrm{CuLi} \xrightarrow[0^\circ\mathrm{C}]{2\ \mathrm{HC{\equiv}CH}} (\mathrm{R{-}CH{=}CH{-}CH{=}CH{-}})_2\mathrm{CuLi} \qquad (20)$$

$$\mathrm{RC{\equiv}CH} \xrightarrow[\mathrm{Et_2O},\ -35\ \text{to}\ -15^\circ\mathrm{C}]{\mathrm{R^1Cu\cdot MgBr_2}} \mathrm{(R)(R^1)C{=}C(H)(Cu\cdot MgBr_2)} \qquad (21)$$

5.2 Zirconium-catalyzed carboalumination

$$\mathrm{RC{\equiv}CZ} \xrightarrow[\mathrm{(CH_2Cl)_2},\ 25^\circ\mathrm{C}]{\mathrm{R^1AlX_2,\ X_2ZrCp_2}} \mathrm{(R)(R^1)C{=}C(Z)(AlX_2)} \qquad (22)$$

X = halogen or C group. Z = H, C, or heteroatom group.

Although acetylene itself reacts with trialkylalanes to give (*Z*)-1-alkenylalanes in good yields,[25] the corresponding reaction of terminal alkynes mainly gives alkynylalanes, and that of internal alkynes tends to produce dimers and oligomers of alkynes.[48] These difficulties can be overcome by using Cl_2ZrCp_2 and other zirconocene derivatives as catalysts (eqn 22). The zirconium-catalyzed carboalumination discovered and developed by Negishi[10,49] proceeds well with alanes containing methyl,[10,49] allyl,[50] and benzyl[50] groups. The use of trialkylalanes containing ethyl and higher alkyl groups in conjunction with Cl_2ZrCp_2 leads to competitive hydroalumination, which can be alleviated by using other reagent combinations, such as R_2AlCl and Cl_2ZrCp_2.[10] Terminal alkynes react with 2 equivalents of Me_3Al and 0.05 to 1 equivalent of Cl_2ZrCp_2 to give high yields of the two *syn*-addition products (95:5 regioisomeric mixture). The stereoselectivity is ⩾98%. Internal alkynes react more slowly. In cases where the two alkyne substituents are different, mixtures of two regioisomers are usually formed.[49] Addition of a small amount of water accelerates the reaction.[51]

The zirconium-catalyzed methylalumination can accommodate various heterofunctional groups, such as OH, $OSiMe_2Bu$-*t*, SPh, I, and $SiMe_3$.[52] Neither conjugated nor isolated alkenyl groups interfere with the reaction. In contrast with carbocupration of propargylic and homopropargylic derivatives containing heterofunctional groups, the zirconium-catalyzed carboalumination of these substrates is not only stereoselective (>98%) but also maintains the normal regiochemistry[52] (eqn 23). On the contrary, the reactions of 1-metallo-ω-halo-1-alkynes with $Me_3Al–Cl_2ZrCp_2$ proceed beyond the expected carbometallation to give cyclic products.[53,54] Some of these reactions involve regiospecific σ-cyclization[55] while the others proceed via π-cyclization processes[54] as in eqn 24.

$$Z(H_2C)_nC\equiv CH \xrightarrow{Me_3Al,\ Cl_2ZrCp_2} Zr(H_2C)_n(Me)C{=}C(H)AlMe_2 \quad (23)$$

Z = OH, $OSiMe_2Bu$-*t*, SPh, I, $SiMe_3$, etc. , *n* = 1 or 2

Me, AlMe2, SiMe3, CD2, Br, Me3Al, Cl2ZrCp2, D, Me, AlMe2Br, SiMe3 (24)

Me, SiMe3, D, D + Me, SiMe3, D, D (1:1)

Protocol 3.
Synthesis of (*E*)-1-iodo-2-phenyl-1-propene. Reaction of phenylacetylene with Me_3Al/Cl_2ZrCp_2[10,49,56]

Caution! Me_3Al is extremely pyrophoric and all organoaluminium compounds are air and moisture sensitive. These compounds must therefore be handled under an inert atmosphere of nitrogen or argon. All procedures should be carried out in a well-ventilated hood, and disposable vinyl or latex gloves and chemical-resistant safety goggles should be worn.

Ph—≡—H → (1. Me_3Al / Cl_2ZrCp_2; 2. I_2) → (*E*)-Ph(H_3C)C=CHI

Equipment

- Magnetic stirrer
- One-necked, round-bottomed flask (100 mL) with a side-arm
- Septum
- Tubing adapter
- Magnetic stirring bar
- Glass syringes with needle-lock Luers (volume appropriate for quantities of solutions to be transferred)
- Six-inch, medium-gauge needles
- Nitrogen or argon line
- Water-jacketed, short-path distillation apparatus

Materials

• Dry 1,2-dichloroethane[a] 25 mL	**flammable, cancer suspect agent**
• Dichlorobis(η^5-cyclopentadienyl)zirconium[b,c] (FW 292.32) 2.92 g, 10 mmol	**irritant, moisture sensitive**
• Trimethylaluminium[b,d] (FW 72.09) 1.44 g, 1.92 mL, 20 mmol	**pyrophoric, moisture sensitive**
• Phenylacetylene[b] (FW 102.14) 1.02 g, 1.11 mL, 10 mmol	**flammable, lachrymator**
• Iodine (FW 253.81) 3.04 g, 12 mmol	**corrosive, toxic**
• Dry THF 15 mL	**flammable, irritant**
• Ether for extraction	**flammable, irritant**

1. Clean glassware, syringes, needles, and stirring bar and dry for at least 4 h in a 120°C electric oven before use.
2. Assemble the round-bottomed flask equipped with the magnetic stirring bar, adapter, and septum while the glassware is still hot and flush the system with nitrogen or argon.
3. Charge the flask with dichlorobis(η^5-cyclopentadienyl)zirconium (2.92 g, 10 mmol).
4. Add 1,2-dichloroethane (25 mL) using a syringe through the septum.
5. Likewise, add trimethylaluminium (1.92 mL, 20 mmol) at 25°C (**caution**! See chapter 1, Protocol 1).
6. Wait until the dichlorobis(η^5-cyclopentadienyl)zirconium dissolves to give a yellow solution (10–15 min).

7. Add the phenylacetylene (1.11 mL, 10 mmol) at 25°C and stir for 24 h.
8. Cool the reaction mixture to 0°C.[e]
9. Add a solution of iodine (3.04 g, 12 mmol) in THF (15 mL).
10. When the iodine color fades, add a mixture of water and ether to the reaction mixture.
11. Transfer the mixture to a separatory funnel and separate the organic layer from the aqueous layer.
12. Wash the organic layer with aqueous $Na_2S_2O_3$, dry ($MgSO_4$), and filter.
13. Concentrate the filtrate under reduced pressure using a rotary evaporator (25°C/20 mmHg).
14. Distil the crude product under reduced pressure to obtain (*E*)-1-iodo-2-phenyl-1-propene (b.p. 72.5°C/0.5 mmHg; 1.79 g, 73% yield; >98% stereoisomeric purity and >97% regioisomeric purity; displays appropriate spectroscopic and analytical data).

[a]Distilled from calcium hydride under an inert atmosphere.
[b]Commercially available.
[c]This may be handled in air. However, it appears desirable to minimize its exposure to air, moisture, and light.
[d]Me_3Al can also be used in solution. It is commercially available in solutions in hexanes, heptane, and toluene.
[e]The shrinkage in volume must be compensated for with nitrogen or argon.

The aluminium-containing groups of alkenylalanes can be readily replaced by various groups containing H, D, halogen, Hg, B, Zr, Cu, C, and so on with essentially complete retention of stereochemistry.[57] These reactions have been applied to the synthesis of various natural products and related organic compounds.[57]

The methylalumination of alkynes with $Me_3Al–Cl_2ZrCp_2$ appears to involve direct addition of a Me–Al bond, rather than a Me–Zr bond, to alkynes via a four-centered concerted process which is facilitated by a $ZrCp_2$ species, as depicted in eqn 25.[10]

$$\mathrm{RC{\equiv}CH} \xrightarrow[\mathrm{Cl_2ZrCp_2}]{\mathrm{Me_3Al}} \underset{\mathrm{Me}}{\overset{\mathrm{RC{\equiv}CH}}{\mathrm{Me{-}\overset{+\delta}{Al}{\cdots}X{\cdots}\overset{-\delta}{Zr}Cp_2ClY}}} \longrightarrow \mathrm{(R)(Me)C{=}C(H)(AlMeX)} + \mathrm{ZrCp_2ClY} \quad \text{X, Y = Me or Cl} \tag{25}$$

6. *anti*-Stereoselective carbometallation. Copper-catalyzed carbomagnesiation of propargyl alcohols

The copper-catalyzed carbomagnesiation reaction of propargyl alcohols discovered by Duboudin and co-workers[14,15] (eqn 26) is synthetically related to

$$RC{\equiv}CCH_2OH \xrightarrow[10\%\ CuI]{R^1MgX} \text{(cyclic alkenylmagnesium alkoxide, Mg–O ring, R, } R^1\text{)} \xrightarrow{E^+} \text{(R)(E)C=C(}R^1\text{)CH}_2\text{OH} \quad (26)$$

R = H, alkyl, aryl, alkenyl, and $SiMe_3$. R^1 = alkyl, aryl, allyl, and benzyl.

the *anti*-addition reaction of $LiAlH_4$ (section 3). Both can be essentially 100% regio- and stereoselective. Whereas the latter produces di- and trisubstituted alkenes, the copper-catalyzed carbomagnesiation provides a selective route to tri- and tetrasubstituted alkenes. The product yields in some cases are at best modest, and further improvement of this potentially valuable methodology is very desirable. Although a mechanism involving bimolecular nucleophilic attack on alkynes has been proposed,[14] unimolecular one-electron transfer and even concerted, i.e. $[\pi 2_s + \sigma 2_a]$, processes cannot be ruled out at this point. Since the products are largely alkenylmagnesium derivatives, a wide variety of groups can be introduced through replacement of magnesium, even though the reactivity of the cyclic alkenylmagnesium derivatives appears to be considerably lower than that of the usual Grignard reagents.

Protocol 4.
Synthesis of (*Z*)-2-allyl-3-iodo-2-buten-1-ol.[58] Reaction of 2-butyn-1-ol with allylmagnesium bromide[14,15]

Caution! All procedures should be carried out in a well-ventilated hood, and disposable vinyl or latex gloves and chemical-resistant safety goggles should be worn.

$$H_3C{-}C{\equiv}C{-}CH_2OH \xrightarrow[2.\ I_2]{1.\ AllylMgBr,\ CuI\ (10mol\%)} \text{(}H_3C\text{)(I)C=C(}CH_2CH{=}CH_2\text{)(}CH_2OH\text{)}$$

Equipment

- Magnetic stirrer
- One-necked round-bottomed flask (250 mL) with a side-arm
- Septum
- Tubing adapter
- Magnetic stirring bar
- Glass syringes with needle-lock Luer (volume appropriate for quantities of solutions to be transferred)
- Six-inch, medium-gauge needles
- Nitrogen or argon line
- Water-jacketed, short-path distillation apparatus

Materials

• Dry ether 45 mL	**flammable, irritant**
• 2-Butyn-1-ol[a] (FW 70.09) 3.15 g, 3.37 mL, 45 mmol	**irritant**
• Copper iodide[a] (FW 190.44) 857 mg, 4.5 mmol	**irritant, light sensitive**
• Allylmagnesium bromide[a] 0.81 M in ether, 139 mL, 112.5 mmol	**flammable, moisture sensitive**
• Iodine (FW 253.81) 20.05 g, 79 mmol	**corrosive, toxic**
• Dry THF 40 mL	**flammable, irritant**
• Ether for extraction	**flammable, irritant**

1. Clean glassware, syringes, needles, and stirring bar and dry for at least 4 h in a 120 °C electric oven before use.
2. Assemble the round-bottomed flask equipped with the magnetic stirring bar, adapter, and septum while the glassware is still hot and flush the system with nitrogen or argon.
3. Charge the flask with the copper iodide (857 mg, 4.5 mmol).
4. Add the ether (45 mL) using a syringe through the septum in the side-arm.
5. Likewise, add 2-butyn-1-ol (3.37 mL, 45 mmol).
6. Cool the mixture to 0°C[b] and add dropwise the allylmagnesium bromide (139 mL, 112.5 mmol) (**caution**! see Chapter 1, Protocol 1).
7. Warm the reaction mixture to 25 °C and stir for 18 h.
8. Cool the reaction mixture to −78 °C.[b]
9. Add a solution of iodine (20.05 g, 79 mmol) in THF (40 mL), stir 1 h at −78 °C, then let warm to 25 °C.
10. Quench the reaction mixture with saturated aqueous NH_4Cl.
11. Add ether, transfer the mixture to a separatory funnel, and separate the organic layer from the aqueous layer.
12. Extract the aqueous layer with ether. Combine the organic layers, wash with aqueous $NaHCO_3$, $Na_2S_2O_3$, and brine, then dry ($MgSO_4$), and filter.
13. Concentrate the filtrate under reduced pressure using a rotary evaporator (25 °C/20 mmHg).
14. Distil the crude product under reduced pressure to obtain (*Z*)-2-allyl-3-iodo-2-buten-1-ol (b.p. 65–70 °C/0.1 mmHg; 7.82 g, 73% yield; displays appropriate spectroscopic data).

[a]Commercially available.
[b]The shrinkage in volume must be compensated for with nitrogen or argon.

A detailed description of various techniques for laboratory operations in an inert atmosphere is available.[1] The use of a one-necked, round-bottomed flask with a side-arm is suggested throughout this chapter, although other arrangements would also be suitable (see Chapter 1).

References

1. Brown, H. C. *Organic Synthesis via Boranes*; Wiley-Interscience: New York, **1975**.
2. Pelter, A.; Smith, K.; Brown, H. C. *Borane Reagents*: Academic: New York, **1988**.
3. Lehmkuhl, H.; Ziegler, K.; Gelbert, H. G. In *Houben–Weyl Methoden der Organischen Chemie*; Müller, G. ed.; Georg Thieme Verlag: Stuttgart, **1970**; Vol. 13, pp. 1–314.
4. Zweifel, G.; Miller, J. A. In *Organic Reactions*; Dauben, W. G. ed.; Wiley: New York, **1984**; Vol. 32, pp. 375–517.

5. Labinger, J. A. In *Comprehensive Organic Synthesis*; Trost, B. M.; Fleming, I., eds.; Pergamon: Oxford, **1991**; Vol. 8, Chapter 3.9.
6. Speier, J. L. *Adv. Organomet. Chem.* **1979**, *17*, 407–447.
7. Normant, J. F.; Alexakis, A. *Synthesis* **1981**, 841–870.
8. Knochel, P. In *Comprehensive Organic Synthesis*; Trost, B. M.; Fleming, I., eds.;Pergamon: Oxford, **1991**; Vol. 4, Chapter 4.4.
9. Negishi, E. *Pure Appl. Chem.* **1981**, *53*, 2333–2356.
10. Negishi, E.; Van Horn, D. E.; Yoshida, T. *J. Am. Chem. Soc.* **1985**, *107*, 6639–6647.
11. Mikhailov, B. M. *Organomet. Chem. Rev. A* **1972**, *8*, 1–65.
12. Corey, E. J.; Katzenellenbogen, J. A.; Posner, G. H. *J. Am. Chem. Soc.* **1967**, *89*, 4245–4247.
13. Corey, E. J.; Kirst, H. A.; Katzenellenbogen, J. A. *J. Am. Chem. Soc.* **1970**, *92*, 6314–6319.
14. Duboudin, J. G.; Jousseaume, B. *J. Organomet. Chem.* **1979**, *168*, 1–11.
15. Duboudin, J. G.; Jousseaume, B.; Bonakdar, A. *J. Organomet. Chem.* **1979**, *168*, 227–232.
16. Leusink, A. J.; Budding, H. A.; Marsman, J. W. *J. Organomet. Chem.* **1967**, *9*, 285–294. Leusink, A. J.; Budding, H. A.; Drenth, W. *J. Organomet. Chem.* **1967**, *9*, 295–306.
17. Wulff, W. D. In *Comprehensive Organic Synthesis*; Trost, B. M.; Fleming, I., eds.;Pergamon: Oxford, **1991**; Vol. 5, Chapter 9.2.
18. Negishi, E. *Acc. Chem. Res.* **1987**, *20*, 65–72.
19. Negishi, E. In *Comprehensive Organic Synthesis*; Trost, B. M.; Fleming, I., eds., Pergamon: Oxford, **1991**; Vol. 5, Chapter 9.5.
20. Negishi, E.; Van Horn, D. E. *J. Am. Chem. Soc.* **1977**, *99*, 3168–3170.
21. Hart, D. W.; Blackburn, T. F.; Schwartz, J. *J. Am. Chem. Soc.* **1975**, *97*, 679–680.
22. Colvin, E. *Silicon in Organic Synthesis*; Butterworths: London, **1981**; pp. 44–96.
23. Brown, H. C.; Zweifel, G. *J. Am. Chem. Soc.* **1959**, *81*, 1512–1512.
24. Pelter, A.; Singaram, S.; Brown, H. C. *Tetrahedron Lett.* **1983**, *24*, 1433–1436.
25. Wilke, G.; Müller, H. *Chem. Ber.* **1956**, *89*, 444–447; *Justus Liebigs Ann. Chem.* **1960**, *629*, 222–240.
26. Wailes, P. C.; Weigold, H.; Bell, A. P. *J. Organomet. Chem.* **1971**, *27*, 373–378.
27. Sun, R. C.; Okabe, M.; Coffen, D. L.; Schwartz, J. *Org. Synth.* **1992**, *71*, 83–88.
28. Hart, D. W.; Schwartz, J. *J. Am. Chem. Soc.* **1974**, *96*, 8115–8116.
29. Buchwald, S. L.; LaMaire, S. J.; Nielsen, R. B.; Watson, B. T.; King, S. M. *Tetrahedron Lett.* **1987**, *28*, 3895–3898; *Org. Synth.* **1992**, *71*, 77–82.
30. Swanson, D. R.; Nguyen, T.; Noda, Y.; Negishi, E. *J. Org. Chem.* **1991**, *56*, 2590–2591.
31. Negishi, E.; Miller, J. A.; Yoshida, T. *Tetrahedron Lett.* **1984**, *25*, 3407–3410.
32. Lipshutz, B. H.; Keil, R.; Ellsworth, E. L. *Tetrahedron Lett.* **1990**, *31*, 7257–7260.
33. Slaugh, L. H. *Tetrahedron* **1966**, *22*, 1741–1746.
34. Zweifel, G.; Steele, R. B. *J. Am. Chem. Soc.* **1967**, *89*, 5085–5086.
35. Bates, E. B.; Jones, E. R. H.; Whiting, M. C. *J. Chem. Soc.* **1954**, 1854–1860.
36. Chan, K. K.; Cohen, N.; De Noble, J. P.; Specian, Jr, A. C.; Saucy, G. *J. Org. Chem.* **1976**, *41*, 3497–3505.
37. Marshall, J. A.; DeHoff, B. S. *J. Org. Chem.* **1986**, *51*, 863–872.
38. Cochrane, J. S.; Hanson, J. R. *J. Chem. Soc., Perkin 1* **1972**, 361–366.

39. Negishi, E.; Sugihara, T. Unpublished results. Marshall, J. A.; Du Bay, W. J. *J. Org. Chem.* **1991**, *56*, 1685–1687.
40. Corey, E. J.; Wollenberg, R. H. *J. Org. Chem.* **1975**, *40*, 2265–2266.
41. Corey, E. J.; Wollenberg, R. H. *J. Am. Chem. Soc.* **1974**, *96*, 5582–5583.
42. Ashe, A. J., III; Shu, P. *J. Am. Chem. Soc.* **1971**, *93*, 1804–1805.
43. Alexakis, A.; Normant, J. F. *Tetrahedron Lett.* **1982**, *23*, 5151–5154.
44. Normant, J. F.; Bourgain, M. *Tetrahedron Lett.* **1971**, 2583–2586.
45. Normant, J. F. *J. Organomet. Chem. Library* **1976**, *1*, 219–256.
46. Alexakis, A.; Cahiez, G.; Normant, J. F. *Org. Synth.* **1984**, *62*, 1–8.
47. Iyer, R. S.; Helquist, P. *Org. Synth.* **1986**, *64*, 1–9.
48. Eisch, J. J.; Fichter, K. C. *J. Am. Chem. Soc.* **1974**, *96*, 6815–6817.
49. Van Horn, D. E.; Negishi, E. *J. Am. Chem. Soc.* **1978**, *100*, 2252–2254.
50. Miller, J. A.; Negishi, E. *Tetrahedron Lett.* **1984**, *25*, 5863–5866.
51. Wipf, P.; Lim, S. *Angew. Chem., Int. Ed. Engl.* **1993**, *32*, 1068–1071.
52. Rand, C. L.; Van Horn, D. E.; Moore, M. W.; Negishi, E. *J. Org. Chem.* **1981**, *46*, 4093–4096.
53. Boardman, L. D.; Bagheri, V.; Sawada, H.; Negishi, E. *J. Am. Chem. Soc.* **1984**, *106*, 6105–6107.
54. Negishi, E.; Boardman, L. D.; Tour, J. M.; Sawada, H.; Rand, C. L. *J. Am. Chem. Soc.* **1983**, *105*, 6344–6346. Negishi, E.; Boardman, L. D.; Sawada, H.; Bagheri, V.; Stoll, A. T.; Tour, J. M.; Rand, C. L. *J. Am. Chem. Soc.* **1988**, *110*, 5383–5396.
55. Negishi, E.; Mohammud, M. M. Unpublished results.
56. Negishi, E.; Van Horn, D. E. *Organomet. Synth.* **1986**, *3*, 467–471.
57. Negishi, E.; Takahashi, T. *Aldrichim. Acta* **1985**, *18*, 31–48.
    Negishi, E. *Pure Appl. Chem.* **1992**, *74*, 323–334.
58. Negishi, E.; Zhang, Y.; Cederbaum, F. E.; Webb, M. B. *J. Org. Chem.* **1986**, *51*, 4080–4082.

8

Catalytic coupling reactions

OLIVER REISER

1. Introduction

Palladium- and nickel-catalyzed coupling reactions have become a powerful method for the functionalization of alkenes and arenes. There is a broad variety of different reactions to choose from, which in many cases can be used to reach the same synthetic end-point. Moreover, in recent years many different procedures for each type of reaction have been published, making it even more difficult to choose the right procedure for a particular synthetic problem. The following chapter will provide the reader with procedures that are applicable to a large number of substrates. Nevertheless, if a protocol should not work for a specific case, it might be worth changing the reaction conditions—even if this change appears to be rather small—since palladium- and nickel-catalyzed reactions are quite sensitive towards parameters such as the palladium source, ligand, base, solvent, temperature, and substrate.

2. Palladium-catalyzed cross-coupling reactions

In the presence of a palladium(0) catalyst, vinyl halides (bromides or iodides) or triflates undergo facile coupling with various organometallic reagents. This way alkenes can be obtained which have been formed by formal nucleophilic substitution, a reaction which is, in contrast to sp^3-centers, not possible at sp^2-centers. Depending on the organometallic reagent, aryl, vinyl, alkynyl, or alkyl groups can be transferred to the alkene, resulting in the synthesis of styrenes, 1,3-dienes, alkenynes, and alkenes. Most widely used have been organometallic reagents based on lithium, magnesium (Grignard reagents), zinc, boron (Suzuki coupling), aluminium, and tin (Stille coupling), which are synthesized either from a halide by halogen metal exchange or metal insertion, or from alkynes or alkenes by hydro- or carbometallation (Chapter 7). Despite the many possible variations in palladium-catalyzed coupling reactions, they can be described by a common mechanism (Scheme 8.1).

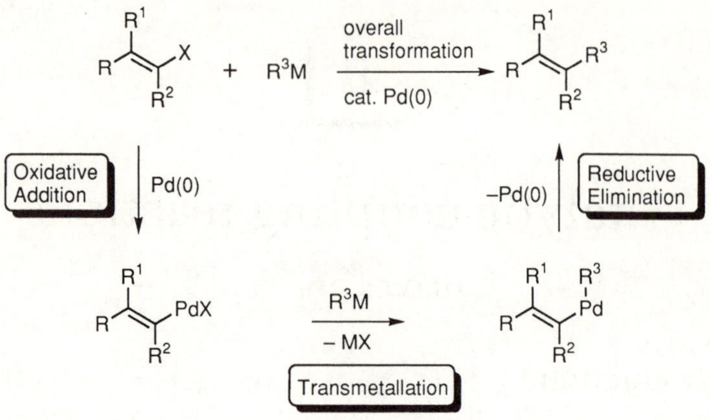

Scheme 8.1

The initial step is the oxidative addition of palladium(0) into the vinyl halide, followed by transmetallation with the organometallic reagent R^3M. Subsequent reductive elimination gives the product and also liberates the palladium(0) catalyst which can enter a new cycle.

2.1 Palladium-catalyzed coupling reactions with organolithium or Grignard reagents[1]

Organolithium and Grignard reagents have been successfully used in the reaction with vinyl bromides or iodides to transfer alkyl and vinyl groups. While organolithium reagents usually require reaction temperatures of about 80 °C (e.g. benzene at reflux) (Protocol 1), the coupling of Grignard reagents has been reported to occur at room temperature (Protocol 2). The reaction is stereospecific in most cases, and therefore the double bond geometry of the substrates is retained in the products. However, the coupling of two *cis*-configurated substrates can result in partial isomerization.[2]

Commonly used catalysts are $Pd(PPh_3)_4$ or a combination of $PdCl_2/PPh_3$, which is reduced *in situ* to palladium(0) by potassium or an alkyllithium compound.

Protocol 1.
Preparation of (*Z*)-β-[2-(*N,N*-dimethylamino)phenyl]styrene 4[3]

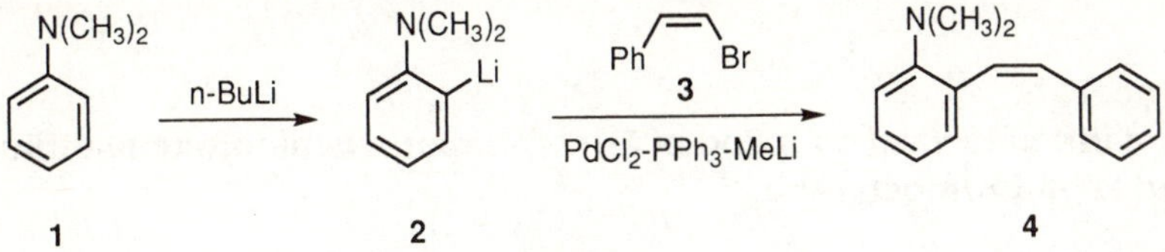

$N(CH_3)_2$ **1** —n-BuLi→ $N(CH_3)_2$, Li **2** —Ph/Br **3**, $PdCl_2$-PPh_3-MeLi→ $N(CH_3)_2$ **4**

Scheme 8.2

Caution! All procedures should be carried out in a well-ventilated hood, and disposable vinyl or latex gloves and chemical-resistant safety goggles should be worn.

Protocol 1a.
Preparation of 2-(*N,N*-dimethylamino)phenyllithium 2[3,4]

Equipment

- Three-necked, round-bottomed flask (500 mL)
- Water-jacketed reflux condenser
- Septum
- Magnetic stirrer
- Magnetic stirring bar
- Glass syringe with needle-lock Luer (50 mL)
- Eight-inch, medium-gauge needle
- Source of dry nitrogen, combined with a vacuum line

Materials

- *N,N*-Dimethylaniline (FW 121.2) 6.05 g, 50 mmol — **toxic, irritant**
- *n*-Butyllithium (FW 64.1) 1.6 M in hexane, 31.3 mL, 50 mmol — **highly flammable, moisture sensitive**

1. Ensure that all glassware is thoroughly clean and has been dried for at least 4 h in a 120 °C electric oven before use. Dry *N,N*-dimethylaniline over sodium and distil it under nitrogen before use.
2. Equip the flask with the magnetic stirring bar, and the reflux condenser. Using a tubing adapter, connect the top of the reflux condenser to the nitrogen/vacuum line, and close all remaining openings with glass stoppers.
3. Alternatively, evacuate, flame dry, and flush the apparatus with nitrogen three times. Ensure that during all the following operations a slow nitrogen stream passes through the apparatus. Whenever you open the reaction vessel, increase the nitrogen stream.
4. After the apparatus has cooled down to room temperature, charge the flask with *N,N*-dimethylaniline (6.05 g, 50 mmol), cap the neck of the flask with a septum, and add slowly via a syringe *n*-butyllithium (31.3 mL, 50 mmol) in hexane (**caution**! see Chapter 1, Protocol 1).

Protocol 1a. *Continued*

5. Stir the solution for 12 h under reflux. The resulting mixture can be used without further purification in Protocol 1b.

Protocol 1b. Palladium-catalyzed coupling of 2-(*N,N*-dimethylamino)phenyllithium 2 and (*Z*)-β-bromostyrene 3[3]

Equipment

- Three-necked, round-bottomed flask (1 L)
- Magnetic stirrer
- Magnetic stirring bar
- Water-jacketed reflux condenser
- Pressure-equalizing addition funnel (500 mL)
- Glass syringes with needle-lock Luers (2 mL and 20 mL)
- Source of dry nitrogen, combined with a vacuum line

Materials

- (*Z*)-β-Bromostyrene **3**[5] (FW 183.1) 9.15 g, 50 mmol — **irritant**
- 2-(*N,N*-Dimethylamino)phenyllithium **2** (obtained from Protocol 1a) — **moisture sensitive**
- Palladium(II) chloride (FW 177.3) 90 mg, 0.5 mmol — **irritant**
- Triphenylphosphine (FW 262.3) 525 mg, 2 mmol — **irritant**
- Methyllithium (FW 22.0) 1 M in ether, 1.1 ml, 1.1 mmol — **flammable, irritant**
- Benzene 400 ml — **flammable, carcinogen**

1. Ensure that all glassware is thoroughly clean and has been dried for at least 4 h in a 120°C electric oven before use. All solvents have to be properly dried and distilled under nitrogen.
2. Equip the flask with the magnetic stirring bar, the reflux condenser, and the addition funnel. Using a tubing adapter connect the top of the reflux condenser to the nitrogen/vacuum line, and close all remaining openings with glass stoppers.
3. Alternatively, evacuate, flame dry, and flush the apparatus with nitrogen three times. Ensure that during all the following operations a slow nitrogen stream passes through the apparatus. Whenever you open the reaction vessel, increase the nitrogen stream.
4. After the apparatus has cooled down, add $PdCl_2$ (90 mg, 0.50 mmol), PPh_3 (525 mg, 2 mmol), and benzene (100 mL) (**caution**!) to the reaction vessel. Close the neck with a septum. Stir the resulting solution under reflux for 1 h.
5. After cooling the solution to room temperature, add methyllithium (1.1 mmol, *c.* 1 M in ether) via a syringe with stirring (**caution**! See Chapter 1, Protocol 1), and heat the solution under reflux for 1 h.
6. After cooling the solution to room temperature, add (*Z*)-β-bromostyrene **3**[5] (9.15 g, 50 mmol) via a syringe in one portion.
7. Dilute the solution of 2-(*N,N*-dimethylamino)phenyllithium **2** (50 mmol)

(Protocol 1a) with benzene (300 mL) and transfer this solution to the addition funnel of the apparatus, via cannula (see Chapter 1, Protocol 2).

8. Add the solution of **2** to the well-stirred, refluxing reaction mixture over 6 h. After the addition is complete, heat the reaction mixture under reflux for an additional 30 min.
9. After cooling to room temperature, slowly add 100 mL of water. Extract the organic layer with water (100 mL) followed by brine (100 mL).
10. Dry the organic layer (Na_2SO_4), filter, and distil off the solvent at normal pressure.
11. Distil the residue under reduced pressure to obtain **4** (8.1 g, 71%); b.p. 100–103 °C (0.05 mmHg).

Whilst it is essential that the organolithium reagent is slowly added to the alkenyl halide, Grignard reagents can be added in one portion to alkenyl halides.

Protocol 2.
Preparation of (3*E*)-1,3-decadiene 7[2]

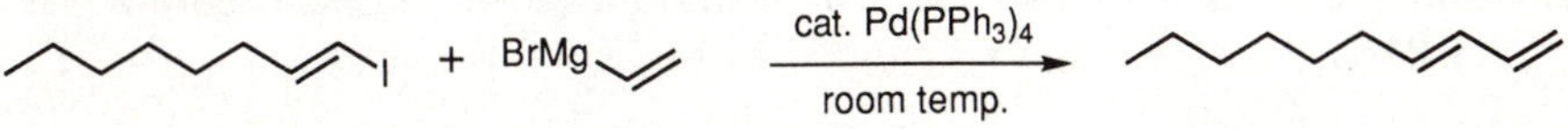

5 + BrMg **6** —(cat. $Pd(PPh_3)_4$, room temp.)→ **7**

Scheme 8.3

Caution! All procedures should be carried out in a well-ventilated hood, and disposable vinyl or latex gloves and chemical-resistant safety goggles should be worn.

Equipment

- One-necked, round-bottomed flask with side-arm (25 mL)
- Magnetic stirrer
- Magnetic stirring bar
- Septum
- Glass syringe with needle-lock Luer (10 mL)
- Column for chromatography
- Source of dry nitrogen, combined with a vacuum line

Materials

- Vinylmagnesium bromide (FW 131.3) 1 M in THF, 4.5 mL, 4.5 mmol — **flammable, moisture sensitive**
- (*E*)-1-Iodo-1-octene **5**[6] (FW 238.1) 715 mg, 3 mmol — **light sensitive, harmful**
- Tetrakis(triphenylphosphine) palladium(0) (FW 1155.6) 173 mg, 0.15 mmol — **light sensitive**
- Benzene 10 mL — **flammable, carcinogenic**
- Hexanes for chromatography — **flammable, irritant**

Protocol 2. *Continued*

1. Ensure that all glassware is thoroughly clean and has been dried for at least 4 h in a 120°C electric oven before use. All solvents have to be properly dried and distilled under nitrogen.
2. Equip the flask with a magnetic stirring bar and the neck with a septum. Using a tubing adapter connect the side-arm of the flask to the nitrogen/vacuum line.
3. Alternatively, evacuate, dry with a heat gun, and flush the flask with nitrogen three times. Ensure that during all the following operations a slow nitrogen stream passes through the apparatus. Whenever you open the reaction vessel, increase the nitrogen stream.
4. After the flask has cooled to room temperature, add $Pd(PPh_3)_4$ (173 mg, 0.15 mmol), (*E*)-1-iodo-1-octene **5** (715 mg, 3 mmol), and benzene (10 mL).
5. Add vinylmagnesium bromide **6** (4.5 mL, 4.5 mmol) in THF via a syringe (**caution**! see Chapter 1, Protocol 1), and stir the resulting mixture for 2 h at room temperature.
6. Add to the mixture saturated aqueous ammonium chloride solution (1 mL), separate the organic layer, dry ($MgSO_4$), filter, and evaporate the solvent under reduced pressure.
7. Purify the residue on a short silica column (hexanes) to obtain pure **7** (335 mg, 81%).

According to this protocol, ethylmagnesium bromide, phenylmagnesium bromide, (*E*)-1-propenyl-1-magnesium bromide, and propynylmagnesium bromide were coupled with (*E*)- or (*Z*)-1-iodo-1-octene in yields ranging between 75 and 87%.

2.2 Palladium-catalyzed coupling reactions with organozinc reagents

Analogous to Grignard reagents, organozinc compounds undergo coupling with aryl or vinyl halides in the presence of a palladium(0) catalyst with similar results. The organozinc reagents are conveniently prepared by transmetallation of Grignard and of organolithium compounds with zinc(II) chloride ($ZnCl_2$). This method is especially advantageous in the case of organolithium precursors, since these compounds often react sluggishly in palladium-catalyzed couplings with vinyl halides.[7]

A particularly valuable application of organozinc reagents is the palladium-mediated coupling of α-metallated enol ethers with vinyl halides, leading after work-up to acylated alkenes (Protocol 3, Scheme 8.4). $Pd(dba)_2$ combined with various phosphines, or alternatively $PdCl_2(PPh_3)_2$, which is reduced *in situ* by DIBAL, have been successfully employed as catalysts.

Protocol 3.
Preparation of *trans*-4-phenyl-3-buten-2-one 11 (Scheme 8.4)[8]

Caution! All procedures should be carried out in a well-ventilated hood, and disposable vinyl or latex gloves and chemical-resistant safety goggles should be worn.

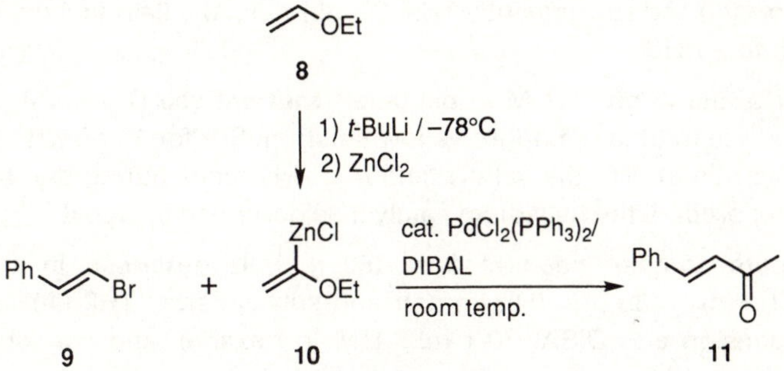

8 → 1) *t*-BuLi / –78°C; 2) $ZnCl_2$ → 10

9 + 10 → cat. $PdCl_2(PPh_3)_2$/DIBAL, room temp. → 11

Scheme 8.4

Equipment

- Two one-necked, round-bottomed, side-arm flasks (50 mL)
- One-necked, round-bottomed, side-arm flask (100 mL)
- Two magnetic stirrers
- Glass syringes with needle-lock Luers (various sizes)
- Two magnetic stirring bars
- Source of dry nitrogen, combined with a vacuum line

Materials

- (*E*)-β-Bromostyrene **9**[9] (FW 183.1) 183 mg, 1 mmol — **irritant**
- Diisobutylaluminium hydride (FW 142.2) 1.0 M in hexane, 0.1 mL, 0.1 mmol — **flammable, moisture sensitive**
- Bis(triphenylphosphine) palladium(II) chloride (FW 701.9) 35 mg, 0.05 mmol — **hygroscopic**
- $ZnCl_2$ (FW 136.3) 850 mg, 6.2 mmol — **toxic, irritant**
- Ethyl vinyl ether (FW 72.1) 0.5 mL, 5.2 mmol — **lachrymator, flammable**
- *t*-Butyllithium (FW 64.06) 1.7 M in pentane 3 mL, 5.1 mmol — **flammable, moisture sensitive**
- Dry THF 40 mL — **flammable, irritant**
- Ether for chromatography — **flammable, irritant**
- Hexane for chromatography — **flammable, irritant**

1. Ensure that all glassware is thoroughly clean and has been dried for at least 4 h in a 120 °C electric oven before use.
2. Connect a side-arm flask (50 mL) to the nitrogen/vacuum line and charge it with $ZnCl_2$ (850 mg). Heat the flask under vacuum ($<$ 1 mmHg) at 50 °C for 2 h. Store the dried $ZnCl_2$ under nitrogen.[10]
3. Alternatively, evacuate, dry with a heat gun, and flush the side-arm flask (50 mL), equipped with a magnetic stirring bar and capped with a septum,

Protocol 3. *Continued*

with nitrogen three times. Ensure that during all the following operations a slow nitrogen stream passes through the apparatus. Whenever you open the reaction vessel, increase the nitrogen stream.

4. To a separate side-arm flask (100 mL), under an atmosphere of nitrogen (see Chapter 1, Figs 1.2–1.4), add ethyl vinyl ether (freshly distilled) (0.5 mL, 5.2 mmol) and subsequently THF (30 mL). Cool the flask in a dry-ice acetone bath to −78°C.
5. Add *t*-BuLi (3 mL, 1.7 M in pentane) (**caution**! see Chapter 1, Protocol 1) dropwise to obtain a bright yellow solution. Stir for 10 min at −78°C, then for 45 min at 0°C; the yellow colour slowly fades during this time. Meanwhile, prepare the palladium catalyst as described in step 6.
6. Prepare another side-arm flask (50 mL) as described in step 3. Add $PdCl_2(PPh_3)_2$ (35 mg, 0.05 mmol) and subsequently THF (10 mL). To this suspension add DIBAL (0.1 mL, 1 M in hexane), and stir for 15 min to obtain the homogeneous, brown catalyst solution.
7. Add the $ZnCl_2$ prepared in step 2 (*c.* 6 mmol) to the metallated enol ether solution prepared in step 5. Stir for 10 min at 0°C, and for 30 min at room temperature. A clear solution should be obtained.
8. Transfer the catalyst solution prepared in step 6 via a syringe to the solution prepared in step 7 (see Chapter 1, Protocol 1 or 2).
9. Add (*E*)-β-bromostyrene **9** (183 mg, 0.13 mL, 1 mmol) and stir the reaction mixture for 6 h at room temperature
10. Add hydrochloric acid (5% solution, 3 mL), and extract the mixture with ether (2 × 30 mL). Wash the organic phase once with brine (30 mL), dry ($MgSO_4$), filter, and remove the solvent under reduced pressure.
11. Purify the residue on a short silica gel column (diethylether/hexane 4:1, R_f = 0.18) to obtain *trans*-4-phenyl-3-buten-2-one **11**[11] (123 mg, 84%).

2.3 Palladium-catalyzed coupling reactions with organotin reagents (Stille coupling)

The coupling of tetraorganotin compounds with organic halides, i.e. aryl, benzyl, and vinyl halides, in the presence of palladium(0) has become a widely used process in organic synthesis. This reaction is named after its discoverer, the late Professor Stille, whose group has made many invaluable contributions regarding mechanistic aspects, and the scope and limitation of this coupling.[12] Initially, many of these tin couplings were carried out in HMPA, but since the great toxicity of HMPA has been realized, alternative procedures have been developed.[13]

$$R^2Li \; (\mathbf{12}) \xrightarrow{XSnR_3} R^2SnR_3$$

$$\underset{\mathbf{13}}{R^1\text{-}X} + \underset{\mathbf{14}}{R^2SnR_3} \xrightarrow{\text{cat. Pd(0)}} \underset{\mathbf{15}}{R^1\text{-}R^2}$$

R^1 = Aryl, Benzyl
R^2 = Alkyl, Vinyl, Aryl

Scheme 8.5

If an unsymmetrical substituted organotin derivative is used for such coupling reactions, the most weakly bound group is selectively transferred, i.e. vinyl and aryl groups are more reactive than alkyl groups (Scheme 8.5). This allows the convenient synthesis of alkenyltributyltin reagents from an alkenyllithium precursor and tributyltin chloride ($SnBu_3Cl$), which can subsequently be used for vinyltransfer to an aryl or benzyl halide or triflate. The most reactive coupling reagents **13** are iodides, while bromides and triflates exhibit somewhat reduced reactivity. However, in the coupling of triflates it is often essential to use lithium chloride (LiCl) as an additive. Moreover, electron-withdrawing groups activate the *ortho-* and *para*-positions of an aryl derivative towards oxidative addition of palladium(0), explaining the excellent selectivity seen in the coupling of **16** with **17** (Protocol 4, Scheme 8.6).

Protocol 4.
Preparation of 2-vinyl-3-nitrophenyl triflate 18 (Scheme 8.6)[13]

Caution! All procedures should be carried out in a well-ventilated hood, and disposable vinyl or latex gloves and chemical-resistant safety goggles should be worn.

OTf, Br, NO_2 (**16**) + $SnBu_3$ (**17**) $\xrightarrow[\text{toluene / reflux}]{\text{cat. } Pd(PPh_3)_4}$ OTf, NO_2 (**18**)

Scheme 8.6

Protocol 4. *Continued*

Equipment

- Three-necked, round-bottomed flask (100 mL)
- Water-jacketed reflux condenser
- Glass syringes with needle-lock Luers (various sizes)
- Septum
- Column for chromatography
- Source of dry nitrogen, combined with a vacuum line

Materials

- 1-Bromo-2-nitrophenyltriflate **16**[13] (FW 350) 3.5 g, 10 mmol — **harmful**
- Tributylvinyltin **17** (FW 317.1) 3.17 g, 10 mmol — **irritant, toxic**
- Tetrakis(triphenylphosphine) palladium(0) (FW 1155.6) 25 mg, 0.02 mmol — **light sensitive**
- Dry toluene 50 mL — **flammable, toxic**
- Acetonitrile for extraction — **flammable, lachrymator**
- Hexane for extraction and chromatography — **flammable, irritant**
- Ethyl acetate for chromatography — **flammable, irritant**

1. Ensure that all glassware is thoroughly clean and has been dried for at least 4 h in a 120°C oven before use. Toluene has to be dried over sodium and distilled under nitrogen.
2. Equip the flask with the magnetic stirring bar and the reflux condenser. Using a tubing adapter, connect the top of the reflux condenser to the nitrogen/vacuum line. Close the remaining openings with a glass stopper and a septum.
3. Alternatively, evacuate, dry with a heat gun, and flush the apparatus with nitrogen three times. Ensure that during all the following operations a slow nitrogen stream passes through the apparatus. Whenever you open the reaction vessel, increase the nitrogen stream.
4. After cooling the apparatus to room temperature, add $Pd(PPh_3)_4$ (25 mg, 0.02 mmol).
5. Add 1-bromo-2-nitrophenyltriflate **16** (3.5 g, 10.0 mmol), toluene (50 mL), and tributylvinyltin **17** (3.17 g, 10.0 mmol), via a syringe through the septum.
6. Stir the reaction mixture under reflux for 48 h.
7. After cooling the mixture to room temperature, evaporate the solvent under reduced pressure.
8. Dissolve the residue in acetonitrile (25 mL), and extract with hexane (25 mL).
9. Evaporate the acetonitrile layer under reduced pressure, and purify the residue on a short silica gel column (EtOAc:hexane 1:4) to obtain **18** (2.7 g, 91%).

2.4 Palladium-catalyzed coupling reactions with organoboron reagents

The palladium-catalyzed reaction of vinyl- and alkyl-boranes with vinyl and aryl halides or triflates, also known as the Suzuki coupling, has become one of the most versatile processes for the synthesis of alkenes.[14] Vinylboranes, both (*E*)- and (*Z*)-isomers, are easily obtained by addition of boranes, such as disiamylborane or catecholborane (Protocol 5, Scheme 8.8), to acetylenes, allowing the synthesis of all geometrical isomers of 1,3-dienes (Scheme 8.7) (see also Chapter 7).

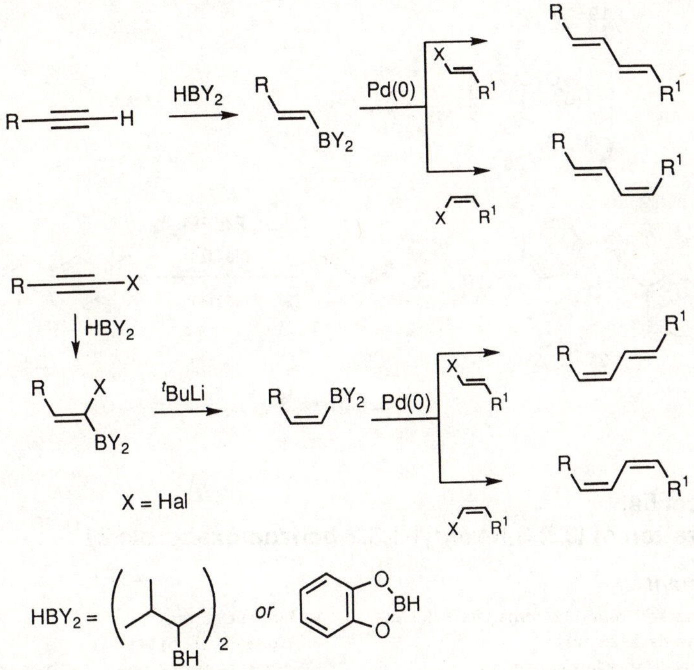

Scheme 8.7

Alkylboranes, easily obtained by the addition of boranes to alkenes, undergo analogous coupling with alkenyl halides to give alkyl-substituted alkenes in generally high yields. Moreover, the couplings of alkylboranes can be often carried out in aqueous media, making the drying of solvents unnecessary.[15]

Protocol 5.
Preparation of (1*E*,3*E*)-3-ethyl-1-phenyl-1,3-hexadiene 22 (Scheme 8.8)[16]

Caution! All procedures should be carried out in a well-ventilated hood, and disposable vinyl or latex gloves and chemical-resistant safety goggles should be worn.

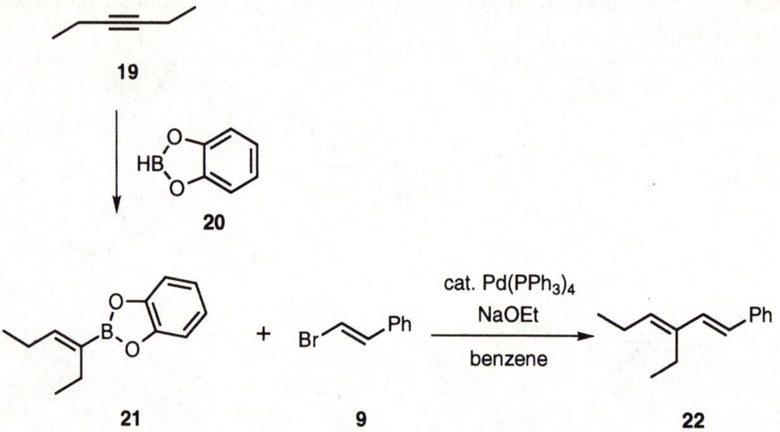

19

HB 20

21 + Br–CH=CH–Ph 9 → cat. $Pd(PPh_3)_4$, NaOEt, benzene → 22

Scheme 8.8

Protocol 5a.
Preparation of (3*Z*)-3-hexenyl-1,3,2-benzodioxaborole 21[17]

Equipment

- Three-necked, round-bottomed flask (50 mL)
- Side arm flask (25 ml)
- Water-jacketed reflux condenser
- Water-jacketed, semi-micro, short-path distillation apparatus
- Magnetic stirrer
- Magnetic stirring bar
- Source of dry nitrogen, combined with a vacuum line

Materials

- 3-Hexyne **19** (FW 82.1) 8.2 g, 100 mmol — **flammable, irritant**
- Catecholborane **20** (FW 119.9) 12.0 g, 100 mmol — **flammable, corrosive**

1. Ensure that all glassware is thoroughly clean and has been dried for at least 4 h in a 120 °C electric oven before use.
2. Equip the three-necked flask with the magnetic stirring bar and the reflux condenser. Using a tubing adapter, connect the top of the reflux condenser to the nitrogen/vacuum line, and close all remaining openings with glass stoppers.

3. Alternatively, evacuate, dry with a heat gun, and flush the apparatus with nitrogen three times. Ensure that during all the following operations a slow nitrogen stream passes through the apparatus. Whenever you open the reaction vessel, increase the nitrogen stream.
4. Charge the three-necked flask with 3-hexyne **19** (8.2 g, 100 mmol) and catecholborane **20** (12.0 g, 100 mmol). Stir the reaction mixture at 70°C for 4 h.
5. Cool the reaction mixture to room temperature.
6. Connect the side-arm flask to the distillation apparatus as a receiver vessel, and connect the side-arm to the nitrogen line, and flush this apparatus with nitrogen.
7. Connect the distillation apparatus to the three-necked flask, while maintaining a nitrogen stream through the apparatus via the side-arm flask.
8. Remove the reflux condenser of the reaction flask and close the opening with a glass stopper.
9. Switch off the nitrogen stream and slowly apply vacuum (0.1–0.2 mmHg) to the apparatus.
10. Distil the product (17.2 g (85%), b.p. 81°C, 0.2 mmHg; 63°C, 0.1 mmHg). After the distillation is complete, close the vacuum line and flush the apparatus with nitrogen. Remove the side-arm flask containing the product, while the flask is still connected to the open nitrogen line, from the distillation apparatus. Close the flask with a glass stopper. Close the nitrogen line to the flask.

Protocol 5b. Palladium-catalyzed reaction of 21 and 9[16]

Equipment

- Three-necked, round-bottomed flask (50 mL)
- Side-arm flask (25 mL)
- Water-jacketed reflux condenser
- Septum
- Magnetic stirrer
- Magnetic stirring bar
- Glass syringes with needle-lock Luers
- Six-inch, medium-gauge needles
- Source of dry nitrogen, combined with a vacuum line

Materials

- (*E*)-β-Bromostyrene **9**[9] (FW 183.1) 915 mg, 5 mmol — **irritant**
- (3*Z*)-3-Hexenyl-1,3,2-benzodioxaborole **21** (Protocol 5a) — **air sensitive**
- Tetrakis(triphenylphosphine) palladium(0) (FW 1155.6) 58 mg, 0.05 mmol — **light sensitive**
- Sodium (FW 23.0) 460 mg, 20 mmol — **flammable, moisture sensitive**
- Dry ethanol 10 mL — **flammable, toxic**
- Dry benzene 15 mL — **flammable, carcinogenic**

Protocol 5b. *Continued*

1. Ensure that all glassware is thoroughly clean and has been dried for at least 4 h in a 120 °C electric oven before use. Dry benzene over sodium and distil it under nitrogen before use.
2. Equip the three-necked flask with the magnetic stirring bar, the reflux condenser, and the septum. Using a tubing adapter, connect the top of the reflux condenser to the nitrogen/vacuum line. Close the remaining opening with a glass stopper.
3. Alternatively, evacuate, dry with a heat gun, and flush the apparatus with nitrogen three times. Ensure that during all the following operations a slow nitrogen stream passes through the apparatus. Whenever you open the reaction vessel, increase the nitrogen stream.
4. After the apparatus has cooled down, add $Pd(PPh_3)_4$ (58 mg, 0.05 mmol).
5. To the three-necked flask, add, via a syringe, benzene (15 ml) and subsequently (*E*)-β-bromostyrene **9** (0.915 g, 5 mmol). Stir the solution for 30 min at room temperature.
6. During this time prepare a 2 M solution of NaOEt in ethanol: connect the side-arm flask to the nitrogen line and pass a slow stream of nitrogen through the flask. Add anhydrous ethanol (10 mL) and subsequently small pieces of freshly cut sodium (460 mg, 20 mmol) (**caution**! hydrogen is evolved). Stir the mixture until the sodium has dissolved.
7. Add (3*Z*)-3-hexenyl-1,3,2-benzodioxaborole **21** (1.11 g, 5.5 mmol, prepared according to Protocol 5a) to the reaction mixture in the three-necked flask (prepared in step 5).
8. To the three-necked flask, add 5 mL of the NaOEt solution which was prepared in step 6, using a syringe. Heat the mixture for 2 h under reflux. A white solid (NaBr) should slowly precipitate.
9. After cooling the reaction mixture to room temperature, add sodium hydroxide (aq.) (3 M, 5 mL). Stir the mixture for 2 h (hydrolysis of the boronic ester).
10. Extract the mixture with a toluene/hexane (1:1) solution (3 × 20 mL), dry ($MgSO_4$), filter, and remove the solvent under reduced pressure. Distil the residue to obtain **22** (750 mg, 81%; b.p. 138–139 °C at 15 mmHg).

---

3. Nickel-catalyzed cross-coupling reactions

Similar to the palladium-catalyzed cross-coupling reactions described in the previous section, Grignard reagents can also be reacted with alkenyl and aryl halides under nickel catalysis.[18] Several dihalodiphosphine nickel derivatives have been successfully employed as catalysts, but bidentate phosphines, especially 1,3-bis(diphenylphosphino)propane (dppp), 1,1′-bis(dimethylphosphino)ferrocene (dmpf), and 1,2-bis(diphenylphosphino)ethane (dppe), have

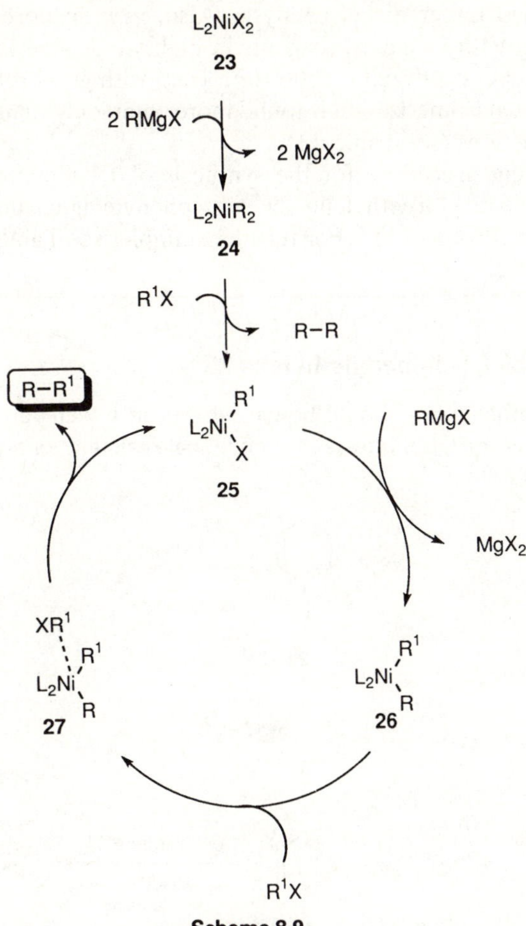

Scheme 8.9

been found to give considerably more reactive nickel complexes than monodentate ligands.[19]

The mechanism of such couplings has been similarly formulated as for the palladium-catalyzed couplings of Grignard reagents. Thus, the nickel complex **23** reacts initially with two molecules of the Grignard compound to form **24**. Reductive elimination of R–R from such nickelorganyl complexes affords a coordinatively unsaturated nickel(0) species which undergoes oxidative addition with a vinyl halide to yield **25**. Reaction of **25** with the Grignard reagent then forms the new complex **26**, from which the cross-coupling product is released by the attack of the vinyl halide, possibly via the pentacoordinated intermediate **27**, and thereby regenerating **25** to complete the catalytic cycle (Scheme 8.9).

Contrary to the palladium-catalyzed reactions, vinyl chlorides can generally be coupled under nickel catalysis. Also, *gem*-dichloroalkenes undergo cross coupling with Grignard reagents in high yields (see Protocol 6). However, while nickel-catalyzed reactions proceed with good retention of alkene geometry, partial isomerization is found more commonly than in the analogous palladium reactions (see Table 8.1).

The following procedure for the synthesis of 1,1-diphenylethylene **28** by reaction of 1,1-dichloroethylene **29** with phenylmagnesium bromide **30** is representative (Protocol 6).[19] For further examples see Table 8.1.

Protocol 6.
Preparation of 1,1-diphenylethylene 28[19]

Caution! All procedures should be carried out in a well-ventilated hood, and disposable vinyl or latex gloves and chemical-resistant safety goggles should be worn.

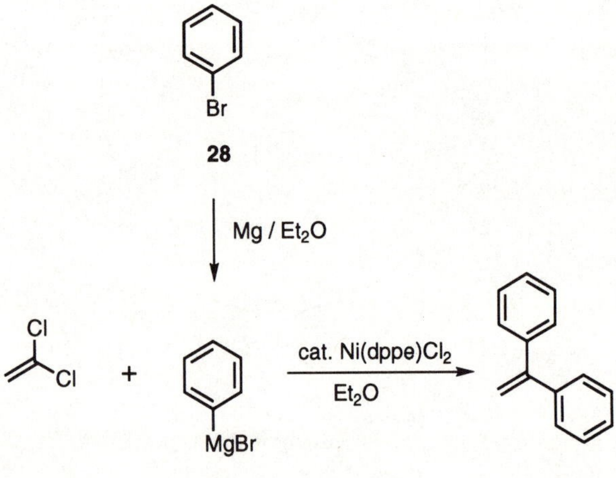

Br

28

Mg / Et_2O

Cl

Cl

+

cat. $Ni(dppe)Cl_2$

Et_2O

MgBr

29 **30** **31**

Scheme 8.10

Equipment

- Three-necked, round-bottomed flask (500 mL)
- Three-necked, round-bottomed flask (250 mL)
- Water-jacketed reflux condenser
- Pressure-equalizing addition funnel (200 mL)
- Mechanical stirrer
- Gas inlet tube
- Drying column (about 40 × 2 cm with phosphorous pentoxide/calcium chloride)
- Source of dry nitrogen, combined with a vacuum line

Materials

- Bromobenzene **28** (FW 157.2) 34.6 g, 0.22 mol — **irritant**
- 1,1-Dichloroethene **29** (FW 96.9) 9.28 g, 95.8 mmol — **flammable, lachrymator**
- Magnesium (FW 24.3) 5.34 g, 0.22 mol — **flammable, moisture sensitive**
- Iodine (FW 253.8) 10 mg — **corrosive, lachrymator**

- [1,2-Bis(diphenylphosphino)ethane]nickel(II) chloride (FW 528.0) 330 mg, 0.62 mmol — **cancer suspect agent, irritant**
- Dry ether 150 mL — **flammable, irritant**

Protocol 6a.
Preparation of phenylmagnesium bromide 30

1. Ensure that all glassware is thoroughly clean and has been dried for at least 4 h in a 120 °C electric oven before use. Bromobenzene **28** should be freshly distilled prior to use.
2. Equip the three-necked flask (250 mL) with the magnetic stirring bar, the reflux condenser, and the addition funnel. Using a tubing adapter, connect the top of the reflux condenser to the nitrogen/vacuum line, and place the system under nitrogen.
3. Charge the flask with magnesium (5.34 g, 0.22 mol), and close the remaining openings with glass stoppers.
4. Alternatively, evacuate, flame dry, and flush the apparatus with nitrogen three times. Ensure that during all the following operations a slow nitrogen stream passes through the apparatus. Whenever you open the reaction vessel, increase the nitrogen stream.
5. Stir the magnesium turnings slowly for 30 min. *If you set up the reaction flask the night before you run the reaction, you should stir the magnesium turnings slowly overnight. However, make sure that the turnings are not exposed to oxygen again.*
6. Charge the addition funnel with bromobenzene **28** (34.6 g, 0.22 mol) in dry ether (100 mL). Add dry ether (50 mL) to the reaction flask.
7. Add about 3 mL of the solution from the addition funnel. Stop the stirring. You should see the reaction starting (mild exothermic reaction, little bubbles on the magnesium surface, some of the magnesium starts to dissolve). If the reaction has not started, add a crystal of iodine in such a way that it lies on the magnesium surface.
8. When the reaction has started, stir the solution and add the rest of the solution from the addition funnel at such a rate that the reaction mixture refluxes gently. After completing the addition, heat the reaction mixture under reflux for about 1 h to ensure that the reaction is complete. The resulting solution of **30** is used without further purification.

Protocol 6b.
Preparation of 1,1-diphenylethylene

1. Ensure that all glassware is thoroughly clean and has been dried for several hours in a 120 °C electric oven before use.
2. Equip the three-necked flask (500 mL) with the mechanical stirrer, the

Protocol 6b. *Continued*

reflux condenser, and the addition funnel. Using a tubing adapter, connect the top of the reflux condenser to the nitrogen/vacuum line, and close all the remaining openings with glass stoppers.

3. Alternatively, evacuate, flame dry, and flush the apparatus with nitrogen three times. Ensure that during all the following operations a slow nitrogen stream passes through the apparatus. Whenever you open the reaction vessel, increase the nitrogen stream.
4. After the apparatus has cooled to room temperature add Ni(dppe)Cl_2 (330 mg, 0.62 mmol), dry ether (100 mL), and subsequently 1,1-dichloroethylene **29** (9.28 g, 95.8 mmol) to the reaction flask.
5. Transfer the Grignard solution prepared in Protocol 6a to the addition funnel, via a cannula (see Chapter 1, Protocol 2).
6. Add the Grignard solution dropwise to the reaction mixture over a 45 min period.
7. Stir the reaction mixture under reflux for 16 h. The reaction mixture will turn black in the course of the reaction.
8. Cool the reaction mixture with an ice bath. Slowly add hydrochloric acid (2 M, 150 mL).
9. Separate the organic layer from the aqueous layer. Extract the aqueous layer with ether (2 × 100 mL).
10. Combine the organic layers and wash them with water (100 mL), saturated sodium hydrogen carbonate solution (100 mL) (**caution**! gas development), and water (100 mL).
11. Dry the organic layer ($MgSO_4$), filter, and evaporate the solvent.
12. Distil the product under reduced pressure to obtain 14.1 g (82%) **31** (b.p. 94–95°C, 11 mmHg).

While primary alkyl Grignard reagents give clean reactions with vinyl or aryl halides, secondary Grignard reagents tend to isomerize to the primary analogs under nickel catalysis, consequently giving rise to product mixtures. However, aryl-substituted secondary Grignard reagents such as **32** can be used in cross-coupling reactions without the danger of isomerization to a primary organomagnesium derivative. Moreover, **32** undergoes rapid racemization, which has been used to develop enantioselective coupling reactions. Thus, vinyl bromide **33** reacts with **32** in the presence of enantiomerically pure nickel catalysts to afford the product **34** with moderate to excellent enantioselectivities[20] (Scheme 8.11).

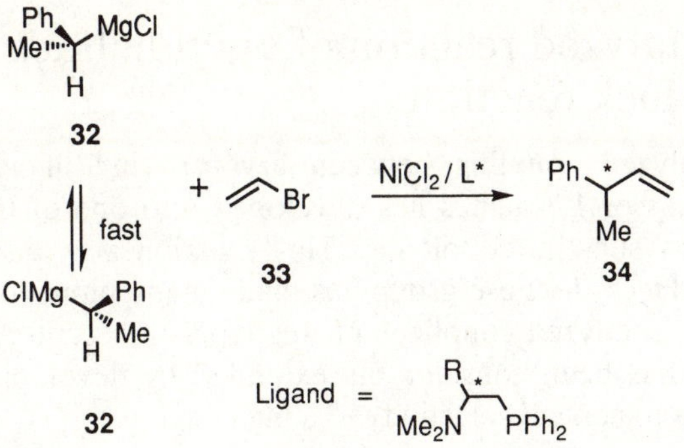

Ligand	R in Ligand	Product
(*S*)-Alaphos	CH_3	38% ee (*S*)
(*S*)-Phephos	Ph	71% ee (*S*)
(*S*)-Valphos	*i*-Pr	81% ee (*S*)
(*R*)-^{t}Leuphos	*t*-Bu	94% ee (*R*)

Scheme 8.11

Table 8.1 Nickel-phosphine-catalyzed Grignard coupling reactions, analogous to Protocol 6[19]

Halide (mmol)	Grignard reagent (mmol)	Catalyst (mmol)^{a}	Time (h)	Product	Yield (%)^{b}
C_6H_5Br (8.4)	$CH_2{=}C(CH_3)MgBr$ (11)^{c}	A (0.08)	20^{d}	$C_6H_5C(CH_3){=}CH_2$	(85)
$CH_2{=}CHCl$ (240)	$4\text{-}ClC_6H_4MgBr$ (200)	B (0.1)	1.5^{e}	$4\text{-}ClC_6H_4CH{=}CH_2$	79^{f}
$CH_2{=}CCl_2$ (50)	$C_6H_5CH_2MgCl$ (130)	C (0.20)	20	$CH_2{=}C(CH_2C_6H_5)_2$	43
(*Z*)-ClCH=CHCl (7.2)	C_6H_5MgBr (12)	C (0.05)	2^{e}	$C_6H_5CH{=}CHC_6H_5$ ((*Z*)/(*E*)–90:10)	(91)
$CH_2{=}C(CH_3)Br$ (12)	$CH_2{=}C(CH_3)MgBr$	B (0.1)	20	$[CH_2{=}C(CH_3)]_2$	(79)
$CH_2{=}C(Cl)CH{=}CH_2$ (108)	C_6H_5MgBr (125)	C (0.43)	20^{e,h}	$CH_2{=}C(C_6H_5)CH{=}CH_2$	(79)
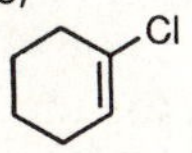 Cl	$n\text{-}C_4H_9MgBr$ (120)	C (0.42)	20	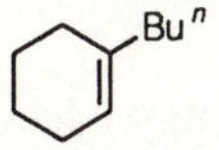 Bu^{n}	67

^{a}Catalysts: A, $Ni(dmpe)Cl_2$; B, $Ni(dppp)Cl_2$; C, $Ni(dppe)Cl_2$.
^{b}Isolated yield based on the halide, unless otherwise noted. Yields determined by quantitative gc are given in parentheses.
^{c}In THF.
^{d}40–50°C.
^{e}Room temperature.
^{f}Based on the Grignard reagent.
^{g}Solution in *m*-xylene.
^{h}A vigorous reaction occurred at 0°C.

4. Palladium-catalyzed reaction of vinyl halides with alkenes: the Heck reaction

The palladium(0)-catalyzed coupling between aryl or vinyl halides and alkenes to form styrenes or 1,3-dienes has developed into one of the most versatile approaches to substituted olefins. This reaction was discovered about 25 years ago by Heck,[21] whose group has made many pioneering contributions to palladium-catalyzed couplings of this type. In recent years the scope of this reaction has been considerably extended by developing mild and selective conditions to control selectivity to a high degree.[22]

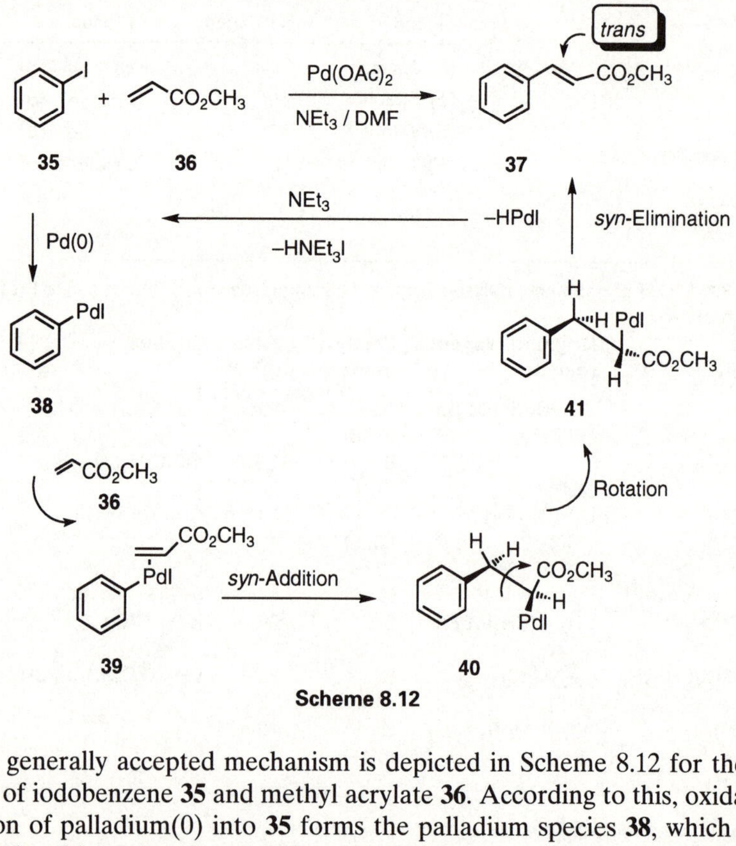

Scheme 8.12

The generally accepted mechanism is depicted in Scheme 8.12 for the reaction of iodobenzene **35** and methyl acrylate **36**. According to this, oxidative addition of palladium(0) into **35** forms the palladium species **38**, which subsequently adds to the alkene **36** in a *syn*-fashion to give **40**, probably via the π-complex **39**. In order to form the product **37**, a *syn*-elimination of HPdX has to take place. Therefore, a rotation around the C–C bond in **40** occurs to **41** in order to make a proton *syn* to the palladium substituent accessible. The substituents Ph and CO_2Me orientate *anti* to each other to minimize steric interactions. Consequently, the Heck reaction shows a high preference for

(*E*)-products. As will be seen later, the rotation around the C–C bond in **40** is a crucial step in this reaction. If this rotation is impossible, e.g. in cyclic systems, the reaction takes a different course (see Scheme 8.17).

Quite a few experimental procedures have been developed for the Heck reaction. Most commonly, palladium(II) acetate is taken as the palladium catalyst which is reduced *in situ* to palladium(0). Polar solvents such as DMF, *N*-methylpyrrolidone (NMP), or acetonitrile have been employed most successfully as solvents. The original reports called for triethylamine as a base (Protocol 7); more recently, phase transfer conditions using a tetrabutylammonium halide (Protocol 8) and potassium carbonate or acetate have been developed which allow some coupling reactions to be run at a temperature of about 30°C. However, the best conditions to apply are quite dependent upon the specific substrate being employed.

Aryl and vinyl iodides have proven to be the most reactive substrates, although bromides and triflates may also be used. Alkenes bearing an electron-withdrawing substituent such as phenyl, carbomethoxy, or cyano are the best coupling partners, although the successful use of ethene, vinylsilane, or alkyl-substituted alkenes has been reported.[23] As the following example shows, aryl iodides can be selectively coupled in the presence of aryl bromides. Note also that free carboxylic acid functionalities are tolerated under the reaction conditions, requiring only the use of additional equivalents of triethylamine (Protocol 7).

Protocol 7.
Preparation of (*E*)-2-bromocinnamic acid[24]

Caution! All procedures should be carried out in a well-ventilated hood, and disposable vinyl or latex gloves and chemical-resistant safety goggles should be worn.

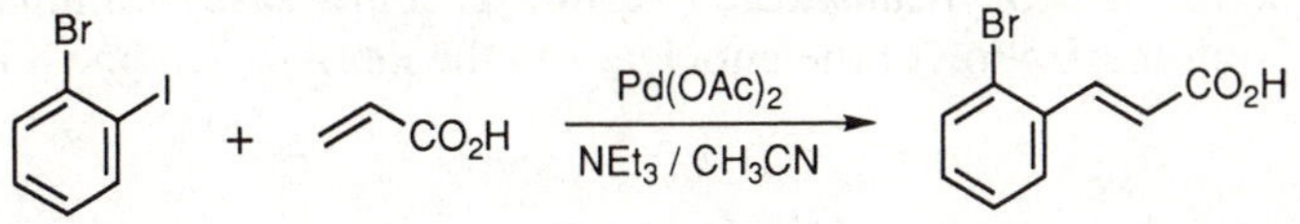

Br I + CO_2H — $Pd(OAc)_2$ / NEt_3 / CH_3CN → Br CO_2H

42 **43** **44**

Scheme 8.13

Equipment

- Thick-walled Pyrex bottle (50 mL) with a self-sealing rubber-lined cap
- Magnetic stirrer
- Magnetic stirring bar
- Beaker (400 mL)
- Protection shield
- Source of dry nitrogen

Protocol 7. *Continued*

Materials

- 2-Bromoiodobenzene **42** (FW 282.9) 2.83 g, 10 mmol — **irritant**
- Acrylic acid **43** (FW 72.06) 900 mg, 12.5 mmol — **corrosive, toxic**
- Palladium(II) acetate (FW 224.5) 22 mg, 0.1 mmol — **harmful**
- Dry triethylamine (FW 101.2) 3.5 mL, 25 mmol — **corrosive, flammable**
- Dry acetonitrile 4 mL — **flammable, lachrymator**

1. Ensure that all glassware is thoroughly clean and has been dried for at least 4 h in a 120°C electric oven before use. Acetonitrile and triethylamine must be dried over calcium hydride and distilled under nitrogen.
2. Equip the Pyrex bottle with the magnetic stirring bar.
3. Using a pipette, flush the open bottle with nitrogen for 5 min.
4. While continuing to flush the Pyrex bottle with a slow flow of nitrogen, add acetonitrile (4 mL), triethylamine (3.5 mL, 25 mmol), 2-bromoiodobenzene (2.83 g, 10 mmol), and acrylic acid (900 mg, 12.5 mmol).
5. Add palladium(II) acetate (22 mg, 0.1 mmol), and cap the bottle with the rubber-lined cap.
6. Place a protection shield in front of the apparatus. *Although we have never experienced any problems, it is possible that the Pyrex bottle could burst.*
7. Stir the mixture at 100°C for 1 h.
8. Cool the reaction mixture to room temperature. Transfer the mixture to the beaker and add hydrochloric acid (10%, 250 mL). A solid should precipitate.
9. Collect the precipitate by filtration. Recrystallize the solid twice from ethanol to obtain **44** (1.86 g, 82%), m.p. 215–216.5°C.

Using Protocol 7, an abundant number of aryl halides and alkenes have been reacted; some examples are given in Table 8.2.

Note that primary and secondary allyl alcohols do not give coupled alkenes under the conditions applied, but instead give saturated ketones, since the proton at the hydroxyl-substituted carbon is more easily eliminated. Subsequent tautomerization of the enol leads to the ketone (Scheme 8.14).

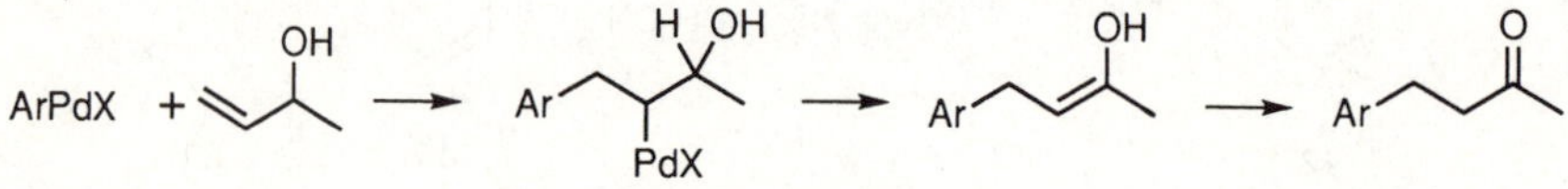

ArPdX + (allyl alcohol, OH) → Ar–CH₂–CH(PdX)–C(H)(OH)– → Ar (enol, OH) → Ar (ketone, O)

Scheme 8.14

Multiple couplings can also be achieved in one step. Thus, 1,2,3-tribromobenzene **45** gives the adducts **47** and **48** in good yields (Scheme 8.15). In these cases it has been found advantageous to use the phase transfer conditions originally suggested by Jeffery,[29] using tetrabutylammonium chloride/

Table 8.2. Heck reaction according to Protocol 7 between aryl halides and alkenes

Halide	Alkene	Time (h)	Product	Yield (%0	Ref.
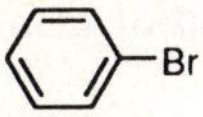 Br	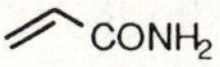 $CONH_2$	1	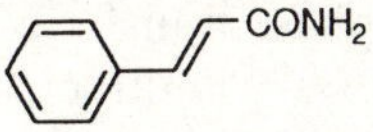 $CONH_2$	70	25
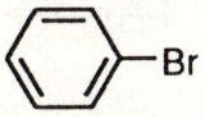 Br	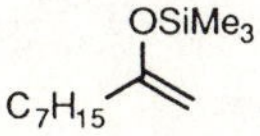 $OSiMe_3$ C_7H_{15}	4[a]	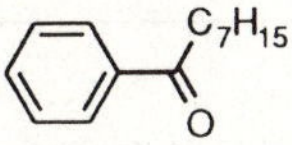 C_7H_{15} O	65	26
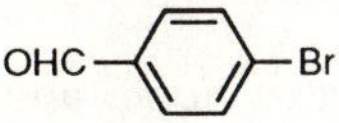 OHC Br	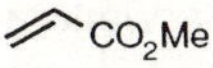 CO_2Me	18	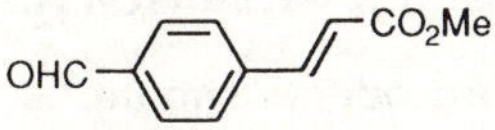 OHC CO_2Me	72	25
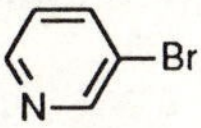 N Br	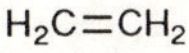 $H_2C{=}CH_2$	66[b]	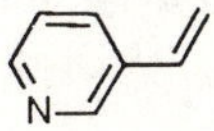 N	52	19a
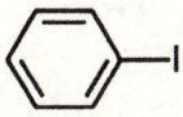 I	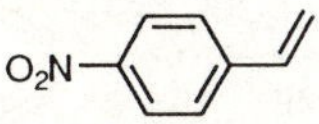 O_2N	2	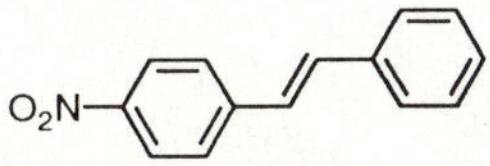 O_2N	85	27
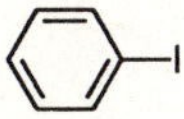 I	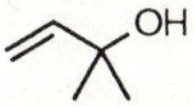 OH	4	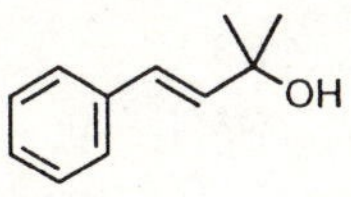 OH	97	28
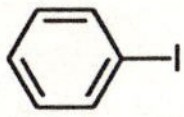 I	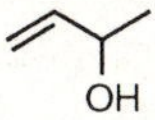 OH	5	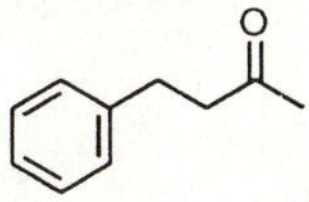 O	86	28
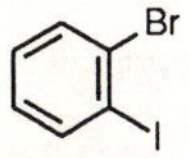 Br I	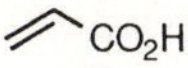 CO_2H	1	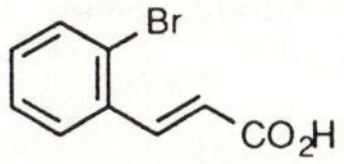 Br CO_2H	82	24
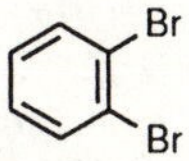 Br Br	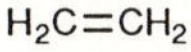 $H_2C{=}CH_2$	15	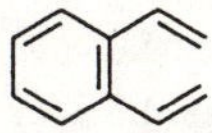	78	23a

[a] 80 °C, Bu_3SnF as base.
[b] 125 °C.

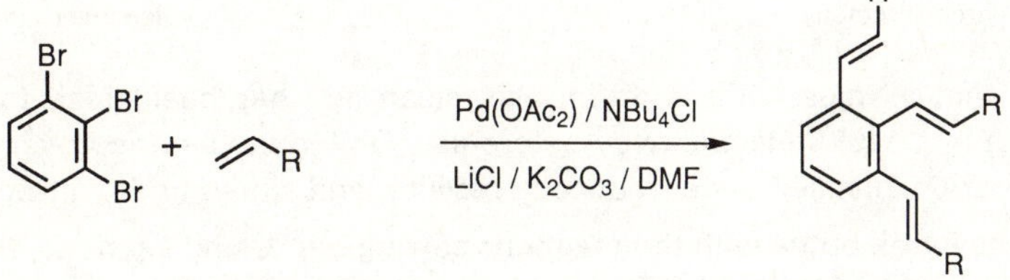

Br, Br, Br + R → [$Pd(OAc_2)$ / NBu_4Cl; $LiCl$ / K_2CO_3 / DMF] → R, R, R

45

36: R = CO_2Me
46: R = Ph

47: R = CO_2Me (58%)
48: R = Ph (71%)

Scheme 8.15

potassium carbonate, modified by also adding lithium chloride to the reaction mixture.[30] Such multiple couplings were unsuccessful using classical conditions with triethylamine as the base.[31]

Analogously, twofold alkenylation of 1,2-dibromobenzene **49** can be readily carried out under phase transfer conditions, although the addition of lithium chloride is not necessary in this case (Protocol 8).

Protocol 8.
Preparation of (*E,E*)-1,2-distyrylbenzene 50 (Scheme 8.16)[26]

Caution! All procedures should be carried out in a well-ventilated hood, and disposable vinyl or latex gloves and chemical-resistant safety goggles should be worn.

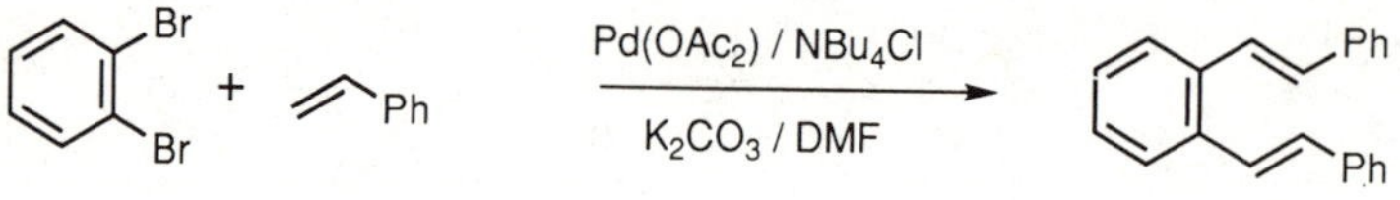

Br Br + Ph — $Pd(OAc_2)$ / NBu_4Cl, K_2CO_3 / DMF → Ph Ph

49 46 50

Scheme 8.16

Equipment

- Thick-walled Pyrex bottle (25 mL) with a self-sealing, rubber-lined cap
- Magnetic stirring bar
- Protection shield
- Short glass column for chromatography
- Source of dry nitrogen

Materials

• 1,2-Dibromobenzene **49** (FW 235.9) 1.0 g, 4.2 mmol	**irritant**
• Styrene **46** (FW 104.1) 1.82 g, 2.0 mL, 17.5 mmol	**flammable, toxic**
• Palladium(II) acetate (FW 224.5) 37 mg, 0.17 mmol	**harmful**
• Tetrabutylammonium chloride (FW 277.9) 1.2 g, 4.3 mmol	**hygroscopic, irritant**
• Potassium carbonate (FW 138.2) 2.9 g, 21 mmol	**hygroscopic, irritant**
• Dry DMF 15 mL	**irritant**
• Ether for extraction and chromatography	**flammable, irritant**
• Hexanes for chromatography	**flammable, irritant**

1. Ensure that all glassware is thoroughly clean and has been dried for at least 4 h in a 120°C electric oven before use. DMF must be dried over calcium hydride, distilled under reduced pressure, and stored under nitrogen.
2. Equip the Pyrex bottle with the magnetic stirring bar. Using a pipette, flush the open bottle with nitrogen for 5 min.
3. While continuing to flush the Pyrex bottle with a slow flow of nitrogen, add K_2CO_3 (2.9 g, 21 mmol), Bu_4NCl (1.2 g, 4.3 mmol), DMF (15 mL), styrene **46** (2.0 mL, 17.5 mmol), and 1,2-dibromobenzene **49** (1.0 g, 4.2 mmol).

4. Add palladium(II) acetate (37 mg, 0.17 mmol), and cap the bottle with the rubber-lined cap.
5. Place a protection shield in front of your apparatus. *Although we have never experienced any problems, it is possible that the Pyrex bottle could burst.*
6. Stir the mixture at 100 °C for 36 h.
7. After cooling to room temperature, dilute the mixture with ether (200 mL), and extract with water (5 × 80 mL).
8. Dry the organic phase ($MgSO_4$), filter, and evaporate the solvent.
9. Dry the residue at 50 °C *in vacuo* (<1 mmHg) to remove as much DMF as possible.
10. Purify the residue on a short silica column (10 g of silica gel, hexanes/ether = 4:1) to obtain crude **50**.
11. Recrystallize the crude product from ethanol to obtain pure **50** (1.1 g, 92%), m.p. 116 °C.

Cyclic alkenes can also be used as coupling components in Heck-type reactions, using the conditions described above.[32] However, there is an important difference in the products obtained compared with acyclic alkenes (Scheme 8.17).

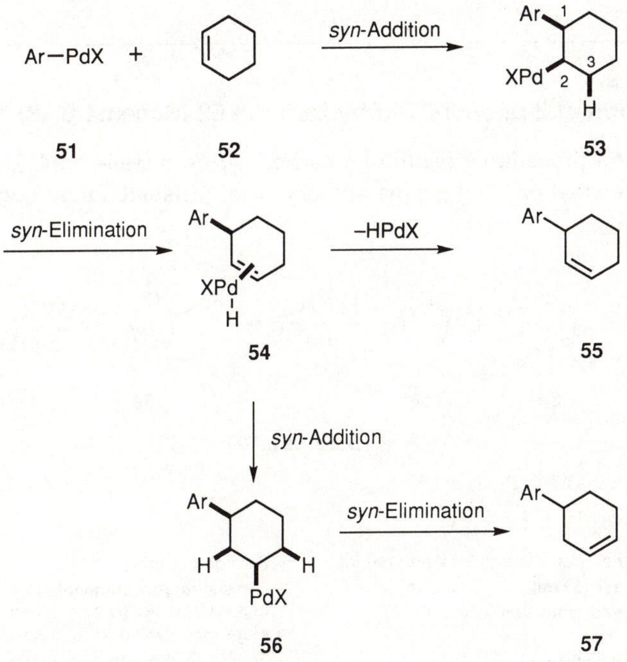

Ar—PdX + **51** **52** *syn*-Addition → **53** (Ar, XPd, H; positions 1, 2, 3)

syn-Elimination → **54** (Ar, XPd–H) –HPdX → **55** (Ar)

54 *syn*-Addition → **56** (Ar, H, H, PdX) *syn*-Elimination → **57** (Ar)

Scheme 8.17

After the *syn*-addition of **51** to the cycloalkene **52**, rotation around the C1-C2 bond in **53** is not possible as it was in the analogous acyclic palladium complex **40** (see Scheme 8.12). Consequently, the proton at C1 in **53** is not accessible for *syn*-elimination but elimination can occur with the *syn*-proton at C3 to yield **55** via **54**, leading to a formal allylation product of the cycloalkene.

The situation, however, can be more complicated since further addition of HPdX to **56** can occur so that further double bond isomers such as **57** can be formed. A number of procedures have been developed to prevent or promote such double bond isomerization. In general, cross-coupling products like **55** can be obtained by using silver(I) salts as additives, while it has been recognized that acetate present can lead to further isomerization.[33]

A good example for the strategies which can be applied is the coupling of iodobenzene and 2,3-dihydrofuran (Protocol 9). Using a catalyst comprising palladium(II) acetate and triphenylphosphine, the thermodynamically favored coupling product **59** is obtained exclusively. However, conducting the reaction with an additional two equivalents of silver(I) carbonate affords the substituted 2,3-dihydrofuran **60** as the sole product (Scheme 8.17).[33a] Electron-rich alkenes such as enol ethers are regioselectively substituted at the α-position, whereas it is the β-position which is substituted for electron-withdrawing alkenes.

Protocol 9.
Preparation of 2-phenyl-2,3-dihydrofuran 59 (Scheme 8.18)[34]

Caution! All procedures should be carried out in a well-ventilated hood, and disposable vinyl or latex gloves and chemical-resistant safety goggles should be worn.

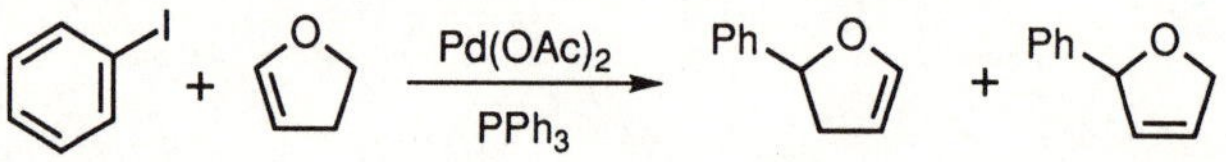

I + O → (Pd(OAc)$_2$ / PPh$_3$) Ph–O + Ph–O

35 58 59 60

Scheme 8.18

Equipment

- Three-necked, round-bottomed flask (100 mL)
- Side-arm flask (50 ml)
- Water-jacketed reflux condenser
- Magnetic stirrer
- Two magnetic stirring bars
- Septum
- Syringe pump
- Glass syringe with needle-lock Luer connected to a syringe needle by a Teflon tube
- Kugelrohr distillation apparatus
- Source of dry nitrogen combined with a vacuum line

Materials

- Iodobenzene **35** (FW 204.0) 5.32 g, 25 mmol — **irritant, light sensitive**
- 2,3-Dihydrofuran **58** (FW 70.1) 8.54 g, 125 mmol — **flammable**
- Palladium(II) acetate (FW 224.5) 130 mg, 0.58 mmol — **harmful**
- Triphenylphosphine (FW 262.3) 310 mg, 1.2 mmol — **irritant**
- Triethylamine (FW 101.2) 3.80 g, 5.25 ml, 37.5 mmol — **corrosive, flammable**
- Dry DMF — **irritant**
- Ether for extraction — **flammable, irritant**

1. Ensure that all glassware is thoroughly clean and has been dried for at least 4 h in a 120°C electric oven before use. Triethylamine must be dried over calcium hydride and distilled under nitrogen. DMF must be dried over calcium hydride, distilled under reduced pressure, and stored under nitrogen.
2. Equip the three-necked flask with the magnetic stirring bar, and the reflux condenser. Using a tubing adapter, connect the top of the reflux condenser to the nitrogen/vacuum line, and close all the remaining openings with glass stoppers.
3. Equip the side-arm flask with a magnetic stirring bar, and connect the flask to the nitrogen/vacuum line. Close the remaining opening with a glass stopper.
4. Alternatively, evacuate, flame dry, and flush both apparatus with nitrogen three times. Ensure that during all the following operations a slow nitrogen stream passes through the apparatus. Whenever you open the reaction vessel, increase the nitrogen stream.
5. After the side-arm flask has cooled down, add palladium(II) acetate (130 mg, 0.58 mmol), triphenylphosphine (310 mg, 1.2 mmol), and DMF (30 mL). Stir this mixture for 30 min at room temperature.
6. After the three-necked flask has cooled down, add iodobenzene **35** (5.32 g, 25 mmol), 2,3-dihydrofuran **58** (8.54 g, 125 mmol), triethylamine (5.25 mL, 37.5 mmol), and DMF (15 mL).
7. Close one neck of the three-necked flask with a septum.
8. Add, via a syringe, 10 mL of the solution prepared in step 5 to the reaction mixture of step 6.
9. Transfer the remaining 20 mL of the solution prepared in step 5 to the 20 mL syringe. Equip the syringe with the Teflon tube and the needle. Place the syringe into the syringe pump, and pierce the needle at the other end of the Teflon tube through the septum of the reaction flask. Adjust the speed of the syringe pump in such a way that the catalyst solution will be added during 12 h to the reaction mixture.
10. Stir the reaction mixture at 80°C for 72 h.
11. After the reaction mixture has cooled down, add ether (100 mL), and extract with water (3×75 mL).

Protocol 9. *Continued*

12. Dry the organic phase ($MgSO_4$), filter, and evaporate the solvent under reduced pressure.
13. Distil the residue with a Kugelrohr distillation apparatus to obtain **59** (2.72 g, 72%, b.p. 40 °C, 0.1–0.15 mmHg).

This reaction has also been carried out using phenyl triflate instead of **35** and Pd-(*R*)-BINAP as catalyst yielding **59** with up to 96% ee.[35]

5. Palladium-catalyzed allylic substitution reactions

Palladium-catalyzed allylic substitution has developed into a versatile method for the functionalization of allyl compounds with a broad variety of nucleophiles. The most common starting materials are allyl acetates such as **62**, but halides, sulfones, carbonates, carbamates, epoxides, or phosphates have also been successfully employed.[36] The alkene moiety is intimately involved in the reaction, hence the inclusion of this reaction.

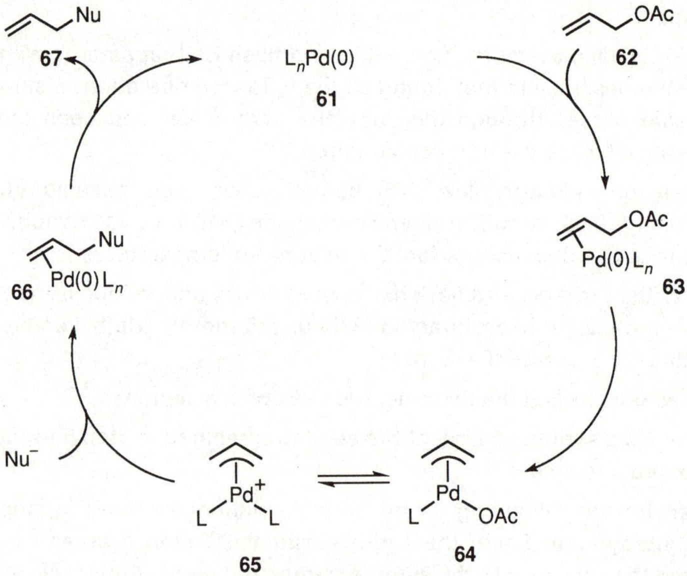

Scheme 8.19

The allylic substrate **62** undergoes oxidative addition with a palladium(0) catalyst **61** to form palladium allyl complexes **64** or **65**, respectively. Attack of a nucleophile liberates via **66** the palladium(0) catalyst and releases the substitution product **67** (Scheme 8.19).

A great range of nucleophiles has been used, with most attention to stabilized anions derived from 1,3-dicarbonyl compounds (e.g. dimethylmalonate or acetylacetone) and nitrogen-based nucleophiles such as amines (Protocol 10).

Protocol 10.
Preparation of 1-(cyclohexylamino)-2,7-octadiene 70[37]

Caution! All procedures should be carried out in a well-ventilated hood, and disposable vinyl or latex gloves and chemical-resistant safety goggles should be worn.

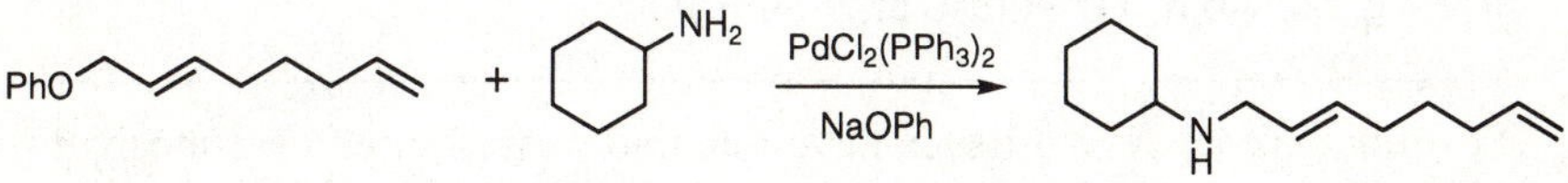

PhO + NH2 → $PdCl_2(PPh_3)_2$, NaOPh → N H

68 **69** **70**

Scheme 8.20

Equipment

- Three-necked, round-bottomed flask (100 mL)
- Water-jacketed reflux condenser
- Magnetic stirrer
- Magnetic stirring bar
- Water-jacketed, semi-micro distillation apparatus
- Source of dry nitrogen combined with a vacuum line

Materials

- 1-Phenoxy-2,7-octadiene **68**[38] 10.1 g, 50 mmol
- Cyclohexylamine **69** (FW 99.2) 9.9 g, 100 mmol — **corrosive, flammable**
- Bis (triphenylphosphine) palladium(II) chloride (FW 701.9) 70 mg, 0.1 mmol — **hygroscopic**
- Sodium phenoxide (FW 116.1) 120 mg, 1 mmol — **corrosive, hygroscopic**

1. Ensure that all glassware is thoroughly clean and has been dried for at least 4 h in a 120°C electric oven before use. Cyclohexylamine **69** (b.p. 134°C) must be dried over calcium hydride and distilled under nitrogen or argon.
2. Equip the three-necked flask with the magnetic stirring bar and the reflux condenser. Using a tubing adapter, connect the top of the reflux condenser to the nitrogen/vacuum line, and close all the remaining openings with glass stoppers.
3. Alternatively, evacuate, flame dry, and flush the apparatus with nitrogen three times. Ensure that during all the following operations a slow nitrogen stream passes through the apparatus. Whenever you open the reaction vessel, increase the nitrogen stream.
4. After the three-necked flask has cooled down to room temperature, charge with 1-phenoxy-2,7-octadiene **68** (10.1 g, 50 mmol) and cyclohexylamine **69** (9.9 g, 100 mmol).

Protocol 10. *Continued*

5. Add sodium phenoxide (120 mg, 1 mmol) and subsequently $PdCl_2(PPh_3)_2$ (70 mg, 0.1 mmol).
6. Stir the reaction mixture at 85 °C for 2 h.
7. After the reaction mixture has cooled to room temperature, equip the flask with the distillation apparatus in such a way that you can collect multiple fractions. Connect the distillation apparatus to the vacuum line, and close all remaining openings. Distil the reaction mixture under reduced pressure to obtain (I) unreacted cyclohexylamine and (II) the desired product **70** (10.4 g, 69%, b.p. 117–118 °C at 2 mmHg).

Several stereochemical issues have been addressed in elegant mechanistic studies, allowing this reaction to be carried out in a regio-, diastereo-, and enantioselective way. With nucleophiles, which do not precoordinate to palladium, e.g. soft nucleophiles, net retention of stereochemistry is observed (Scheme 8.21).

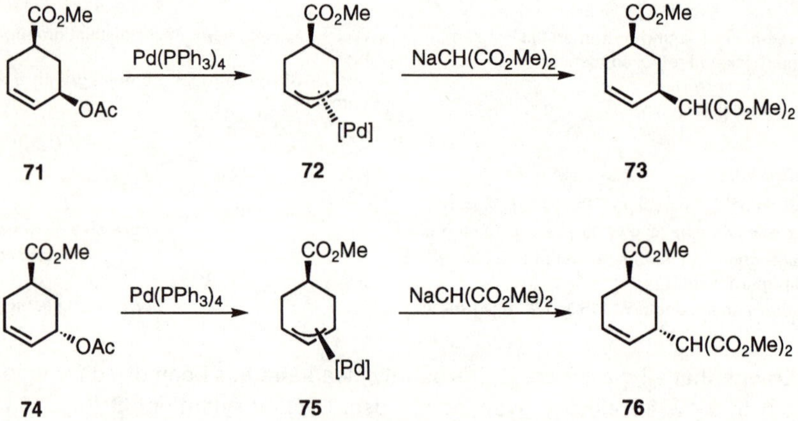

CO_2Me / OAc **71** —$Pd(PPh_3)_4$→ CO_2Me / [Pd] **72** —$NaCH(CO_2Me)_2$→ CO_2Me / $CH(CO_2Me)_2$ **73**

CO_2Me / OAc **74** —$Pd(PPh_3)_4$→ CO_2Me / [Pd] **75** —$NaCH(CO_2Me)_2$→ CO_2Me / $CH(CO_2Me)_2$ **76**

Scheme 8.21

The palladium allyl complex **72** is formed by displacing the leaving group with inversion; subsequent attack of the nucleophile from the opposite side of the palladium fragment results again in inversion, accounting for the overall retention in the product **73**. In agreement with the depicted mechanism, the *trans*-substituted cyclohexene **74** gives **76** as the sole product.[39]

However, certain nucleophiles such as hydride[40] or tributyltin[41] give inversion of stereochemistry in the reaction with **71**. This can be explained by a precoordination of the nucleophile to the palladium, and subsequent deliverance from there to the allyl system.

Another important factor is the geometry of the allyl component. It has been generally observed that the double bond geometry in (*E*)-allyl acetates is preserved, since *syn*-configurated π-allyl complexes such as **78** are favored, in general, for steric reasons. However, *anti*-π-allyl complexes **81**, derived from (*Z*)-configurated allyl acetates, generally rearrange by a π–σ–π mechanism to the *syn*-complex **78** leading also to (*E*)-configurated products **79**. The (*Z*)-derivative **80** affords inversion rather than retention of stereochemistry in the reaction with nucleophiles[42] (Scheme 8.22).

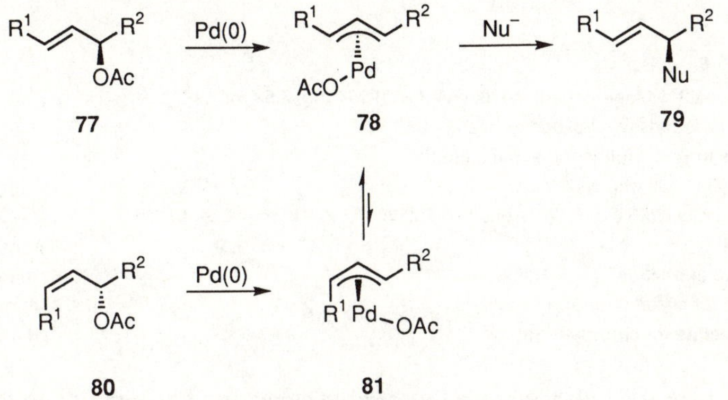

Scheme 8.22

Another issue to be addressed is that attack of the nucleophile at unsymmetrically substituted allyl termini can lead to regioisomers (see Scheme 8.22; the nucleophile could attack either at R^1C or R^2C). Here, the general rule is that soft nucleophiles will attack the sterically less-hindered position of the allyl system. The stereochemical outcome in the synthesis of **84** (Scheme 8.23) can be understood following these rules (Protocol 11).[43]

Protocol 11.
Preparation of ((*E*)-3-trimethylsilyl-2-propenyl)-2-methyl-1,3-cyclopentanedione 84[37]

Caution! All procedures should be carried out in a well-ventilated hood, and disposable vinyl or latex gloves and chemical-resistant safety goggles should be worn.

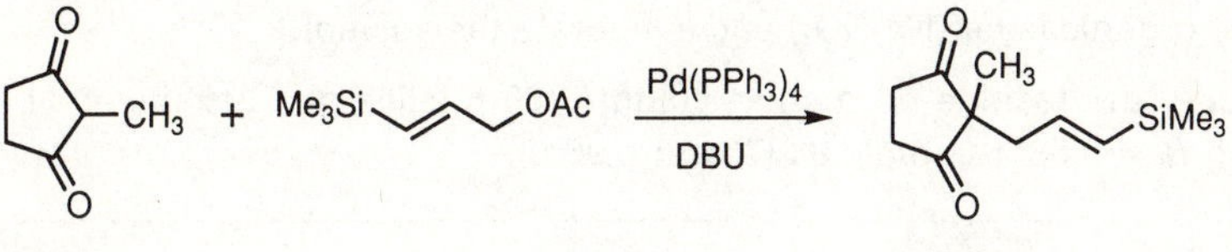

Scheme 8.23

Protocol 11. *Continued*

Equipment

- Side-arm flask (25 mL)
- Magnetic stirrer
- Magnetic stirring bar
- Glass stopper
- Septum
- Glass syringes
- Six-inch, medium-gauge needles
- Short glass column for chromatography
- Source of dry nitrogen combined with a vacuum line

Materials

- 2-Methyl-1,3-cyclopentanedione **82** (FW 112.1) 394 mg, 3.52 mmol
- (*E*)-3-(Trimethylsilyl)-2-propenyl acetate **83**[44]
- Tetrakis (triphenylphosphine) palladium(0) (FW 1155.6) 135 mg, 0.12 mmol — **light sensitive**
- 1,8-Diazabicyclo[5.4.0]undec-7-ene (DBU) (FW 152.2) 640 mg, 4.21 mmol — **corrosive**
- Dry THF — **flammable, irritant**
- Ether for extraction — **flammable, irritant**
- Hexane for chromatography — **flammable, irritant**
- Ethyl acetate for chromatography — **flammable, irritant**

1. Ensure that all glassware is thoroughly clean and has been dried for at least 4 h in a 120°C electric oven before use. DBU should be distilled under reduced pressure (b.p. 80–83°C, 0.6 mmHg) prior to use.
2. Equip the side-arm flask with a magnetic stirring bar, and connect the flask to the nitrogen/vacuum line. Close the remaining opening with a glass stopper.
3. Alternatively, evacuate, flame dry, and flush the apparatus with nitrogen three times. Ensure that during all the following operations a slow nitrogen stream passes through the apparatus. Whenever you open the reaction vessel, increase the nitrogen stream.
4. After the flask has cooled down to room temperature, add $Pd(PPh_3)_4$ (135 mg, 0.12 mmol) and 2-methyl-1,3-cyclopentanedione **82** (394 mg, 3.52 mmol). Close the flask with a septum.
5. Add, via a syringe, THF (10 mL), and subsequently DBU (0.63 mL, 4.21 mmol) and (*E*)-3-(trimethylsilyl)-2-propenyl acetate **83** (759 mg, 4.41 mmol).
6. Stir the mixture for 34 h at room temperature.
7. Add ether (100 mL) and wash the organic layer with brine (2 × 20 mL). Dry the organic layer (Na_2SO_4) and evaporate the solvent.
8. Purify the residue on a short column (30 g silica gel, hexane/ethyl acetate 5:1, $R_f = 0.58$) to obtain **84** (754 mg, 96%).

Recently, a remarkable catalytic system has been discovered, which allows the (*Z*)-geometry of allylic acetates to be retained in the substitution products. In the presence of 2,9-dimethyl-1,10-phenanthroline (neocuproine **91**) as a ligand for palladium(0) and DMF as solvent, the formation of an *anti*-configurated palladium–allyl complex is thermodynamically favored, and the *syn–anti* isomerization is slow compared with the attack of the nucleophile (Scheme 8.24, Protocol 12).[45]

Protocol 12.
Preparation of diethyl(4-methoxy-(*Z*)-2-butenyl) methylmalonate 92[45]

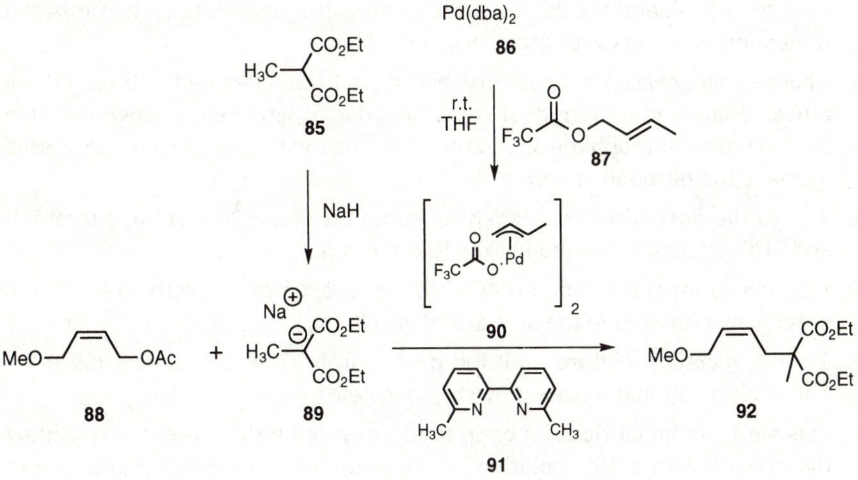

Pd(dba)$_2$ **86**, r.t. THF, F_3C–C(=O)–O–CH$_2$CH=CHCH$_3$ **87** → [(F_3CCO$_2$)Pd(η^3-butenyl)]$_2$ **90**

H_3C–CH(CO_2Et)$_2$ **85** —NaH→ $Na^{\oplus}$ H_3C–C$^{\ominus}$(CO_2Et)$_2$ **89**

MeO–CH$_2$CH=CHCH$_2$–OAc **88** + **89** —(**90**, **91**)→ MeO–CH$_2$CH=CHCH$_2$–C(CH_3)(CO_2Et)$_2$ **92**

Scheme 8.24

Protocol 12a.
Preparation of bis(μ-trifluoroacetato)bis[(1,2,3-η)2-butenyl]dipalladium 90

Caution! All procedures should be carried out in a well-ventilated hood, and disposable vinyl or latex gloves and chemical-resistant safety goggles should be worn.

Equipment

- Side-arm flask (50 mL)
- Side-arm flask (25 mL)
- Side-arm flask (10 mL)
- Magnetic stirrer
- Magnetic stirring bar
- Glass syringes with needle-lock Luers
- Septum
- Six-inch, medium-gauge needles
- Source of dry nitrogen combined with a vacuum line

Protocol 12a. *Continued*

Materials

- [Bis(dibenzylideneacetone) palladium (0) **86**[46] 1.15 g, 1.0 mmol
- (*E*)-2-Butenyltrifluoroacetate **87**[47] 370 mg, 2.2 mmol — **assume toxic**
- Dry THF 15 mL — **flammable, irritant**
- Dry acetonitrile 2 mL — **flammable, irritant**
- Acetonitrile for extraction — **flammable, irritant**

1. Ensure that all glassware is thoroughly clean and has been dried for at least 4 h in a 120°C electric oven before use. Acetonitrile must be dried over calcium hydride and distilled under nitrogen or argon.
2. Equip the side-arm flask (50 mL) with a magnetic stirring bar, and connect both the 50 mL and the 25 mL flasks to the nitrogen/vacuum line. Close the remaining openings with glass stoppers.
3. Alternatively, evacuate, flame dry, and flush both flasks with nitrogen three times. Ensure that during all the following operations a slow nitrogen stream passes through the apparatus. Whenever you open the reaction vessel, increase the nitrogen stream.
4. Add to the flask bis(dibenzylideneacetone) palladium(0) (1.15 g, 1.0 mmol) and THF (16 mL). Close the flask with a septum.
5. Add, via syringe, a solution of **87** (370 mg, 2.2 mmol) in acetonitrile (2 mL), previously prepared in the side-arm flask (25 mL).
6. Stir the reaction mixture until the deep purple color disappears (30 min). The solution should have a greyish-green color.
7. This step can be carried out open to air: evaporate the solvent, and extract the residue with a 10% solution of acetonitrile in water (4 × 5 mL). Filter, and subsequently evaporate the aqueous layer to obtain **90** (400 mg, 73%) as a pale yellow solid.
8. Store the complex in the side-arm flask (10 mL) in a freezer; it can be used without further purification. The complex is stable to air, but it should be stored under nitrogen if kept for extended periods of time.

Protocol 12b.
Preparation of diethyl(4-methoxy-(*Z*)-2-butenyl)methylmalonate 92[45]

Caution! All procedures should be carried out in a well-ventilated hood, and disposable vinyl or latex gloves and chemical-resistant safety goggles should be worn.

Equipment

- Side-arm flask (25 ml)
- Two side-arm flasks (10 mL)
- Magnetic stirrer
- Magnetic stirring bar

- Glass syringes, with needle-lock Luers.
- Short glass, column for chromatography
- Six-inch, medium-gauge needle
- Source of dry nitrogen combined with a vacuum line

Materials

• 1-Acetoxy-4-methoxy-(*Z*)-2-butene **88**[48] 288 mg, 2 mmol	
• Diethylmethylmalonate **85** (FW 174.20) 700 mg, 4 mmol	
• **90** (prepared in Protocol 12a) 5.4 mg, 0.01 mmol	**assume toxic**
• 2,9-Dimethyl-1,10-phenanthroline **91** (FW 226.3) 21 mg, 0.1 mmol	**irritant**
• Sodium hydride (80% in mineral oil) (FW 24.0) 120 mg, 4 mmol	**flammable, moisture sensitive**
• Dry DMF	**irritant**
• Pentane for chromatography	**flammable, irritant**
• Ether for chromatography	**flammable, irritant**

1. Ensure that all glassware is thoroughly clean and has been dried for at least 4 h in a 120°C electric oven before use. DMF must be dried over calcium hydride and distilled under nitrogen or argon. Diethylmethylmalonate **85** should be distilled prior to use. Dissolve the 2,9-dimethyl-1,10-phenanthroline **91** in THF, dry the solution ($MgSO_4$), filter, and evaporate the solvent.
2. Equip the 25 mL side-arm flask and one of the 10 mL side-arm flasks with magnetic stirring bars. Connect all three flasks to the nitrogen/vacuum line. Close the remaining openings with glass stoppers.
3. Alternatively, evacuate, flame dry, and flush the flasks with nitrogen three times. Ensure that during all the following operations a slow nitrogen stream passes through the apparatus. Whenever you open the reaction vessel, increase the nitrogen stream.
4. Charge the 10 mL flask (with stirrer bar) with NaH (120 mg, 4 mmol) and DMF (4 mL). Close the opening with a septum.
5. Whilst stirring, slowly add a solution of **85** (700 mg, 4 mmol) in DMF (1 mL), previously prepared in the other side-arm flask (10 mL). When the NaH has dissolved and the gas evolution has subsided, cool the solution to 0°C.
6. Add to the side-arm flask (25 mL) **91** (21 mg, 0.1 mmol), **90** (5.4 mg, 0.01 mmol), and DMF (2 mL), then cool the solution to 0°C.
7. Add **88** (288 mg, 2 mmol) and subsequently the solution of **89**, prepared in step 5, to the reaction mixture.
8. Stir the solution for 90 min at 0°C.
9. Add 30 mL of water to the mixture. Extract with ether (4 × 25 mL). Wash the combined organic layers with brine (25 mL), dry ($MgSO_4$), filter, and evaporate the solvent.
10. Purify the residue by filtration on a short column (silica gel, pentane/ether 3:1, $R_f = 0.19$) to obtain **92** (463 mg, 90%).

Most of the asymmetric allylic substitution processes start from racemic allylic compounds such as *rac*-**93**, which in the absence of chiral ligands form *meso*-complexes of the type **94** with palladium(0). Since a nucleophile can attack at either of the two ends of the allylic component, the enantiomers **95** and *ent*-**95** are formed (Scheme 8.25). The degree of the enantioselectivity of a reaction depends on how well a chiral ligand in **94** can direct the attack of the nucleophile to one of the two allylic termini.

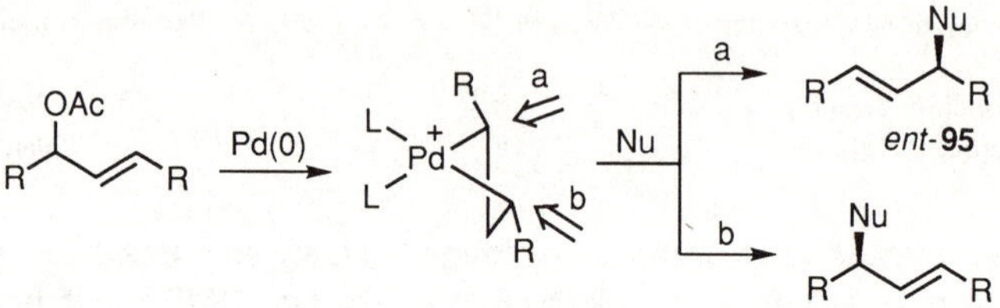

Scheme 8.25

There has been a number of ligands which are able to carry out the reaction, especially in the case of **93** with R = Ph, with very high optical induction (ee > 95%).[49]

The reactions introduced in this chapter on palladium- and nickel-catalyzed coupling processes could only be a small part of the great number of current developments in this fast-growing field of organic catalysis. All of these reactions have found wide use in organic synthesis, since the excellent stereocontrol allows the selective synthesis of numerous products. I hope that this chapter can be taken as an inspiration to explore further the fascinating world of palladium- and nickel-catalyzed reactions.

Acknowledgement

This work was supported by the Deutsche Forschungsgemeinschaft (Habilitationsstipendium) and Fonds der Chemischen Industrie.

References

1. Heck, R. F. *Palladium Reagents in Organic Syntheses;* Academic Press: London, **1985,** p. 191ff.
2. Dang, H. P.; Linstrumelle, G. *Tetrahedron Lett.* **1978**, 191.
3. Murahashi, S.-I.; Yamamura, M.; Yanagisawa, K.-I.; Mita, N.; Kondo, K. *J. Org. Chem.* **1979**, *44*, 2408–2417.
4. Lepley, A. R.; Khan, W. A.; Giumanini, A. B.; Giumanini, A. G. *J. Org. Chem.* **1966**, *31*, 2047–2051.
5. Christol, S. J.; Norris, W. R. *J. Am. Chem. Soc.* **1953**, *75*, 2645.

6. Zweifel, G.; Whitney, C. C. *J. Am. Chem. Soc.* **1967**, *89*, 2753.
7. (a) Negishi, E.; Valente, L. F.; Kobayashi, M. *J. Am. Chem. Soc.* **1980**, *102*, 3298. (b) Miller, R. B.; Al-Hassan, M. I. *J. Org. Chem.* **1985**, *50*, 2121.
8. Russell, C. E.; Hegedus, L. S. *J. Am. Chem. Soc.* **1983**, *105*, 943–949.
9. Whitesides, G. M.; Fischer, Jr, W. F.; Filippo, Jr, J. S.; Basle, R. W.; House, H. O. *J. Am. Chem. Soc.* **1969**, *91*, 4871.
10. Negishi, E.; Luo, F.-T. *J. Org. Chem.* **1983**, *48*, 1560–1562.
11. The Aldrich Library of NMR and IR spectra: NMR, **6**, 42A; IR, 745 G.
12. Milstein, D.; Stille, J. K. *J. Am. Chem. Soc.* **1979**, *101*, 4992.
13. Krolski, M. E.; Renaldo, A. F.; Rudisill, D. E.; Stille, J. K. *J. Org. Chem.* **1988**, *53*, 1170–1176.
14. Suzuki, A. *Pure Appl. Chem.* **1991**, *63*, 419.
15. Miyaura, N.; Ishiyama, T.; Sasaki, H.; Ishikawa, M.; Satoh, M.; Suzuki, A. *J. Am. Chem. Soc.* **1989**, *111*, 314–321.
16. Miyaura, N.; Yamada, K.; Suginome, H.; Suzuki, A. *J. Am. Chem. Soc.* **1985**, *107*, 972–980.
17. Brown, H. C.; Gupta, S. K. *J. Am. Chem. Soc.* **1972**, *94*, 4370–4373.
18. Knight, D. W. In *Comprehensive Organic Synthesis*; Trost, B. M.; Fleming, I., eds; Pergamon Press: Oxford, **1991**; Vol. 3, p. 441 ff.
19. Tamao, K.; Sumitani, S.; Kiso, Y.; Zembayashi, M.; Fujioka, A.; Kodama, S.; Nakajima, I.; Kumada, M. *Bull. Chem. Soc. Jpn.* **1976**, *49*, 1958–1969.
20. (a) Hayashi, T.; Konishi, M.; Fukushima, K.; Kanehira, K.; Hioki, T.; Kumeda, M. *J. Org. Chem.* **1983**, *48*, 2195. (b) Hayashi, T.; Hagihara, T.; Katsuro, Y.; Kumuda, M. *Bull. Chem. Soc. Jpn.* **1983**, *56*, 363.
21. (a) Heck, R. F. *Palladium Reagents in Organic Syntheses*; Academic Press: London, **1985**. (b) Heck, R. F. *Org. React.* **1983**, *27*, 1.
22. For a recent review see: de Meijere, A.; Meyer, F. *Angew. Chem. Int. Ed. Engl.* **1994**, *33*, 2379.
23. (a) Plevyak, J. E.; Heck, R. F. *J. Org. Chem.* **1978**, *43*, 2454. (b) de Meijere, A.; Gerson, F.; Knig, B.; Reiser, O.; Wellauer, T. *J. Am. Chem. Soc.* **1990**, *112*, 6827. (c) Reiser, O.; Reichow, S.; de Meijere, A. *Angew. Chem.* **1987**, *99*, 1285–1286; *Angew. Chem., Int. Ed. Engl.* **1987**, *26*, 1277.
24. Plevyak, J. E.; Dickerson, J. E.; Heck, R. F. *J. Org. Chem.* **1979**, *44*, 4078–4080.
25. Patel, B. A.; Kao, L.-C.; Cortese, N. A.; Minkiewicz, J. V.; Heck, R. F. *J. Org. Chem.* **1979**, *44*, 918.
26. Kuwajima, I.; Urabe, H. *J. Am. Chem. Soc.* **1982**, *104*, 6831.
27. Heck, R. F.; Nolley, J. P. J. *J. Org. Chem.* **1972**, *37*, 2320–2322.
28. Melpolder, J. B.; Heck, R. F. *J. Org. Chem.* **1976**, *41*, 265.
29. Jeffery, T. *Tetrahedron Lett.* **1985**, *26*, 2667–2670.
30. Lansky, A.; Reiser, O.; de Meijere, A. *Synlett* **1990**, *1*, 405.
31. Yao, W.; Nesbitt, S.; Heck, R. *J. Org. Chem.* **1990**, *55*, 63.
32. Larock, R. C.; Yum, E. K.; Yang, H. *Tetrahedron* **1994**, *50*, 323–334.
33. (a) Larock, R. C.; Gong, W. H.; Baker, B. E. *Tetrahedron Lett.* **1989**, *30*, 2603–2606. (b) Larock, R. C.; Gong, W. H. *J. Org. Chem.* **1989**, *54*, 2047–2050.
34. Hillers, S.; Reiser, O. Unpublished results. See also reference 33a.
35. Hayashi, T.; Kubo, A.; Ozawa, F. *Pure Appl. Chem.* **1992**, *64*, 421–427.
36. Reviews: (a) Frost, C. G.; Howarth, J.; Williams, J. M. J. *Tetrahedron: Asymmetry* **1992**, *3*, 1089–1122. (b) Godleski, S. A. In *Comprehensive Organic Synthesis*;

Trost, B. M., ed., Pregamon Press; Oxford, **1991**; Vol. 4, p. 585. (c) Trost, B. M. *Angew. Chem.* **1989**, *101*, 1199–1219; *Angew. Chem., Int. Ed. Engl.* **1989**, *28*, 1173. (d) Consiglio, G.; Waymouth, R. *Chem. Rev.* **1989**, *89*, 257.

37. Takahashi, K.; Miyake, A.; Hata, G. *Bull. Chem. Soc. Jpn.* **1972**, *45*, 230–236.
38. Smutny, E. J. *J. Am. Chem. Soc.* **1967**, *89*, 6793.
39. Trost, B. M.; Verhoeven, T. R. *J. Org. Chem.* **1976**, *41*, 3215.
40. Greenspoon, N.; Keinan, E. *J. Org. Chem.* **1988**, *53*, 3723.
41. Trost, B. M.; Herndon, J. W. *J. Am. Chem. Soc.* **1984**, *106*, 6835.
42. Hayashi, T.; Yamamoto, A.; Hagihara, T. *J. Org. Chem.* **1986**, *51*, 723.
43. Mori, M.; Isono, N.; Kaneta, N.; Shibasaki, M. *J. Org. Chem.* **1993**, *58*, 2972–2976.
44. Jones, T. K.; Denmark, S. E. *Org. Synth.* **1985**, *64*, 182.
45. (a) Akermark, B. Personal communication. (b) Sjrgen, M.; Hansson, S.; Norrby, P.-O.; Akermark, B.; Cucciolito, M. E.; Vitagliano, A. *Organometallics* **1992**, *11*, 3954–3964.
46. Rettig, M. F.; Maitlis, P. M. *Inorg. Synth.* **1977**, *17*, 135.
47. Prepared from crotyl alcohol and trifluoroacetic anhydride.
48. Prepared by monomethylation and subsequent acetylation of 2-butene-1,4-diol: see Mc Dougal, P. G.; Rice, J. G.; Oh, Y.-I.; Condon, B. D. *J. Org. Chem.* **1986**, *51*, 3388–3390.
49. Review: Reiser, O. *Angew. Chem.* **1993**, *105*, 576–578; *Angew. Chem., Int. Ed.* **1993**, *32*, 547–549.

9

Summary of alkene syntheses

ANDREW D. WESTWELL and JONATHAN M. J. WILLIAMS

1. Introduction

Each previous chapter in this book has been devoted to particular methods for the synthesis of alkenes, and has included experimental protocols which provide details for the preparation of alkenes. These chapters have broadly been arranged into the types of starting material which are required. There have been alkene syntheses from two components, from single C–C bonds, from triple C–C bonds, and from other alkenes.

No further protocols are provided in this chapter. However, this last chapter is divided into alkene category, and provides a commentary on the approaches available for each class of alkene. There are many cross-references to the individual protocols which have been described in earlier chapters.

The first sections deal with monosubstituted, 1,1-disubstituted, (*E*)- and (*Z*)-1,2-disubstituted, trisubstituted, and tetrasubstituted alkenes. Later sections deal with cyclic alkenes, dienes and related compounds, heteroatom-substituted alkenes, and unusual alkenes (such as allenes and ketenes). These categories are summarised in Scheme 9.1.

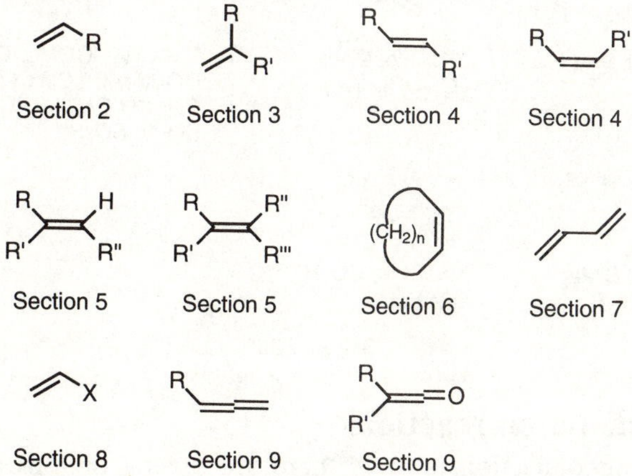

R
Section 2
R R'
Section 3
R R'
Section 4
R R'
Section 4
R H R' R''
Section 5
R R'' R' R'''
Section 5
$(CH_2)_n$
Section 6
Section 7
X
Section 8
R
Section 9
R R' O
Section 9

Scheme 9.1

Clearly, some alkenes may fall into two or more categories. Thus, in Section 8, heteroatom-substituted alkenes may be substituted with additional groups. We have discussed particular alkenes in the category which we believe to be the most appropriate. In this chapter, we have paid particular attention to methods leading to the stereo- and regioselective synthesis of alkenes. One general class of reaction that is not considered in detail in this chapter is the rearrangement of one alkene to another alkene. There are cases when this is a valuable method, and detailed reviews are available in this area.[1]

2. Monosubstituted alkenes

Many of the methods applicable to the synthesis of more highly substituted alkenes are also suitable for the preparation of monosubstituted alkenes. All that is required is the use of less highly substituted starting materials.

Simple monosubstituted alkenes are often commercially available, and hence their synthesis is unnecessary. Scheme 9.2 includes some of the more common commercially available alkenes, although this is a representative list, not an exhaustive one.

The synthesis of a monosubstituted alkene can be envisioned from the precursors shown in Scheme 9.3. In principle, disconnections more remote from the alkene moiety are also possible.

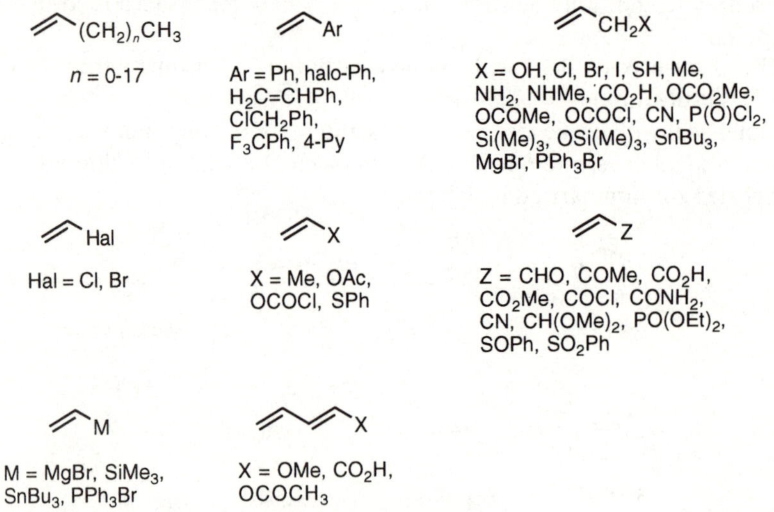

$(CH_2)_nCH_3$

n = 0-17

Ar

Ar = Ph, halo-Ph, H_2C=CHPh, $ClCH_2Ph$, F_3CPh, 4-Py

CH_2X

X = OH, Cl, Br, I, SH, Me, NH_2, NHMe, CO_2H, OCO_2Me, OCOMe, OCOCl, CN, $P(O)Cl_2$, $Si(Me)_3$, $OSi(Me)_3$, $SnBu_3$, MgBr, PPh_3Br

Hal

Hal = Cl, Br

X

X = Me, OAc, OCOCl, SPh

Z

Z = CHO, COMe, CO_2H, CO_2Me, COCl, $CONH_2$, CN, $CH(OMe)_2$, $PO(OEt)_2$, SOPh, SO_2Ph

M

M = MgBr, $SiMe_3$, $SnBu_3$, PPh_3Br

X

X = OMe, CO_2H, $OCOCH_3$

Scheme 9.2

2.1 Methylenation reactions

Considering disconnection ‘a’ first, there are several options available. An aldehyde precursor may be converted into the corresponding alkene[2] by Wittig

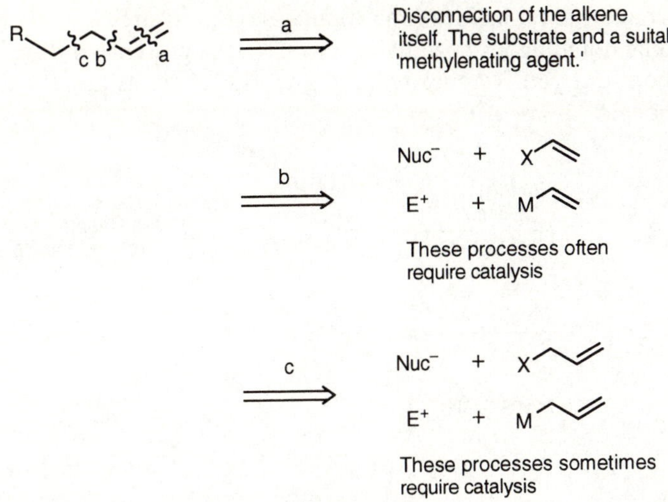

Disconnection of the alkene itself. The substrate and a suitable 'methylenating agent.'

These processes often require catalysis

These processes sometimes require catalysis

Scheme 9.3

reaction (eqn 1),[3] Peterson reaction[4] (eqn 2), Johnson methylenation[5] (eqn 3), or the use of Tebbe and related reagents (eqn 4).[6]

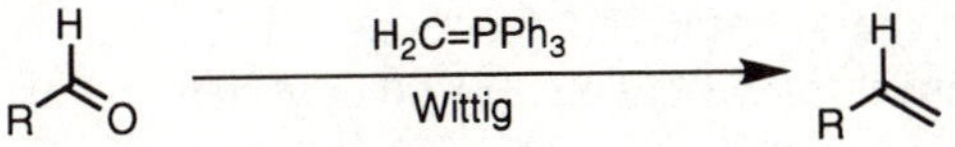

$$\mathrm{RCHO} \xrightarrow[\text{Wittig}]{H_2C{=}PPh_3} \mathrm{RCH{=}CH_2} \qquad (1)$$

see Chapter 2, Protocol 6

$$\mathrm{RCHO} \xrightarrow[\text{Peterson}]{\text{(i) } Me_3SiCH_2MgCl\text{; (ii) } SOCl_2} \mathrm{RCH{=}CH_2} \qquad (2)$$

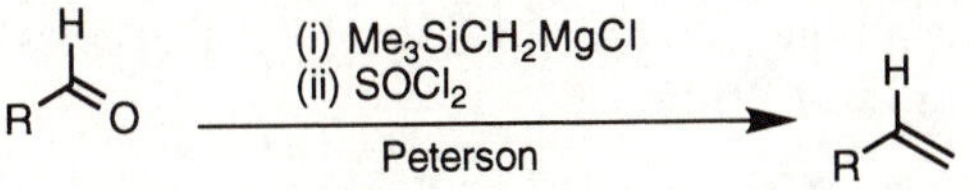

see Chapter 3, Protocol 1

$$\mathrm{RCHO} \xrightarrow[\text{(ii) Al / Hg}]{\text{(i) } H_2^{-}C{-}S(=O)(=NMe){-}Ph} \mathrm{RCH{=}CH_2} \qquad (3)$$

Johnson methylenation

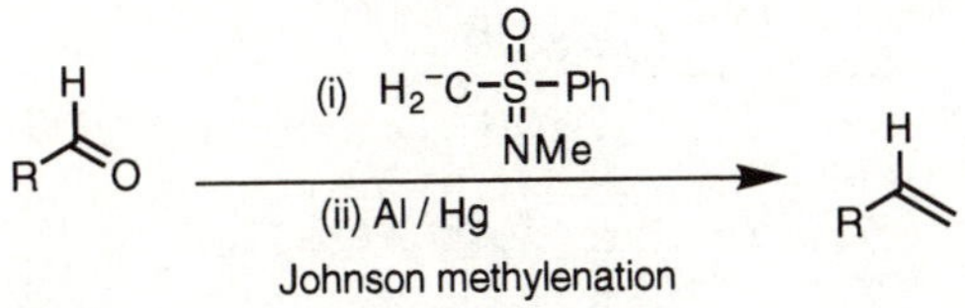

see Chapter 3, Protocol 5

$$\mathrm{RCHO} \xrightarrow[\text{or } Zn{-}CH_2Br_2{-}TiCl_4 \text{ (Oshima-Lombardo)}]{Cp_2Ti(\mu\text{-}CH_2)(\mu\text{-}Cl)AlMe_2 \text{ (Tebbe)}} \mathrm{RCH{=}CH_2} \qquad (4)$$

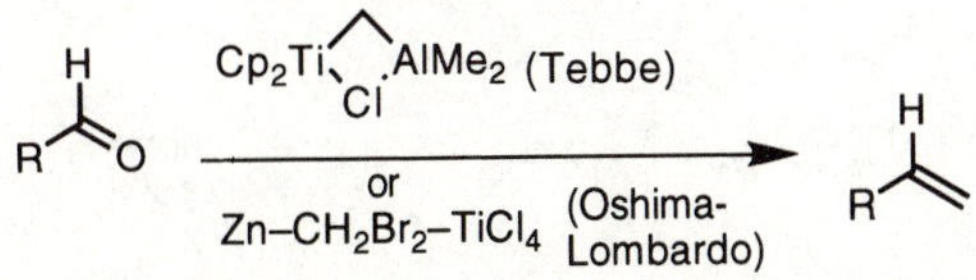

see Chapter 4, Protocol 2

Alternative precursors for methylenation reactions are alkyl halides (or tosylates). Again, the Wittig reaction may be employed for this transformation. Conversion of the alkyl bromide into the corresponding phosphonium salt, deprotonation, and trapping with formaldehyde provides one method (eqn 5). Recently, the use of dimethylsulfonium methylide has been demonstrated to convert alkyl halides (eqn 6) and alkyl mesylates (eqn 7) into the corresponding homologous terminal alkenes.[7] In the presence of three equivalents of the sulfur ylid, ketones react to give allyl alcohols (i.e. the addition of two carbons).[8]

R–Br —(i) PPh_3, (ii) Base, (iii) $H_2C{=}O$, Wittig→ R–CH=CH$_2$ see Chapter 2, Protocols 1 and 5 (5)

R–CH_2Br —Me_2S^+–CH_2^-→ R–CH=CH$_2$ (6)

R–CH_2OMs —Me_2S^+–CH_2^-→ R–CH=CH$_2$ (7)

2.2 Vinyl nucleophiles

Simple vinyl nucleophiles such as vinylmagnesium bromide or vinyllithium are able to add a vinyl group to reactive electrophiles such as aldehydes or ketones (eqn 8). Lithium divinyl cuprate adds in a conjugate addition fashion to enals (eqn 9).[9] Transition metal catalysed reactions can be employed to couple vinylmagnesium bromide to aryl halides (or alkenyl halides) as a convenient synthesis of styrenes (or dienes, see Section 7) (eqn 10). The Stille reaction can employ tributylvinylstannane as the 'nucleophilic' coupling partner (eqn 11),[10,11] and the Suzuki coupling can employ a vinylboron species as the 'nucleophilic' coupling partner (eqn 12).[12]

CH$_2$=CH–MgBr + RCHO → CH$_2$=CH–CH(OH)–R (8)

H–C(=O)–CH=CH–R + (CH$_2$=CH)$_2$CuLi —Me_3SiCl→ H–C(=O)–CH$_2$–CH(CH=CH$_2$)–R (9)

CH$_2$=CH–MgBr + Ar–Hal —Pd cat.→ Ar–CH=CH$_2$ see Chapter 8, Protocol 2 (10)

CH$_2$=CH–SnBu$_3$ + Ar–Hal —Pd cat.→ Ar–CH=CH$_2$ see Chapter 8, Protocol 4 (11)

vinyl catecholborane —Pd cat., Ar–Hal→ Ar–CH=CH$_2$ see Chapter 8, Protocol 5 (12)

2.3 Vinyl electrophiles

Whilst nucleophilic substitution reactions of vinyl halides are not generally possible, there are exceptions. Thus, cuprates are able to react with vinyl bromide to give the substitution product, as seen in eqn 13. In cases where the vinyl bromide is substituted, it has been shown that the reaction proceeds with retention of configuration.

In the presence of transition metal catalysts, vinyl halides are able to function as coupling partners with aromatic and aliphatic Grignard reagents (eqn 14). In the presence of a suitable enantiomerically pure ligand, racemic Grignard reagents have been used to give the coupled product with high asymmetric induction (eqn 15). The use of an aryllithium coupling with a vinyl bromide is discussed in Chapter 8, Protocol 1.

Furthermore, vinyl halides are also able to participate in Stille (eqn 16) and Suzuki (eqn 17) coupling reactions. Vinyl triflates are also able to act as 'electrophiles' in coupling reactions (eqn 18).

$$CH_2{=}CH\text{–}Br \xrightarrow{R_2CuLi} CH_2{=}CH\text{–}R \qquad (13)$$

$$CH_2{=}CH\text{–}Hal \xrightarrow[\text{Ni cat.}]{RMgBr} CH_2{=}CH\text{–}R \qquad (14)$$

see Chapter 8, Table 8.1 + Protocol 6

$$CH_2{=}CH\text{–}Br \xrightarrow[\text{Ni cat. L}^*]{PhCH(Me)MgCl} CH_2{=}CH\text{–}{}^{*}CH(Me)Ph \qquad (15)$$

see Chapter 8, Scheme 8.11

$$CH_2{=}CH\text{–}Hal \xrightarrow[\text{Pd cat.}]{ArSnBu_3} CH_2{=}CH\text{–}Ar \qquad (16)$$

$$CH_2{=}CH\text{–}Hal \xrightarrow[\text{Pd cat.}]{\text{Ar–B(catechol)}} CH_2{=}CH\text{–}Ar \qquad (17)$$

see Chapter 8, Protocol 5

$$CH_2{=}CH\text{–}OTf \xrightarrow[\text{Pd cat.}]{ArSnBu_3} CH_2{=}CH\text{–}Ar \qquad (18)$$

2.4 Allyl nucleophiles

Allylmetal reagents are sometimes found to be more reactive than otherwise similar organometallic species. This is particularly true in the addition of allylmetals to carbonyl compounds, where a six-membered cyclic transition state can be invoked.[13] A straightforward example would be the addition of allylmagnesium bromide to an aldehyde to give the corresponding homoallylic alcohol (eqn 19). However, more subtle variations of this basic process have been achieved. Thus, allylchromium compounds (formed *in situ* from an allyl bromide and chromous chloride) react with high diastereoselectivity with aldehydes (eqn 20).[14] Similar chemistry is observed for allylboronates, where the *syn*-product is obtained if the (*Z*)-precursor is employed (eqn 21). Enantiomerically pure allylboronates have proved to be very popular for the synthesis of homoallylic alcohols.[15,16] Allylstannanes also add to aldehydes (eqn 22),[17] and this reaction can be catalysed by rhodium complexes[18] and by enantiomerically pure titanium catalysts.[19] In the latter case, highly enantioselective reactions have been observed.

RCHO + allyl-MgBr → RCH(OH)CH₂CH=CH₂ (19)

PhCHO + crotyl bromide —$CrCl_2$, 96% yield→ anti-homoallylic alcohol (20)

RCHO + (Z)-crotyl pinacol boronate —then $N(CH_2CH_2OH)_3$→ syn-homoallylic alcohol (21)

RCHO + allyl-$SnBu_3$ —catalyst→ RCH(OH)CH₂CH=CH₂ (22)

2.5 Allyl electrophiles

Allyl bromide is an example of a simple allyl electrophile. It is more reactive as an electrophile than propyl bromide, due to the activating nature of the alkene. A nucleophile such as dimethylmalonate or an alkoxide will react with allyl bromide readily (eqn 23). When issues of regio- and stereochemistry are being considered, palladium-catalysed allylic substitution may be employed.[20] Thus, a less reactive electrophile such as an allyl acetate becomes activated to nucleophilic attack in the presence of a palladium catalyst (eqn 24). Various nucleophilic coupling partners have been employed, including nitrogen-based (Chapter 8, Protocol 10) and carbon-based (Chapter 8, Protocol 11) nucleophiles.[21]

allyl-Br + PhO^- —EtOH, reflux→ allyl-OPh (23)

allyl-OAc + Nuc^- —Pd cat.→ allyl-Nuc (24)

3. 1,1-Disubstituted alkenes

3.1 Methylenation reactions

With 1,1-disubstituted alkenes, there is no issue of controlling the alkene geometry, and therefore, as with monosubstituted alkenes, synthesis is often relatively straightforward.

The methylenation reactions discussed in Section 2.1 can be extended to the methylenation of ketones (eqn 25). These methods are widely used, and have been discussed in detail in the relevant chapters, as indicated.[22]

R(R')C=O —'methylenation'→ R(R')C=CH_2 (25)

Wittig – Chapter 2, Protocol 6
Peterson – Chapter 3, Protocol 1
Johnson – Chapter 3, Protocol 5
Boron-Wittig – Chapter 3, Protocol 8
Tebbe, Oshima-Lombardo – Chapter 4, Protocol 2

3.2 Transition metal-catalysed coupling reactions

Transition metal-catalysed coupling reactions provide an effective means for the synthesis of 1,1-disubstituted alkenes. Thus, the use of an appropriate 2-haloalkene under standard coupling conditions leads to the corresponding product (eqn 26).[23] The synthetic accessibility of the vinyl halide is the main synthetic challenge with this approach. When 1,1-dichloroethene is employed as the precursor, the synthesis of symmetrical 1,1-disubstituted alkenes is possible (eqn 27). Palladium-catalysed coupling reactions are also high yielding processes (eqn 28).[24] Depending on the availability of the vinylmetal reagent, transition metal-catalysed coupling reactions can be conducted in the opposite sense (eqn 29).[25]

ArMgBr; $NiCl_2(dppp)$ (0.005-0.01eq.) (26)

ArMgBr; Ni cat. — see Chapter 8, Protocol 6 (27)

$PhCH_2ZnBr$, $PdCl_2(MeCN)_2$ (5 mol%); DMF, 25°C, 12 h → $PhCH_2$ … OH (28)

M = MgBr, ZnBr, $SnBu_3$, BR_2, etc.; ArBr; Pd cat. (see Chapter 8, Protocols 3 and 8) (29)

3.3 Elimination reactions

Elimination reactions can be used to obtain 1,1-disubstituted alkenes (see Chapter 5, Protocols 3 and 4). This may be achieved in one of two basic ways, illustrated by eqns 30 and 31. The first of these approaches is generally more reliable, since there may be alternative elimination reactions possible, if R or R′ contain α-hydrogens.

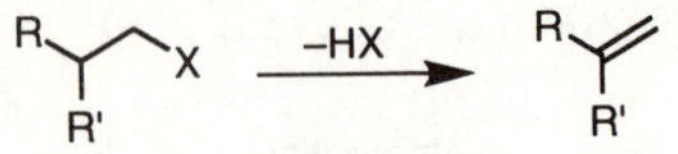

R, R′, X; –HX; R, R′ (30)

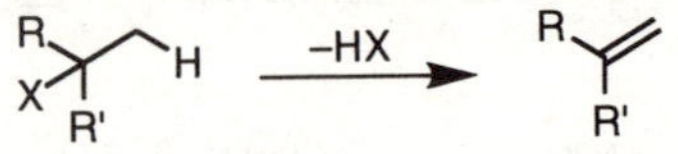

R, X, R′, H; –HX; R, R′ (31)

Useful synthetic methods based on elimination strategies include the conversion of lactones into α-methylene lactones (eqn 32),[26] the use of α-lithioselenoxides as nucleophiles which eliminate after addition (eqn 33),[27] and the elimination of H–X (eqn 34, see Chapter 5).

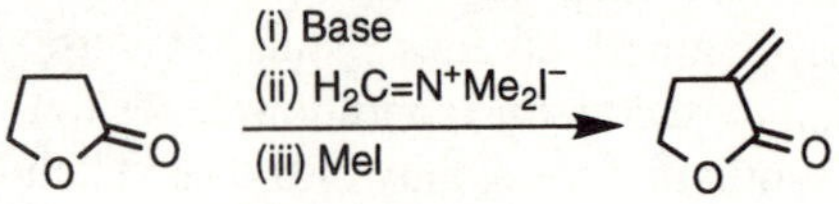

O, O; (i) Base; (ii) $H_2C{=}N^+Me_2I^-$; (iii) MeI; O, O (32)

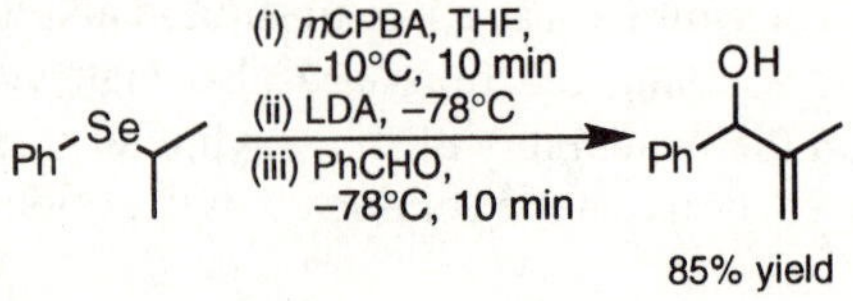

Ph, Se; (i) *m*CPBA, THF, –10°C, 10 min; (ii) LDA, –78°C; (iii) PhCHO, –78°C, 10 min; OH, Ph; 85% yield (33)

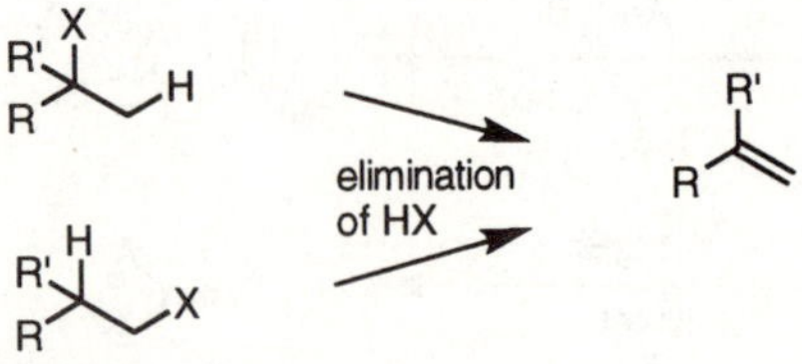

X, R′, R, H; H, R′, R, X; elimination of HX; R′, R (34)

3.4 Baylis–Hillman reaction

The Baylis–Hillman reaction provides a useful means for the conversion of monosubstituted alkenes attached to an electron-withdrawing group into 1,1-disubstituted alkenes.[28] Thus, the reaction of acrylonitrile with benzaldehyde, catalysed by an amine (usually DABCO), affords the Baylis–Hillman adduct (eqn 35).[29]

$$\text{CH}_2{=}\text{CHCN} \xrightarrow[\text{DABCO}]{\text{PhCHO}} \text{CH}_2{=}\text{C(CN)CH(OH)Ph} \qquad (35)$$

4. 1,2-Disubstituted alkenes

1,2-Disubstituted alkenes may exist as (*E*)- and (*Z*)-isomers. Therefore, the synthesis of 1,2-disubstituted alkenes is usually concerned with the control of the geometry of the alkene. There are many methods available to choose from for the stereoselective synthesis of 1,2-disubstituted alkenes. It is often a matter of personal preference which method to employ, although the following sections offer some guidelines as to the most suitable methods available.

4.1 Reduction of alkynes

Chapter 6 considers the synthesis of 1,2-disubstituted alkenes from the corresponding alkyne. Typically, hydrogenation of an alkyne over a heterogeneous catalyst affords the (*Z*)-alkene (eqn 36), whereas treatment of an alkyne with sodium/ammonia affords the (*E*)-alkene (eqn 37). Chapter 6 discusses the details and procedures behind these basic methods, as well as other stereoselective alkyne reduction methods. Reduction of alkynes to alkenes is particularly useful for the preparation of either (*E*)- or (*Z*)-1,2-disubstituted alkenes which do not contain too much other functionality. However, the reduction of alkynes cannot be applied to the synthesis of more highly substituted alkenes, and other methods are needed.

$$\text{R}{-}{\equiv}{-}\text{R'} \xrightarrow[\text{catalyst}]{\text{H}_2} (Z)\text{-RCH}{=}\text{CHR'} \qquad (36)$$

see Chapter 6, Protocol 2

$$\text{R}{-}{\equiv}{-}\text{R'} \xrightarrow{\text{Na/NH}_3} (E)\text{-RCH}{=}\text{CHR'} \qquad (37)$$

see Chapter 6, Protocol 4

Alkynes may also be converted into alkenes mediated by boranes. Boranes add across alkynes to afford vinylboranes. The carbon–boron bond may be replaced by a carbon–carbon bond (eqn 38)[30] or by a carbon–hydrogen bond (eqn 39).[31] By suitable choice of starting material, (*Z*)-alkenes are produced, although a variation of this strategy can be employed to afford the (*E*)-alkene.[32]

$$\text{R}_2\text{BH} + \text{H}{-}{\equiv}{-}\text{R'} \longrightarrow \left[\text{R}_2\text{B(H)C}{=}\text{C(H)R'}\right] \xrightarrow[\text{I}_2]{\text{NaOH}} (Z)\text{-RCH}{=}\text{CHR'} \qquad (38)$$

$$\text{R}_2\text{BH} + \text{R'}{-}{\equiv}{-}\text{R''} \longrightarrow \left[\text{R}_2\text{B(R')C}{=}\text{C(H)R''}\right] \xrightarrow{\text{AcOH}} (Z)\text{-R'CH}{=}\text{CHR''} \qquad (39)$$

see Chapter 6, Protocol 3

4.2 Wittig and related reactions

The factors which affect the geometrical outcome of the Wittig reaction are discussed in Chapter 2. In short, there are Wittig methods available for the synthesis of either (*E*)- or (*Z*)-alkenes, as shown in Scheme 9.4.[33] The Wittig reaction and its variants are particularly effective for the stereoselective synthesis of (*E*)- and (*Z*)-1,2-disubstituted alkenes, but selectivity is generally poor in the preparation of more highly substituted alkenes. On occasions when a Wittig reaction affords only poor selectivity, the thermodynamic ratio of (*E*)- and (*Z*)-isomers has been obtained by either subsequent[34] or *in situ*[35] irradiation in the presence of diphenyldisulfide.

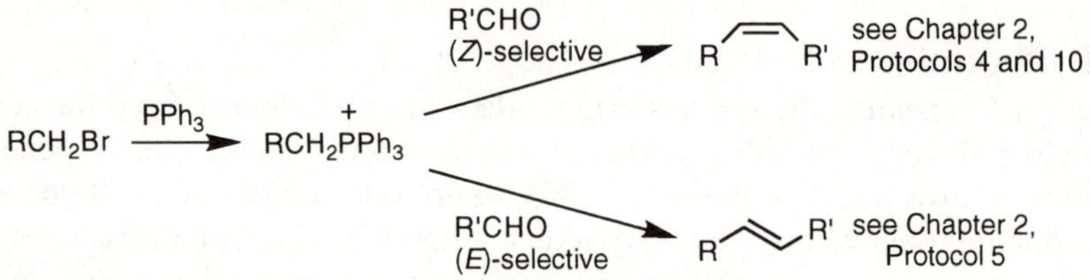

$RCH_2Br \xrightarrow{PPh_3} RCH_2\overset{+}{P}Ph_3$

→ R'CHO, (*Z*)-selective → (*Z*)-RCH=CHR' — see Chapter 2, Protocols 4 and 10

→ R'CHO, (*E*)-selective → (*E*)-RCH=CHR' — see Chapter 2, Protocol 5

Scheme 9.4

The Wadsworth–Horner–Emmons procedure is generally (*E*)-selective (eqn 40), although there are are examples when (*Z*)-selectivity can be obtained (eqn 41). The Horner–Wittig reaction of phosphine oxide anions

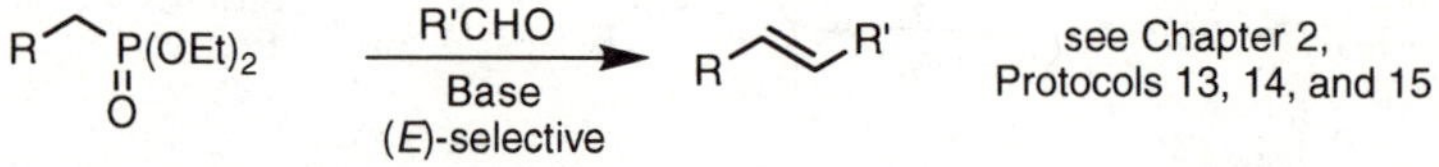

$RCH_2P(O)(OEt)_2$ —R'CHO, Base, (*E*)-selective→ (*E*)-RCH=CHR' — see Chapter 2, Protocols 13, 14, and 15 (40)

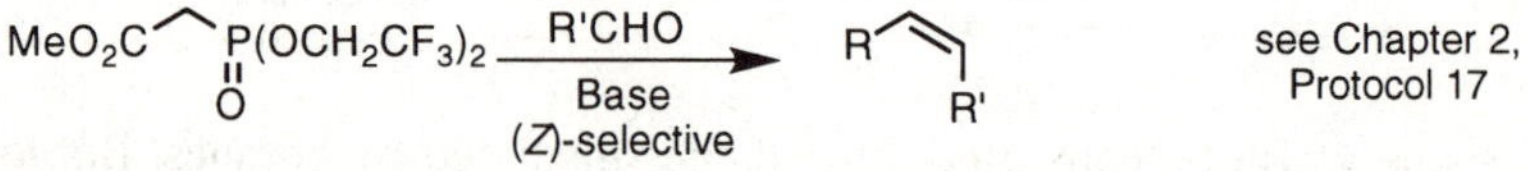

$MeO_2CCH_2P(O)(OCH_2CF_3)_2$ —R'CHO, Base, (*Z*)-selective→ (*Z*)-RCH=CHR' — see Chapter 2, Protocol 17 (41)

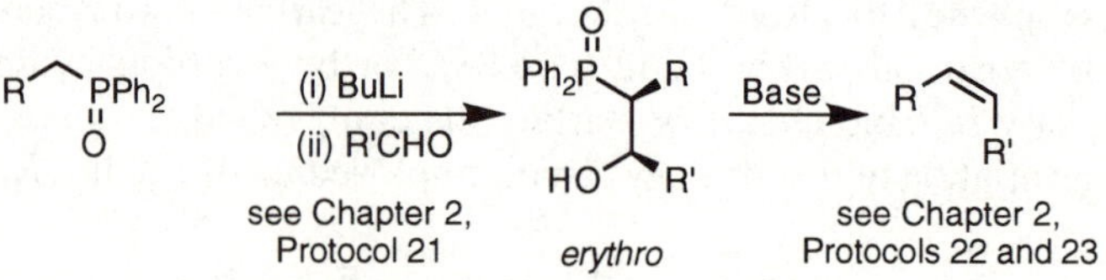

$RCH_2P(O)Ph_2$ —(i) BuLi, (ii) R'CHO (see Chapter 2, Protocol 21)→ $Ph_2P(O)CH(R)CH(OH)R'$ (*erythro*) —Base (see Chapter 2, Protocols 22 and 23)→ (*Z*)-RCH=CHR' (42)

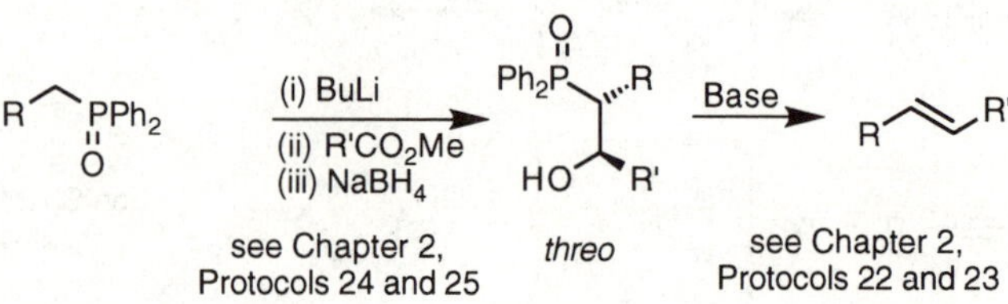

$RCH_2P(O)Ph_2$ —(i) BuLi, (ii) $R'CO_2Me$, (iii) $NaBH_4$ (see Chapter 2, Protocols 24 and 25)→ $Ph_2P(O)CH(R)CH(OH)R'$ (*threo*) —Base (see Chapter 2, Protocols 22 and 23)→ (*E*)-RCH=CHR' (43)

offers the advantage that the diastereomeric *erythro*- and *threo*-adducts are isolable and may be purified to diastereomeric purity, and are stereospecifically converted into the (*Z*)- and (*E*)-alkenes respectively (eqns 42 and 43).[36,37] The factors and reasoning for the selectivities are more fully elaborated in Chapter 2.

4.3 Peterson reaction

The Peterson reaction has been used in the stereoselective synthesis of 1,2-disubstituted alkenes. Like the Horner–Wittig reaction, the alkene precursors may be isolated, and can be prepared in a variety of ways (synthetically, the problem is the diastereoselective preparation of the β-hydroxysilane; see Chapter 3, Section 2). However, a single diastereomer of the precursor may be stereoselectively converted into either the (*E*)- or the (*Z*)-alkene, depending upon the elimination conditions employed (Scheme 9.5).

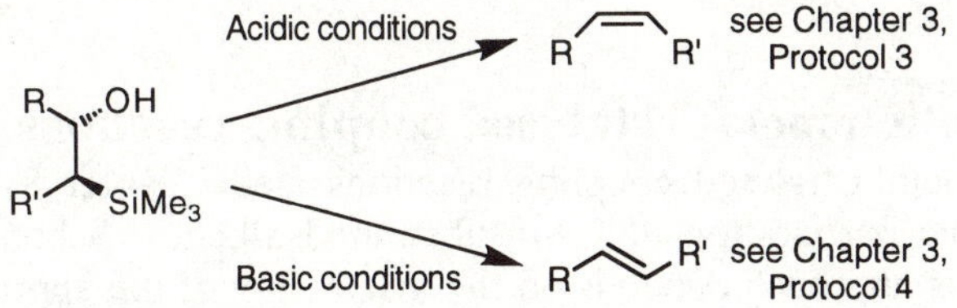

R, OH; R', $SiMe_3$ — Acidic conditions → (Z)-R–CH=CH–R' (see Chapter 3, Protocol 3)

— Basic conditions → (E)-R–CH=CH–R' (see Chapter 3, Protocol 4)

4.4 Julia olefination

The Julia olefination reaction (see Chapter 3, Section 4) has been less widely used than the Wittig and Peterson reactions, although it is useful for the synthesis of 1,2-disubstituted alkenes with high levels of (*E*)-selectivity.[38] The reaction involves the addition of an α-lithiosulfone to an aldehyde, followed by hydroxyl activation and reductive elimination (eqn 44). The main disadvantages are the length of the sequence and that the elimination of the hydroxysulfone requires the use of sodium amalgam. However, other reducing agents, including samarium diiodide/HMPA[39] have recently been used effectively.

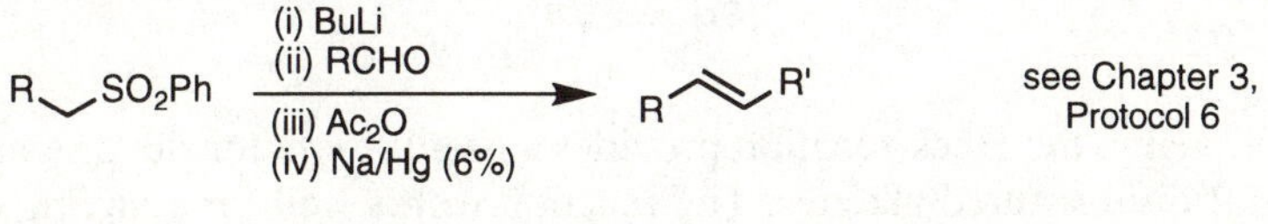

$$R\text{CH}_2SO_2Ph \xrightarrow[\text{(iii) }Ac_2O\text{ (iv) Na/Hg (6\%)}]{\text{(i) BuLi (ii) RCHO}} R\text{CH=CH}R' \qquad (44)$$

see Chapter 3, Protocol 6

4.5 Aldol condensation and related processes

The aldol condensation and the family of closely related reactions are amongst the simplest alkene-forming reactions from a practical viewpoint (see Chapter 5, Section 4).

These reactions provide a means for the synthesis of 1,2-disubstituted alkenes which are almost invariably formed as the thermodynamic product (generally the (*E*)-alkene).

Thus, treatment of a ketone and an aldehyde with a base such as sodium hydroxide affords the (*E*)-α,β-unsaturated ketone (eqn 45). As discussed in more detail in Chapter 5, a mixture of products will arise unless the two aldol components are carefully chosen.

The Reformatsky reaction and its variants[40] also provide a related method for the preparation of (*E*)-α,β-unsaturated carbonyls (eqn 46).

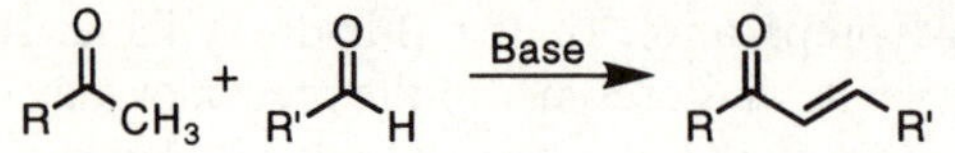

$$\mathrm{RC(O)CH_3} + \mathrm{R'CHO} \xrightarrow{\text{Base}} \mathrm{RC(O)CH{=}CHR'} \qquad (45)$$

see Chapter 5, Protocol 6
cf. Chapter 5, Protocol 7

$$\mathrm{RC(O)CH_2Br} + \mathrm{R'CHO} \xrightarrow{\text{Zn}} \mathrm{RC(O)CH{=}CHR'} \qquad (46)$$

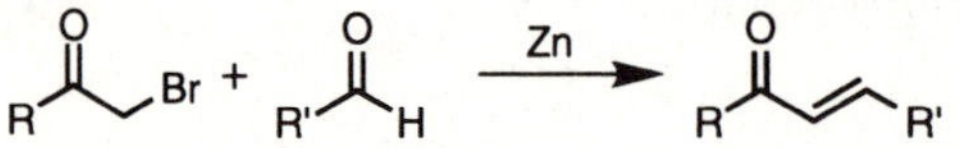

see Chapter 5, Section 4.6

4.6 Transition metal-catalysed coupling reactions

Transition metal-catalysed coupling reactions (see Chapter 8) are readily applied to the preparation of 1,2-disubstituted alkenes (Scheme 9.6). The success of this approach depends on the availability of the appropriate vinyl halide or vinylmetal (see Section 8).[41]

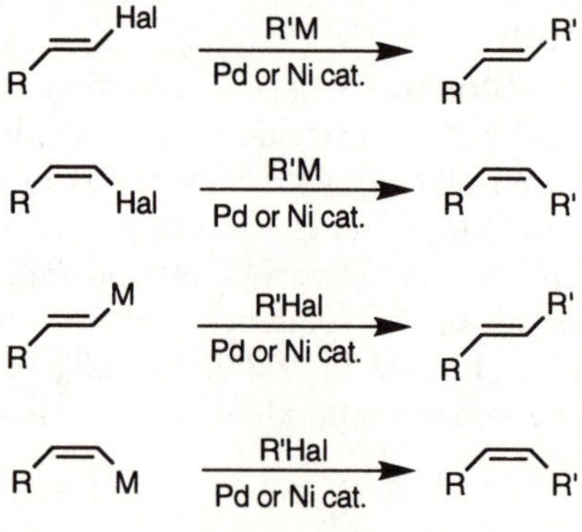

$$(E)\text{-RCH{=}CHHal} \xrightarrow[\text{Pd or Ni cat.}]{\text{R'M}} (E)\text{-RCH{=}CHR'}$$

$$(Z)\text{-RCH{=}CHHal} \xrightarrow[\text{Pd or Ni cat.}]{\text{R'M}} (Z)\text{-RCH{=}CHR'}$$

$$(E)\text{-RCH{=}CHM} \xrightarrow[\text{Pd or Ni cat.}]{\text{R'Hal}} (E)\text{-RCH{=}CHR'}$$

$$(Z)\text{-RCH{=}CHM} \xrightarrow[\text{Pd or Ni cat.}]{\text{R'Hal}} (Z)\text{-RCH{=}CHR'}$$

Scheme 9.6

Additionally, the Heck reaction provides a useful route for the preparation of (*E*)-1,2-disubstituted alkenes. The reaction works well for coupling either aryl or vinyl halides with alkenes attached to electron-withdrawing groups (eqn 47). The Heck reaction has an advantage over other transition metal-catalysed procedures since there is no need for a metal-containing component. When the R-group contains α-protons, consideration of the regioselectivity of the product needs to be made, as discussed in Chapter 8, Section 4.

$$\text{Ar-Hal} + \text{CH}_2\text{=CH-R} \xrightarrow[\text{Base (Et}_3\text{N)}]{\text{cat. Pd(OAc)}_2} \text{Ar-CH=CH-R} \quad \text{see Chapter 8, Protocols 7 and 8} \tag{47}$$

4.7 Miscellaneous methods

There are many other stereoselective syntheses of 1,2-disubstituted alkenes. Hydrometallation of an alkyne (see Chapter 7) and subsequent functionalisation are perhaps most useful in the preparation of tri-substituted alkenes, but eqns 48[42] and 49[43] show that the alkenylzirconocene complex obtained by hydrometallation of the corresponding alkyne may be converted into synthetically useful 1,2-disubstituted alkenes. The chromous chloride promoted coupling of *gem*-dihalides[44] and 1-acetoxy-1-bromides[45] with aldehydes provides an alternative which is discussed in Chapter 4, Section 4. Generally, this is most useful in the preparation of vinyl halides, silanes, and stannanes, but other (*E*)-1,2-disubstituted alkenes can be prepared in this way (eqn 50). Decarboxylation of β-lactones proceeds with preservation of stereochemistry to give the corresponding alkenes (eqn 51).[46] Furthermore, many sequences have been devised for the stereospecific interconversion of one alkene geometry into another. This topic has been reviewed elsewhere.[47]

$$\text{R-CH=CH-ZrCp}_2\text{Cl} \xrightarrow[\text{(ii) R'CHO}]{\text{(i) Me}_2\text{Zn}} \text{R-CH=CH-CH(OH)-R'} \tag{48}$$

$$\text{R-CH=CH-ZrCp}_2\text{Cl} \xrightarrow[\text{(ii) 4-RO-cyclopent-2-enone}]{\text{(i) MeLi/Me}_3\text{ZnLi, cat. Me}_2\text{Cu(CN)Li}_2} \text{3-(R-CH=CH)-4-RO-cyclopentanone} \tag{49}$$

$$\text{RCHO} \xrightarrow[\text{CrCl}_2]{\text{R'CHHal}_2} \text{R-CH=CH-R'} \tag{50}$$

$$\beta\text{-lactone (R, R')} \xrightarrow[-\text{CO}_2]{\text{heat}} \text{R-CH=CH-R'} \tag{51}$$

5. Tri- and tetrasubstituted alkenes

In the previous section, it is apparent that there are many successful strategies for the highly stereoselective synthesis of 1,2-disubstituted alkenes. There are fewer methods available for the stereoselective synthesis of trisubstituted alkenes, and even fewer for tetrasubstituted alkenes. One problem with more highly substituted alkenes is that there is generally less thermodynamic difference between the isomeric products. It is therefore easier to first consider the synthesis of tri- and tetrasubstituted alkenes which do not have isomeric forms.

5.1 Non-stereoselective methods

Since this section describes non-stereoselective methods, only substrates which afford alkenes where stereochemistry is not an issue have been described. The Wittig reaction (eqn 52), Wadsworth–Emmons reaction (eqn 53), and Peterson reaction (eqn 54) are suitable for the synthesis of trisubstituted alkenes, but for steric reasons, these reactions are not very successful for the preparation of tetrasubstituted alkenes. Aldol-type procedures (eqn 55) and elimination reactions (eqn 56)[48] are also suitable.

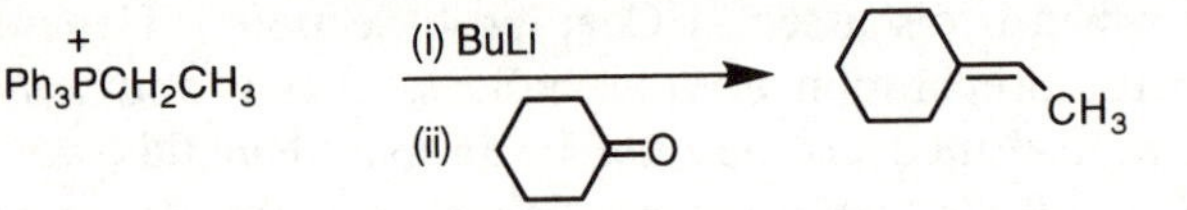

$Ph_3\overset{+}{P}CH_2CH_3$ —(i) BuLi; (ii) cyclohexanone→ ethylidenecyclohexane (CH_3) (52)

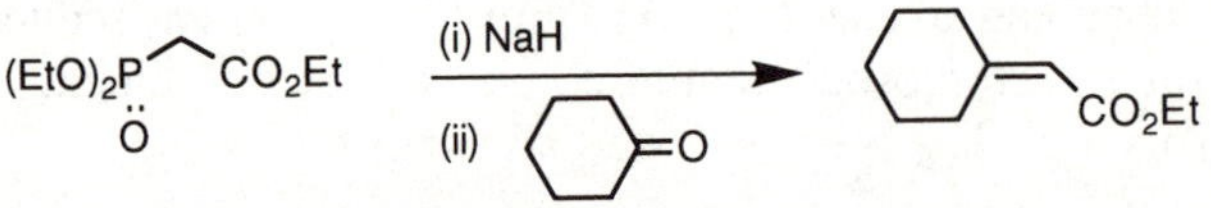

$(EtO)_2P(O)CH_2CO_2Et$ —(i) NaH; (ii) cyclohexanone→ cyclohexylidene–$CH\,CO_2Et$ (53)

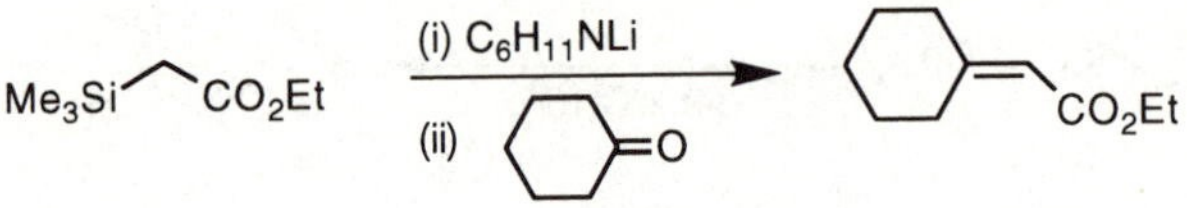

$Me_3SiCH_2CO_2Et$ —(i) $C_6H_{11}NLi$; (ii) cyclohexanone→ cyclohexylidene–$CH\,CO_2Et$ (54)

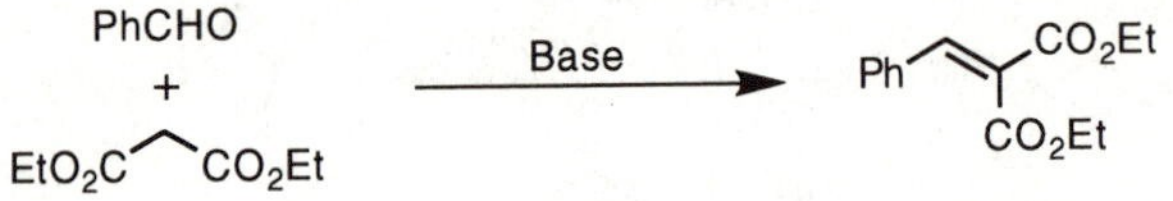

PhCHO + $EtO_2CCH_2CO_2Et$ —Base→ $PhCH{=}C(CO_2Et)_2$ (55)

see Chapter 5, Protocol 8

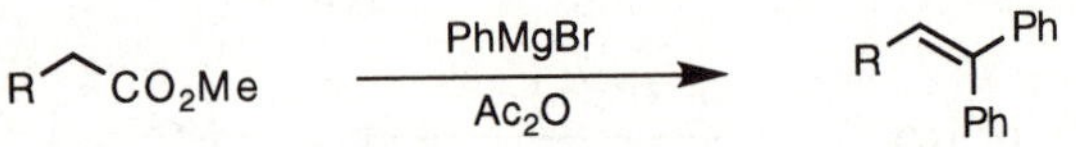

RCH_2CO_2Me —PhMgBr; Ac_2O→ $RCH{=}CPh_2$ (56)

Tetrasubstituted alkenes may be prepared by the McMurry coupling reaction[49] (eqn 57) and by the Barton–Kellogg method (eqn 58),[50] which involves extrusion of sulfur and dinitrogen from 1,3,4-thiadiazolines, which in turn can be prepared in two steps from ketones[51] or thioketones. This last method is particularly useful for the synthesis of very hindered alkenes.

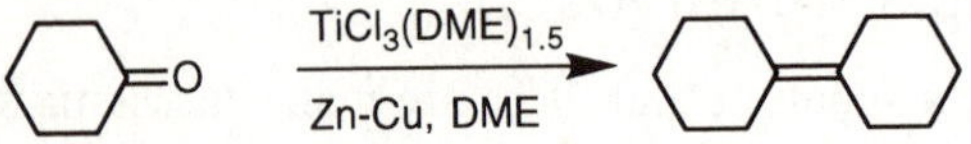

cyclohexanone —$TiCl_3(DME)_{1.5}$; Zn-Cu, DME→ cyclohexylidenecyclohexane (57)

see Chapter 4, Protocol 1

spiro-thiadiazoline (S, N=N) —$P(OEt)_3$; heat→ cyclohexylidenecyclohexane (58)

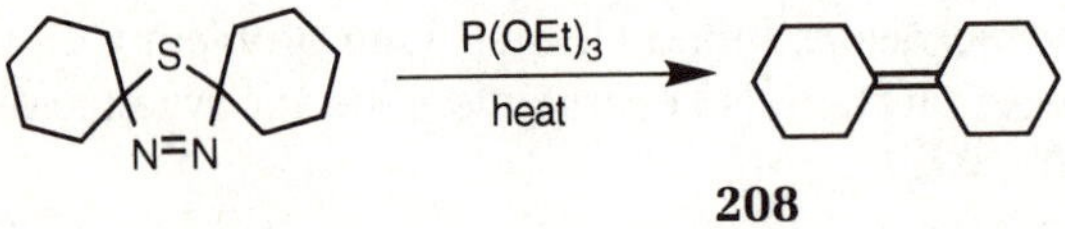

5.2 Transition metal-catalysed synthesis of trisubstituted alkenes

Many of the transition metal-catalysed syntheses of trisubstituted alkenes are made possible by using precursors which are synthesised from alkynes. Thus the vinyl halide (eqn 59) and vinyl metal (eqn 60) precursors are often derived by an addition process to an appropriate alkyne.

R, R', Hal → (R"M; Pd or Ni cat.) → R, R', R" (59)

R, R', M → (R"Hal; Pd or Ni cat.) → R, R', R" (60)

The methods for preparation of these precursors are discussed in Chapter 7 and also in Section 8 of this chapter. Some examples of the stereoselective synthesis of trisubstituted alkene syntheses are given in eqns 61,[52] 62,[53] 63,[54] 64,[55] 65,[56] 66,[57] and 67.[58] These examples are merely representative of the types of process which are possible. The transition metal-catalysed reactions are remarkably tolerant of other functional groups, and many researchers choose to construct alkenes using transition metal-catalysed reactions.

C_8H_{17}, H, bis(pinacolboronate) → (PhI, 3 mol% $Pd(PPh_3)_4$; aq. KOH, 91%) → C_8H_{17}, H, Ph, Ph (61)

Me, H, I, OH → ((i) EtZnCl; (ii) $PhCH_2ZnCl$, 5 mol% $((Ph_3P)_2PdCl_2$ + 2BuLi), 90%) → Me, H, PhH_2C, OH (62)

C_3H_7, C_3H_7, ZnX → (PhI; cat. $Pd(PPh)_3$) → C_3H_7, C_3H_7, Ph (63)

Ph, pinacolboronate, $ZrCp_2Cl$ → ((i) ArI, $Pd(PPh_3)_4$, THF, 0°C; (ii) Ar'Br, $Pd(PPh_3)_4$, EtONa, benzene, reflux, 71-82%) → Ph, Ar', Ar (64)

TfO, Ph (cyclohexene) → ($NaBPh_4$; cat. $Pd(PPh_3)_4$, 73%) → Ph, Ph (cyclohexene) (65)

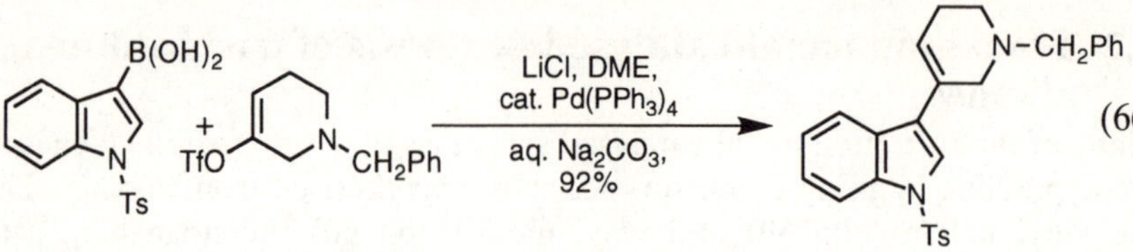

(66)

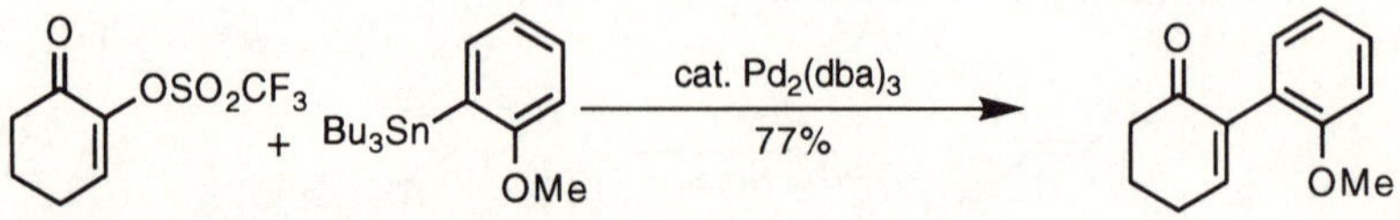

(67)

5.3 Strategies using cyclic precursors

The geometry of an alkene contained within a small ring is thermodynamically defined and fixed. Indeed, even in larger rings ($n>8$), although both (E)- and (Z)-isomers are stable, there may be large differences in energy if other substituents within the ring place conformational constraints upon it. The geometry inherent in cyclic alkenes may be exploited in the synthesis of acyclic alkenes, as the following examples show. In eqn 68, the enforced (Z)-geometry of the alkene is retained after reduction to the (Z)-acyclic alkene.[59] Kocienski *et al.*[60] have demonstrated the use of dihydrofuran in the transition metal-catalysed synthesis of stereodefined trisubstituted alkenes. Again the geometry of the cyclic alkene is employed to control the geometry of the acyclic products (eqn 69).[60] Evans and Carreira[61] have used a strategy involving a tether to control the geometry of an alkene via an intermediate macrocycle. The macrocyclisation via a Horner–Wadsworth–Emmons reaction provides a single alkene geometry, which is preserved upon removal of the tether (eqn 70).[61]

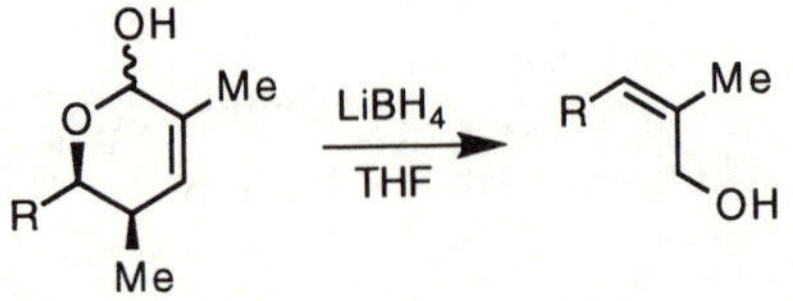

(68)

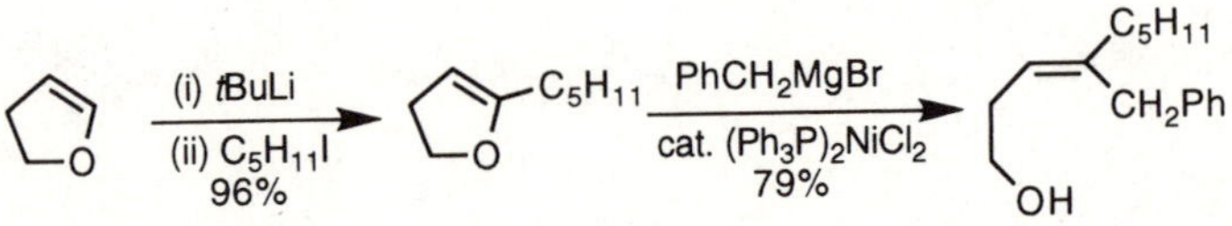

(69)

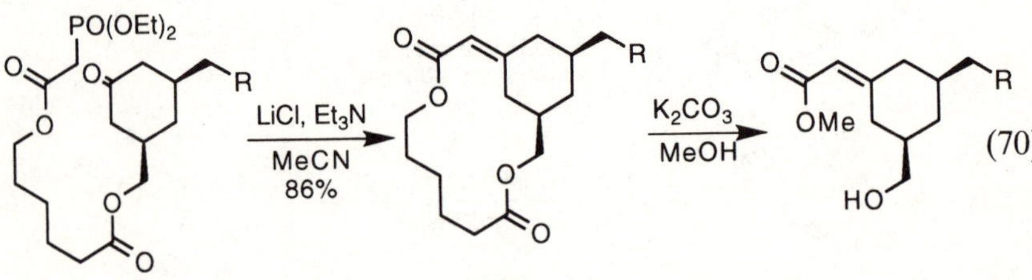

(70)

5.4 Miscellaneous methods

Aldol-type reactions can give good levels of selectivity in the synthesis of trisubstituted alkenes when the product has a strong thermodynamic preference, which is the case for (*E*)-α,β-unsaturated carbonyl compounds. Thus, treatment of methyl acrylate and benzaldehyde with sodium methoxide affords an (*E*)-α,β-unsaturated acid after hydrolysis (eqn 71; this reaction is assumed to proceed by conjugate addition of methoxide to the acrylate to form an ester enolate which then adds to the benzaldehyde).[62]

Alkynylborates are formed by the reaction of boranes with lithium acetylides. The alkynylborate may be quenched with electrophiles to afford an intermediate vinylborane, which upon treatment with acetic acid affords a trisubstituted alkene.[63] Equation 72 represents the reaction using carbon dioxide as the electrophilic quench, an example which proceeds to give exclusively the (*Z*)-alkene product.[64]

Baylis–Hillman adducts are 1,1-disubstituted alkenes, but they are readily converted into trisubstituted alkenes upon treatment with magnesium bromide (on the derived acetates)[65] or with *N*-bromosuccinimide (eqn 73).[66]

Enantioselective desymmetrisation of an achiral ketone has been achieved by conversion into a trisubstituted alkene using an enantiomerically pure phosphonamide reagent (eqn 74).[67] Disubstituted epoxides can be reduc-

PhCHO + $CH_2{=}CHCO_2Me$ —(i) NaOMe; (ii) hydrolysis; 44%→ Ph–CH=C(CO_2H)(CH_2OMe) (71)

R_3B + R'C≡CLi ⟶ $R_3\bar{B}C{\equiv}CR'$ Li^+ —(i) CO_2; (ii) CH_3CO_2H; 71-84%→ (R)(H)C=C(CO_2H)(R') (72)

R–CH(OH)–C(=CH_2)CO_2Me —N-Bromosuccinimide, Me_2S→ R–CH=C(CH_2Br)CO_2Me (73)

Phosphonamide (Me, N, P=O, N, Me, CH2Ph on cyclohexane-diamine) + 4-*t*Bu-cyclohexanone —(i) *n*BuLi, -78°C; (ii) H_2O; (iii) AcOH; >99% e.e.; 91% yield→ PhCH= (4-*t*Bu-cyclohexylidene) (74)

nC_4H_9-epoxide-*n*C_4H_9 —*s*BuLi; –70°C to r.t.; 90 min→ *n*C_4H_9–CH=C(*s*C_4H_9)(*n*C_4H_9) (75)

tively alkylated to give trisubstituted alkenes (eqn 75).[68] Epoxides may also be isomerised to allylic alcohols by bases.[69]

6. Cyclic alkenes

There are additional methods available for the preparation of cyclic alkenes which are generally less applicable to the synthesis of acyclic alkenes. In this section, it is not possible to give a detailed treatment of each of these processes, but to remind the reader of their existence, and to provide appropriate references.

One of the most effective ways of synthesising cyclic structures is via the Diels–Alder reaction. This process does result in the synthesis of cyclic alkenes (eqn 76);[70] however, three double bonds are converted into one double bond, and so this reaction cannot be primarily considered as an alkene synthesis.

The Pauson–Khand reaction provides a widely used and synthetically useful route to cyclopentenones (eqn 77).[71] This reaction and related cycloaddition reactions involving alkynes have been the topic of a review.[72] Catalytic versions of the Pauson–Khand reaction have also been reported.[73]

Grubbs and co-workers[74–76] have developed a synthetically powerful ring-

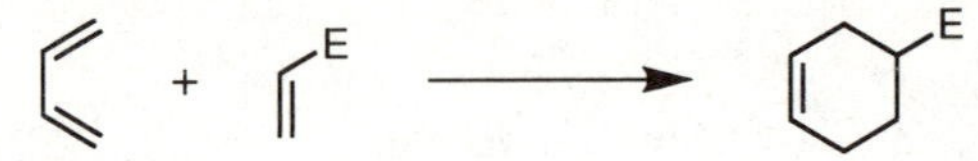

(76)

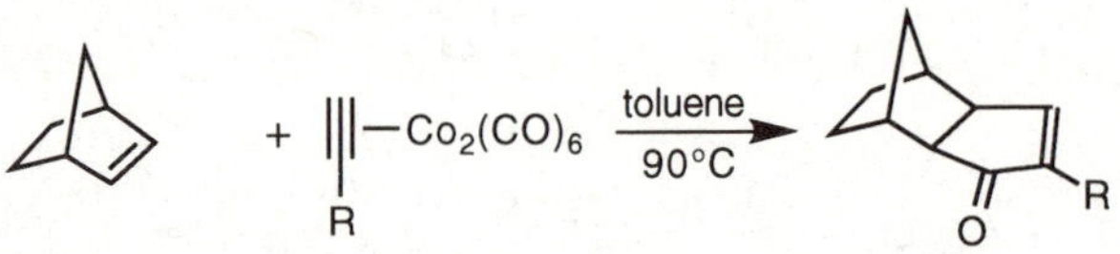

(77)

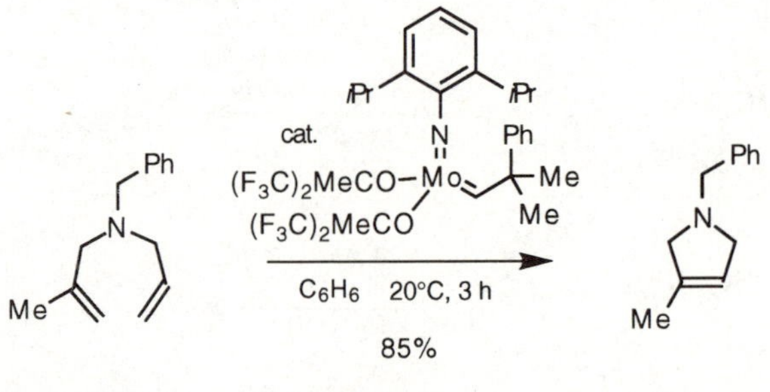

(78)

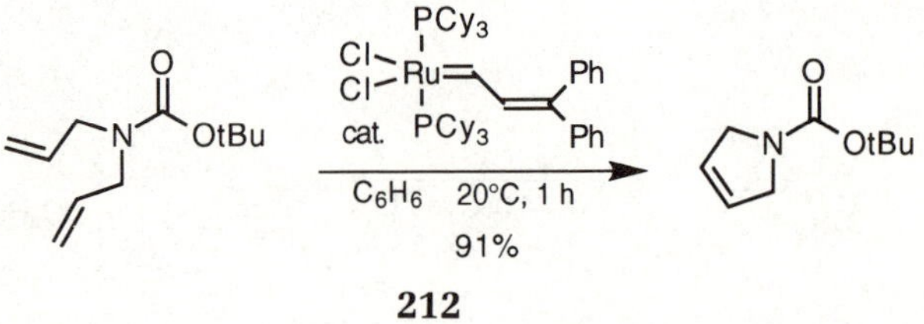

(79)

closing metathesis process. The alkene metathesis process can be catalysed by molybdenum- (eqn 78)[74] and ruthenium- (eqn 79)[75] based catalysts. Already, this strategy has been applied to the synthesis of a number of natural products.[76]

7. Dienes, polyenes, and related systems

Many of the modern syntheses of complex conjugated polyene systems require the use of transition metal-catalysed coupling reactions, and this section deals mainly with such reactions. However, this is by no means exclusively true, especially in the synthesis of dienes, and hence the first section considers these reactions.

7.1 Synthesis of dienes without transition metal catalysis

Diene syntheses via Wittig and related reactions have been widely used. The reaction between an α,β-unsaturated aldehyde and a phosphonium ylid affords the corresponding diene (eqn 80), with the usual selectivity issues associated with the Wittig reaction (see Chapter 2). Phosphine-stabilised allyl anions have been reported to afford mainly the (*Z*)-diene (eqn 81), whereas the corresponding phosphine oxides can be used to give mainly the (*E*)-diene.[77] (*E*,*E*)-Dienes conjugated to esters have been prepared using an allylic Wadsworth–Emmons reagent (eqn 82).[78] (*E*,*Z*)-Dienes conjugated to esters can also be prepared (see Chapter 2, Protocol 17).

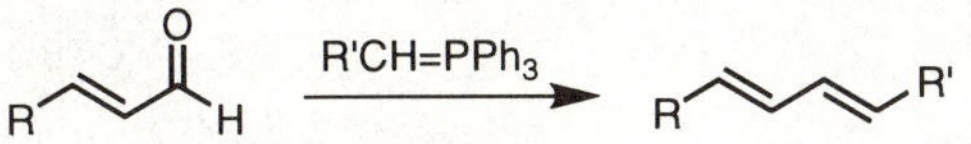

R'CH=PPh$_3$ (80)

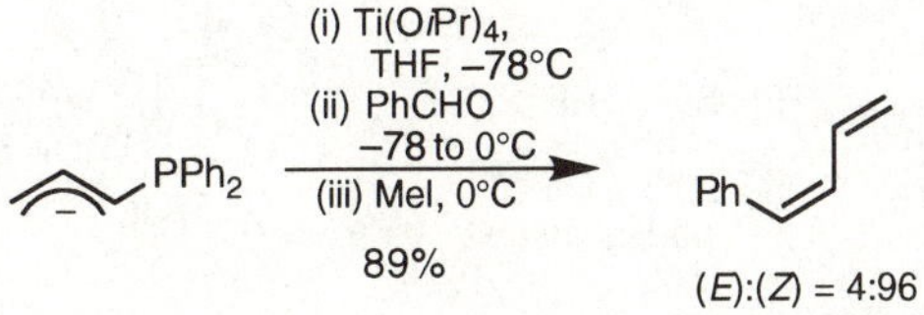

(i) Ti(O*i*Pr)$_4$, THF, –78°C
(ii) PhCHO –78 to 0°C
(iii) MeI, 0°C
89%
(*E*):(*Z*) = 4:96 (81)

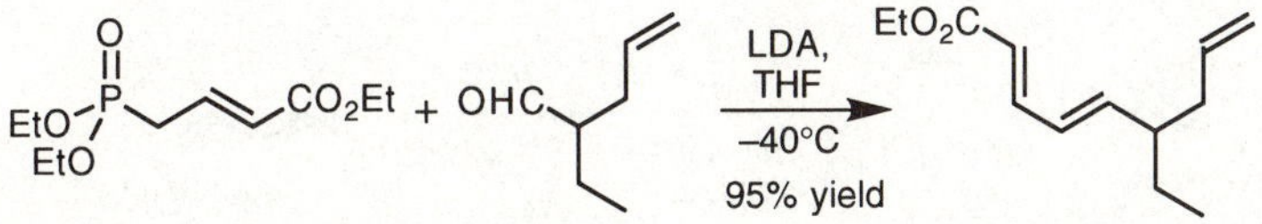

LDA, THF
–40°C
95% yield (82)

An unusual Peterson-type diene construction was reported for the preparation of the diene portion of the marine metabolite discodermolide.[79] Equation 83 shows the application of the addition of an allylchromium species to an aldehyde to form a β-hydrcxysilane (83:17 mixture of isomers, both with an *anti*-relationship between the oxygen and silicon substituents). Treatment of the mixture with potassium hydride afforded the (*Z*)-diene.

The Julia coupling has been used in the preparation of dienes and polyenes.[80] For example, Hanessian *et al.*[81] have demonstrated the use of the Julia coupling in the preparation of a diene moiety contained within a multitude of other functional groups, but the overall yield is low (eqn 84).

Dienes have been prepared by aldol-type reactions, as shown in the preparation of a dienyl sulfoxide (eqn 85).[82] Enantiomerically pure 2-sulfinyldienes have also been prepared by an elimination strategy.[83] Dienes can be prepared by the thermolysis of 3-sulfolenes (eqn 86).[84] Dienes may be prepared stereoselectively by reduction of the corresponding enynes, but the use of this method clearly depends upon the availability of the enyne.[85]

R, O, O, Si, tBu, tBu, H, O + L_nCr $SiMe_3$ → R, O, O, Si, tBu, tBu, OH, $SiMe_3$ —KH, 94%→ R, O, O, Si, tBu, tBu (83)

TBDMSO, Me, Me, SO_2Ph, O, O, Me, Me, Me, OTBDMS + OHC, OR, CO_2Me, O, H, Me, OTBDMS, R = TMS —(i) *n*BuLi, THF, –78°C, 47% (ii) $SOCl_2$, pyr., Na-Hg, MeOH, 35%→ TBDMSO, Me, Me, O, O, Me, Me, Me, OTBDMS, OR, CO_2Me, O, H, Me, OTBDMS, R = TMS (84)

Ph CHO + Me S^+ O^- Tol —(i) LDA, THF –78°C (ii) NaH, THF 0°C (iii) MeI→ Ph S^+ O^- Tol, 82% yield (85)

R, S, O, O —heat, $-SO_2$→ R (86)

7.2 Synthesis of dienes using transition metal catalysis

Identified in Chapter 8 are the strategies required for the preparation of (*E*,*E*)-, (*E*,*Z*)-, (*Z*,*E*)- and (*Z*,*Z*)-dienes via the hydroboration of alkynes and coupling with an appropriately substituted vinyl halide. In fact these strategies can be extended to include the use of other vinylmetal reagents in the coupling process, Scheme 9.7.

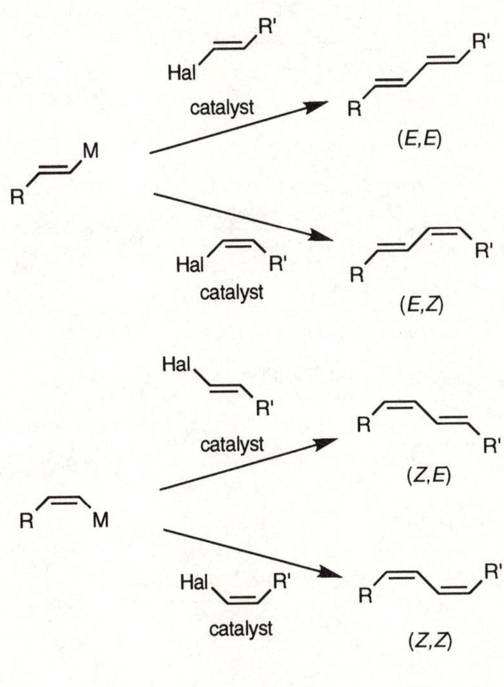

Scheme 9.7

The transition metal-catalysed construction of dienes in a highly stereoselective manner is indeed a versatile and synthetically important strategy. Equations 87,[86] 88,[87] 89,[88] 90,[89] and 91[90] demonstrate some of the applications for which this reaction has been used. Dienamides[91] and dienylsulfoxides[92] are amongst the many other dienes which have been prepared in this manner.

As well as the above coupling reactions using a vinylmetal coupling partner, the Heck reaction may also be employed in the formation of dienes, by coupling together a vinyl halide with an alkene. This is exemplified by the formation of dienes tethered to protected amino acid units (eqn 92)[93] and by the synthesis of taxane analogues (eqn 93).[94]

1,4-Dienes (skipped dienes) can be prepared using transition metal-catalysed methods[95] and by reduction of skipped diynes.[96]

R–CH=CH–I + BrMg–CH=CH$_2$ —(cat. $Pd(PPh_3)_4$)→ R–CH=CH–CH=CH$_2$ see Chapter 8, Protocol 2 (87)

Ph–CH=CH–Br + (catecholboronate) O–B(R)=CH–R —(cat. $Pd(PPh_3)_4$, NaOEt)→ Ph diene R, R see Chapter 8, Protocol 5 (88)

nC_5H_{11}, H, H, Al(iBu)$_2$ + I, H, H, nC_4H_9 —(cat. Pd(PPh)$_4$, ether/hexane, 74%)→ nC_5H_{11}, H, H, H, H, nC_4H_9 (89)

Ph–CH=CH–B(OH)$_2$ + Br, CHO, Ph —(cat. $Pd(PPh_3)_4$, K_2CO_3, THF-H_2O, reflux, 50%)→ Ph, CHO, Ph (90)

OMe, MeO, 4, I, SiMe$_3$ + (HO)$_2$B, 3CO$_2$Me —(cat. $Pd(PPh_3)_4$, TlOH, THF, H_2O, 71%)→ OMe, MeO, 4, SiMe$_3$, 3CO$_2$Me (91)

CO$_2$Et, I, NHAc + O —(5 mol% $PdCl_2(PPh_3)_2$, Et_3N, 78%)→ CO$_2$Et, NHAc, O (92)

I, HO, O —($Pd(PPh_3)_2(OAc)_2$, Et_3N, THF, 70°C, 52%)→ HO, O (93)

7.3 Enynes and enediynes

There is particular interest in the preparation of enediynes due to the ability of these compounds to cleave DNA and hence act as anti-tumour agents.[97] Transition metal-catalysed methods have been exploited for the preparation of enynes and enediynes with great effect.

Thus an enyne can be prepared by coupling an acetylide with a suitable vinyl substrate, such as a vinyl carbamate (eqn 94). Enynes are also available from the catalysed cross coupling of allenes and alkynes,[98] as well as from other transition metal-catalysed routes.[99] An effective strategy for enediyne synthesis is shown in eqn 95.[100] An alternative and particularly striking approach was demonstrated by Danishefsky and co-workers who constructed the enediyne via the cyclisation strategy shown (eqn 96).[101]

$$\text{R-CH=CH-OCON}i\text{Pr}_2 + \text{R}'\text{-C}\equiv\text{C-MgBr} \xrightarrow[\text{C}_6\text{H}_6,\ 67\%]{\text{10 mol\% (Ph}_3\text{P)}_2\text{NiCl}_2,\ \text{25 mol\% EtMgBr}} \text{R-CH=CH-C}\equiv\text{C-R}' \quad (94)$$

$$\text{Cl-CH=CH-Cl} \xrightarrow[\text{(ii) cat. PdCl}_2\text{(PhCN)}_2,\ \text{piperidine},\ \equiv\text{-SiMe}_3,\ 72\%]{\text{(i) C}_5\text{H}_{11}\text{-}\equiv\ \text{cat. Pd(PPh}_3)_4,\ \text{cat. CuI, BuNH}_2} \text{C}_5\text{H}_{11}\text{-C}\equiv\text{C-CH=CH-C}\equiv\text{C-SiMe}_3 \quad (95)$$

(96): Me$_3$Sn–CH=CH–SnMe$_3$ + diiodo-diyne substrate (MeO$_2$C–N, O bridge, OAc, OAc, OTBS) $\xrightarrow[\text{1h, 80\%}]{\text{10 mol\% Pd(PPh}_3)_4,\ \text{DMF, 60°C}}$ cyclic enediyne (MeO$_2$C–N, O bridge, OAc, OAc, OTBS)

7.4 Trienes and polyenes

Conjugated trienes and polyenes are generally prepared by coupling together smaller alkenyl and dienyl subunits. The enediyne prepared by the principles shown in eqn 95 and the corresponding (*E*)-isomer are converted into trienes by reduction with activated zinc in methanol–water (eqns 97 and 98).[100]

A typical example of a more complex coupling is provided by the Sonogashira coupling[102] between a terminal alkyne and a vinyl iodide (eqn 99).[103]

Nicolaou *et al.*[104] employed a remarkable stitching–cyclisation as the very last step in the total synthesis of rapamycin (eqn 100).

Polyenes do not have to be constructed via transition metal-catalysed reactions, as exemplified by the Julia coupling reaction to form a pentaene (eqn 101).[105]

8. Heteroatom-substituted alkenes

Throughout this chapter, the need for the availability of heteroatom-substituted alkenes has been made apparent. Therefore, it is appropriate to

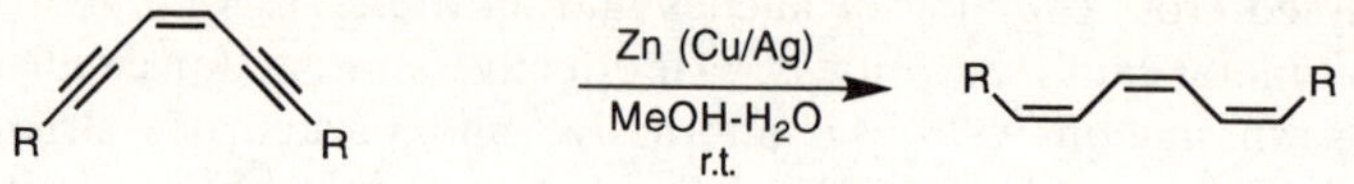

Zn (Cu/Ag)
MeOH-H_2O
r.t.
(97)

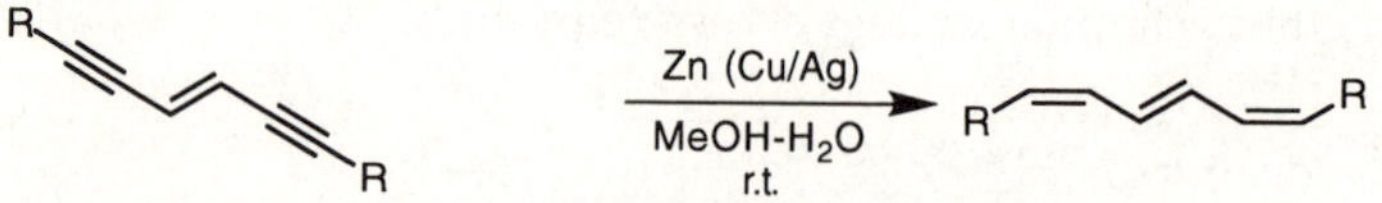

Zn (Cu/Ag)
MeOH-H_2O
r.t.
(98)

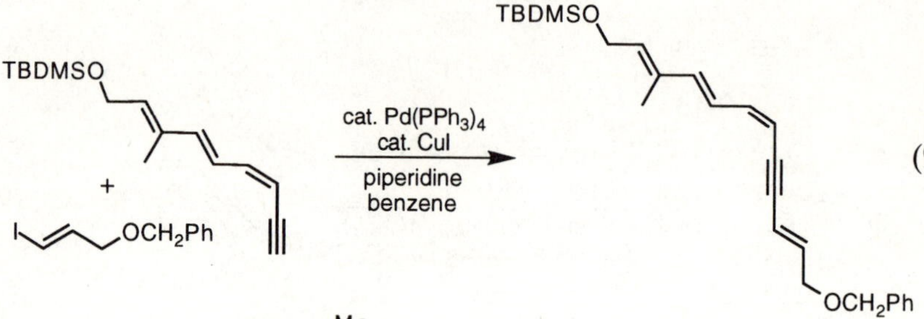

TBDMSO
+
I OCH$_2$Ph
cat. Pd(PPh$_3$)$_4$
cat. CuI
piperidine
benzene
TBDMSO
OCH$_2$Ph
(99)

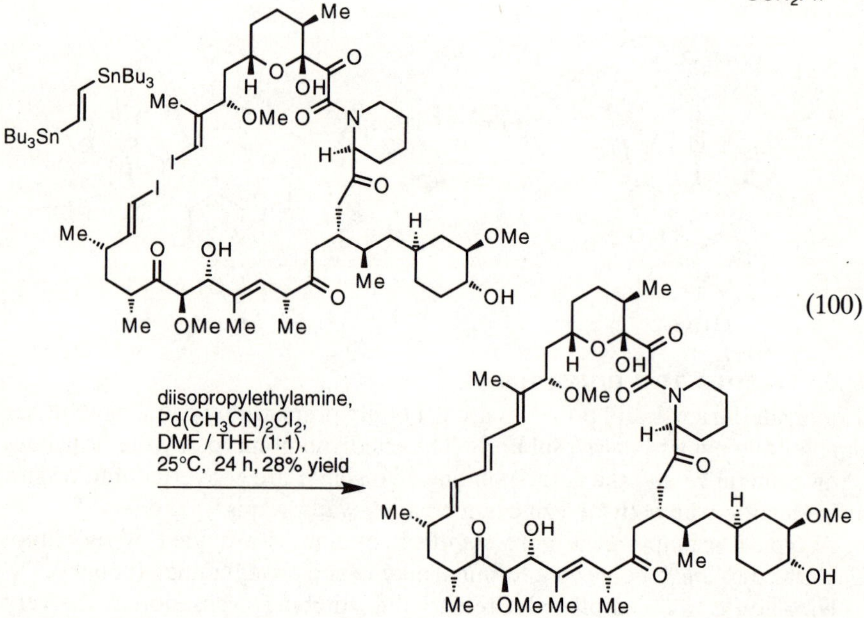

SnBu$_3$
Bu$_3$Sn
diisopropylethylamine,
Pd(CH$_3$CN)$_2$Cl$_2$,
DMF / THF (1:1),
25°C, 24 h, 28% yield
(100)

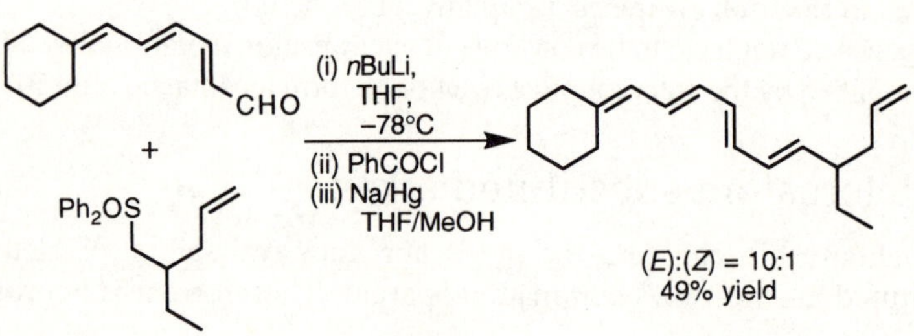

CHO
+
Ph$_2$OS
(i) *n*BuLi,
THF,
−78°C
(ii) PhCOCl
(iii) Na/Hg
THF/MeOH
(*E*):(*Z*) = 10:1
49% yield
(101)

consider the preparations of these compounds. Whilst particular emphasis is placed on heteroatom-substituted alkenes which are able to participate in transition metal-catalysed coupling reactions, other alkene categories have also been discussed.

8.1 Halogen-substituted alkenes

There are many routes for the preparation of halogen-substituted alkenes (vinyl or alkenyl halides). Amongst the simplest are the additions of a halide to an alkyne. Depending on the conditions, either the (*E*)- or the (*Z*)-product[106,107] can be formed preferentially (eqs 102 and 103). The addition of a halide to phenylacetylene can also be achieved (eqn 104).[108] Cyclohexenone can be converted into the 2-iodo derivative by treatment with iodine in pyridine (eqn 105).[109]

NaI, TFA; r.t., 30 min; 95% (102)

NaI, HOAc; 70-80°C (103)

$Ph{-}C{\equiv}C{-}H + Et_3\overset{+}{N}H\ HCl_2^-$ → 70°C, 73% yield (104)

I_2, pyridine; CCl_4, 67% (105)

Other methods for the synthesis of alkenyl halides from alkynes involve the reduction of the halo-alkyne[110] (eqn 106 and see Chapter 6, Protocol 6) and a sequence involving hydroboration of the alkyne to give an alkenylborane which is dibrominated. Suitable choice of elimination conditions leads to either geometry of the vinyl bromide (eqn 107).[111]

In Chapter 7, several methods for the stereoselective preparation of substituted alkenyl iodides are discussed. These methods arise from the hydrometallation[112] (eqn 108) or carbometallation[113] (eqn 109) of an alkyne with a subsequent conversion of the so-formed alkenylmetal into an alkenyl iodide. The experimental details and scope of these reactions are discussed in Chapter 7.

Other procedures discussed in earlier chapters include the preparation of alkenyl halides via the Wittig reaction[114] (eqn 110) and also the chromium-promoted homologation of aldehydes (eqn 111).

$$nC_6H_{13}-C\equiv C-Cl \xrightarrow[\text{(ii) } CH_3OH \text{, 80\% yield}]{\text{(i) } LiAlH_4\text{, THF, } -30 \text{ to } 0°C} nC_6H_{13}(H)C{=}C(H)Cl \tag{106}$$

(i) Sia_2BH, CCl_4, 0-5°C; (ii) Br_2: $nC_5H_{11}-C\equiv C-H$ → nC_5H_{11}, H, Br / BSia$_2$, Br, H ⇌ Br, BSia$_2$, H, Br, nC_5H_{11}, H (107)

H_2O → H, H, nC_5H_{11}, Br — 67% yield

76°C → H, Br, nC_5H_{11}, H — 75% yield

R—≡—R → (i) tBuMgCl/Cl_2ZrCp_2; (ii) I_2 → R, R, I — see Chapter 7, Protocol 1 (108)

R—≡—H → (i) Me_3Al/Cl_2ZrCp_2; (ii) I_2 → R, Me, I — see Chapter 7, Protocol 3 (109)

Alkenyl bromides may be obtained by the oxidative decarboxylation of α, β-unsaturated carboxylic acids,[115] and also by the regioselective transition metal-catalysed mono cross-coupling reaction of dibromoalkenes.[116] Fluoroalkenes have been prepared in a number of ways including the use of Horner–Wadsworth–Emmons methodology,[117] aldol-type reactions,[118] and transition metal-catalysed cross-coupling reactions involving fluoro-substituted alkenylstannanes.[119]

$R_2C{=}O$ → $Ph_3P{=}CH$–Hal → $R_2C{=}CH$–Hal — see Chapter 2, Protocol 7 cf. Chapter 2, Protocol 12 (110)

RCHO → CHI_3, $CrCl_2$, THF → R–CH=CH–I — see Chapter 4, Protocol 4 (111)

8.2 Oxygen-substituted alkenes

In many ways, oxygen-substituted alkenes are best thought of as derivatives of carbonyl compounds, and indeed, many syntheses of these compounds are derived from carbonyl compounds, Scheme 9.8.[120] Enol ethers can be prepared in other ways such as elimination from acetals.[121]

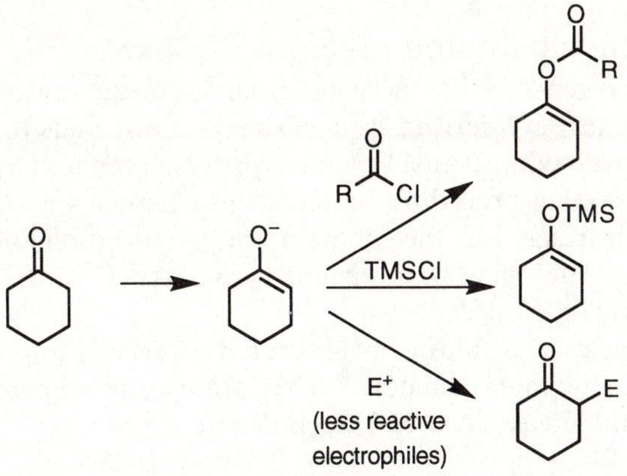

Scheme 9.8

Suitable oxygen-substituted alkenes may also participate in Wittig reactions (see Chapter 2, Protocol 8), Wittig–Horner reactions (eqn 112),[122] and in transition metal-catalysed coupling reactions (eqns 113 and 114).[123,124] Enol triflates are able to function as the coupling partner with displacement of the triflate and the formation of a new carbon–carbon bond (for examples, see eqns 18, 65, 66, and 114).

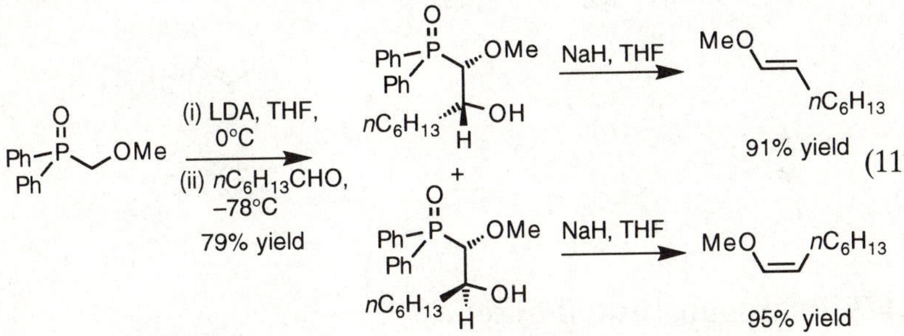

(112)

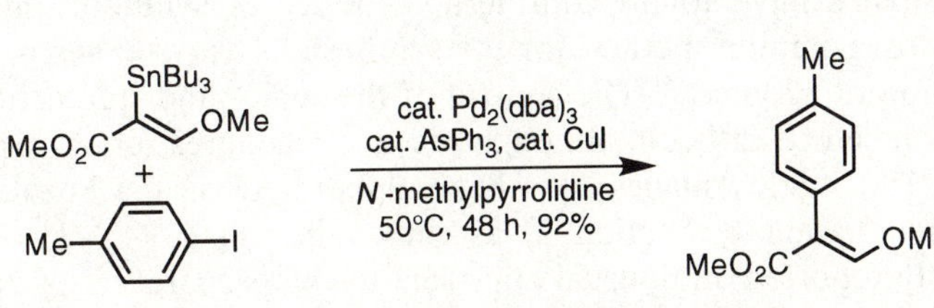

(113)

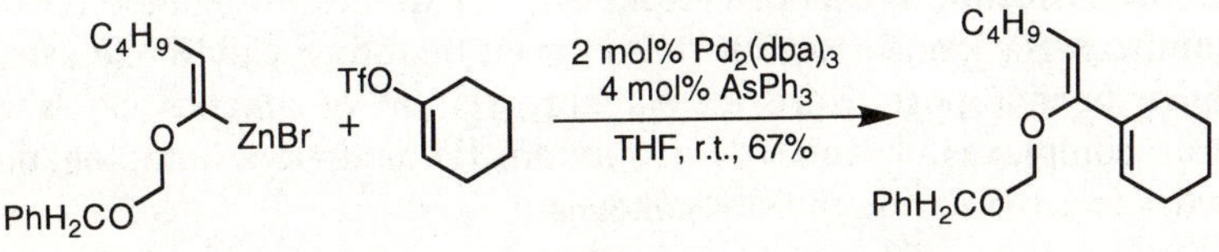

(114)

8.3 Nitrogen-substituted alkenes

Like their oxygen-substituted counterparts, many nitrogen-substituted alkenes (enamines) are derived from carbonyl compounds (eqn 115),[125] and function as enolate equivalents. Enamines are also prepared by the use of the Tebbe methylenating procedure described in Chapter 4.[126] Enamido esters are useful substrates for the preparation of enantiomerically enriched α-amino esters. They may be prepared by a Horner–Wadsworth–Emmons procedure (eqn 116).[127]

An important class of nitrogen-substituted alkenes is nitroalkenes, which are useful synthetic intermediates.[128] They are usually prepared via a Henry reaction[129] (eqn 117, see Chapter 5, Section 4.5).

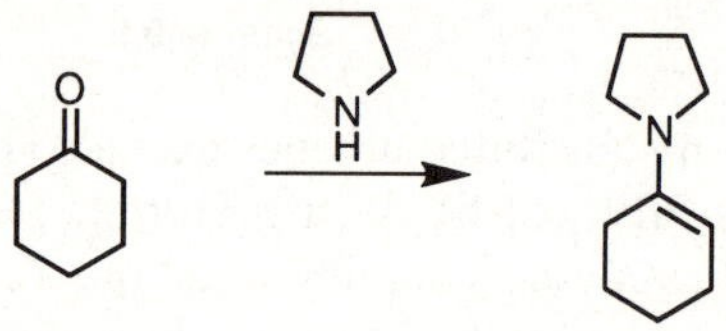

(115)

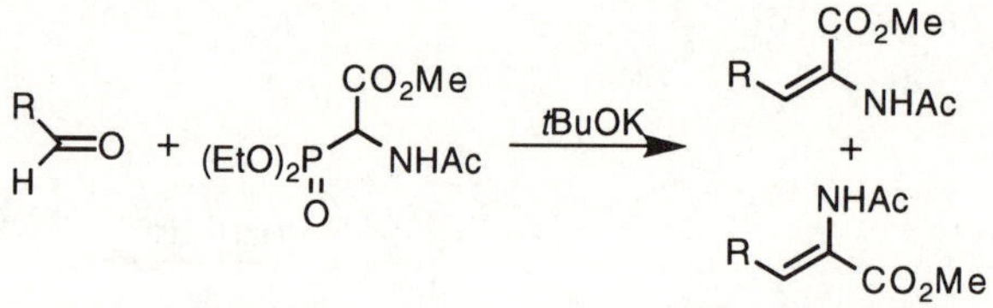

(116)

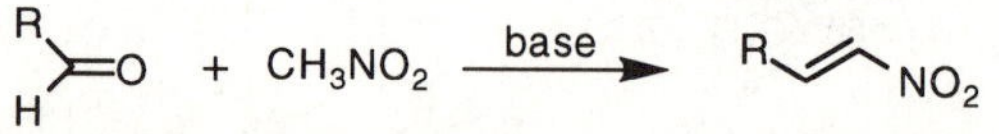

(117)

8.4 Silicon-substituted alkenes

Vinylsilanes have found considerable utility as synthetic intermediates.[130] The most common method for the synthesis of vinylsilanes is by the hydrosilylation of alkynes.[131] The control of the regio- and stereochemistry of the addition can be difficult, although recent procedures report high selectivities (eqn 118).[132] Vinylsilanes have been elaborated via the Suzuki coupling reaction (Chapter 8, Section 2) as shown in eqn 119.[133] Two groups have recently reported an unusual approach to the preparation of vinylsilanes via episulfone substitution reactions (eqn 120)[134] in an interesting modification of the Ramberg–Bäckland reaction.[135] Amongst the other vinylsilane syntheses that have been reported are the silylformylation of alkenes catalysed by rhodium complexes,[136] and the chromium(II)-mediated homologation of aldehydes to 1,1-bis(trimethylsilyl)alkenes.[137]

$HSiEt_3$, cat. $[Rh(COD)_2]BF_4$, cat. PPh_3; acetone, r.t., 30 min, 83-95% (118)

RBr, cat. $Pd(PPh_3)_4$; NaOH, 58-89% (119)

LDA (3 eq), Me_3SiCl (15 eq), 85%; toluene, reflux (120)

8.5 Phosphorus-substituted alkenes

There have been reports of the preparation and uses of vinylphosphine oxides, vinylphosphinates,[138] vinylphosphonium salts,[139] vinylphosphine oxides,[140] and vinylphosphonates.[141] This last category has been the subject of review. Some procedures for the preparation of these compounds are given in eqns 121,[142] 122,[143] and 123.[144]

$pClC_6H_4CHO$, toluene; 110°C, 48 h; 84% yield (121)

(i) *n*BuLi, THF, hexane, −78°C, 20 min; (ii) RCHO, THF, −78 to 20°C; 60-85% yield (122)

$Pd(PPh_3)_4$ (5 mol%), NEt_3, toluene; 90°C, 0.2 h; 93% yield (123)

8.6 Sulfur-substituted alkenes

Vinyl sulfides, vinyl sulfoxides, and vinyl sulfones have been prepared in a number of ways, and the preparation and uses of vinyl sulfones have been the most widely investigated.[145] Vinyl sulfones have been prepared by conjugate addition to acetylenic sulfones (eqn 124),[146] by the Peterson reaction (eqn 125)[147] and the Horner–Wadsworth–Emmons reaction (eqn 126).[148]

Vinyl sulfoxides can be prepared without paying any regard to the stereochemistry of the sulfoxide (eqn 127),[149] or when stereochemistry is important, alternative methods are available (eqn 128).[150]

Vinyl sulfides may be prepared by the palladium-catalysed reaction between vinyl halides and either arylthiolates[151] or triisopropylsilanethiolate

PhSO$_2$—≡—H → [R_2CuLi, Et_2O; (*E*)-isomer, 80-90%yield] → PhSO$_2$(H)C=C(H)R (124)

Cyclohexanone + Me$_3$SiCH$_2$SO$_2$Ph → [*n*BuLi, DME; –78°C to r.t., 92% yield] → cyclohexylidene–CH–SO$_2$Ph (125)

(EtO)$_2$P(O)CH$_2$SO$_2$R → [(i) *n*BuLi, THF, –78°C; (ii) R'R"CO; >72% yield] → R'R"C=CH(SO$_2$R), (*E*)-selective (126)

MeO$_2$CCH$_2$S$^+$(O$^-$)Ph + PhCHO → [piperidine, CH$_3$CN; 60°C, 6 h; 85% yield] → PhCH=C(CO$_2$Me)S$^+$(O$^-$)Ph (127)

epoxy sulfoxide (S$^+$(O$^-$)Ph) → [NaOH, H$_2$O; r.t. 93% yield] → allylic alcohol (OH) vinyl sulfoxide S$^+$(O$^-$)Ph, (*E*) : (*Z*) = 17:1 (128)

Me$_2$C=CHBr → [KSTIPS, Pd(PPh$_3$)$_4$, benzene,THF; reflux, 36 h; 84% yield] → Me$_2$C=CH(STIPS) (129)

C$_6$H$_{11}$—≡—SCH$_3$ + catecholborane (HB) → [PdCl$_2$(dppf), benzene; r.t.] → catecholboryl–C(C$_6$H$_{11}$)=CH(SCH$_3$) → [PhCH=CHBr; PdCl$_2$(dppf), 3M NaOH, reflux, 85%] → PhCH=CH–C(C$_6$H$_{11}$)=CH–SCH$_3$ (130)

(eqn 129).[152] Suzuki and co-workers have reported an interesting sequence involving the hydroboration of an acetylenic sulfide followed by a Suzuki coupling reaction (eqn 130). The palladium-catalysed addition of both thiols[153] and selenols[154] to alkynes has also been reported.

8.7 Tin-substituted alkenes

Vinylstannanes are particularly useful synthetic intermediates because they are amenable to transformation into other alkenes, as already demonstrated. Chapter 7, Section 4 discusses the hydrostannation of alkynes, and the factors controlling the regio- and stereoselectivity of this process (eqn 131). The reaction proceeds either thermally or with palladium catalysis.[155]

As well as the hydrostannation of alkynes, vinylstannanes can be produced by an electrophilic quench of a vinyllithium, as shown in eqn 132[156] (although vinylstannanes may also be used as precursors to vinyllithium reagents.[157]). The chromium-mediated homologation of aldehydes to vinylstannanes is discussed in Chapter 4 (eqn 133). The reaction between lithiated dihydrofuran with a stannylcuprate and an electrophilic quench (note the position of the addition of the electrophile) affords a vinylstannane (eqn 134).[158]

R —(HSnR'3)→ R, SnR'3 + R'3Sn, R (131)

R, OMe, H —(tBuLi, −78 to 20°C)→ [R, OMe, Li] —(Me3SnCl)→ R, OMe, SnMe3 (132)

RCHO —(Bu3SnCHBr2, LiI, CrCl2; DMF, THF, 25°C)→ R, SnBu3 (133)

Chapter 4, Protocol 5

O, Li —(1. (Bu3Sn)2CuCNLi2; 2. MeI)→ HO, Me, SnBu3 (134)

8.8 Boron-substituted alkenes

The main synthetic route to vinylboranes involves the hydroboration of alkynes.[159] This process is described in Chapter 7, Section 2.2 (see also Chapter 6, Section 4 and Chapter 8, Scheme 8.7 and Protocol 5). Addition of a borane to an alkyne affords the (*E*)-vinylborane (eqn 135). Hydroboration of haloalkynes is also a *syn*-addition, although because of the nomenclature rules, this is classified as a (*Z*)-alkene product. The adducts may be converted into products where the boron and alkyl substituent are on the same face of the alkene (eqn 136).[160]

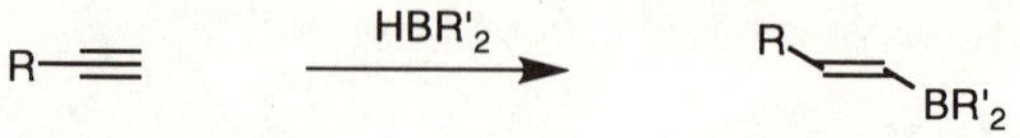

(135)

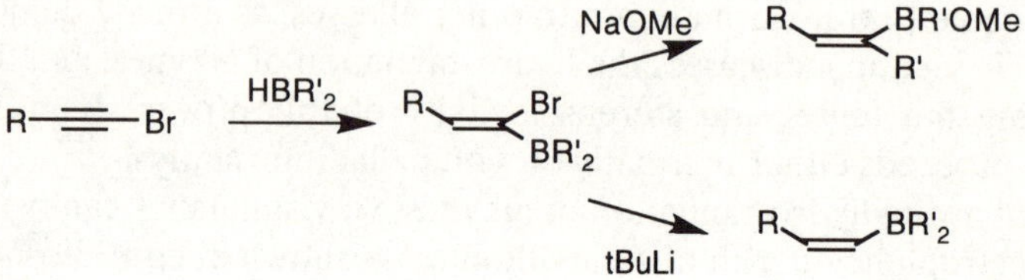

(136)

The hydrozirconation of alkynylboranes leads to a boryl zirconocene 1,1-dimetallic species, which can be functionalised in many ways (see eqn 64) including hydrolysis to the (*Z*)-alkenyl boronate (eqn 137).[161] Palladium-catalysed thioboration (eqn 138)[162] and diboration (eqn 139)[52] procedures have also been reported.

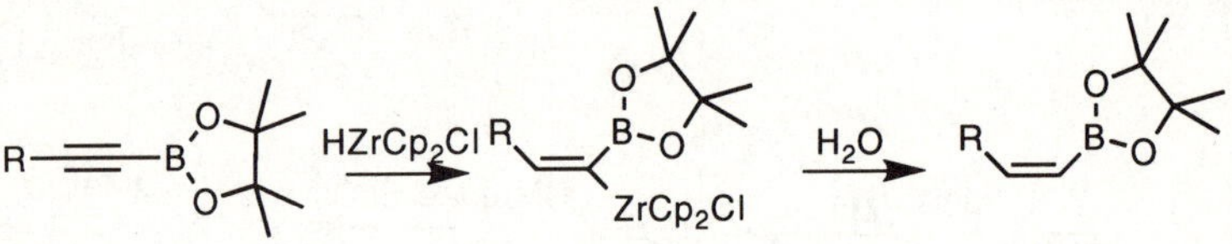

(137)

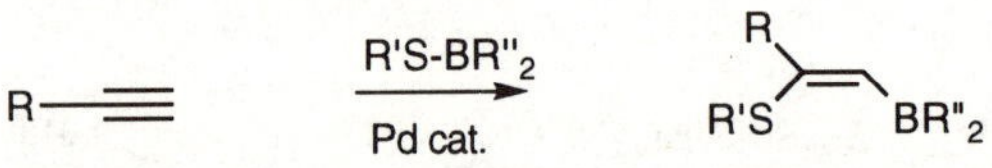

(138)

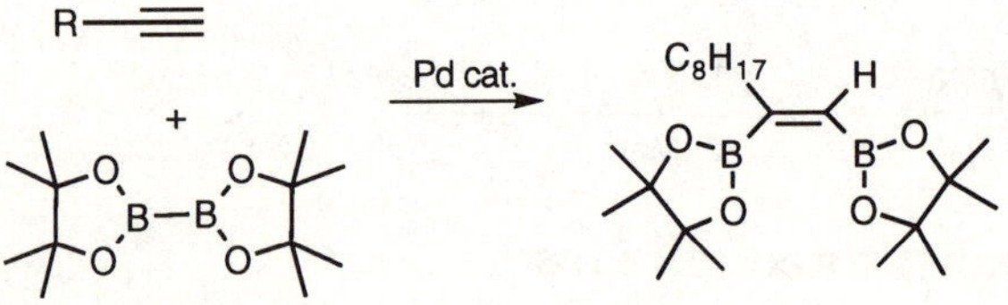

(139)

9. Unusual alkenes

In this final section, a brief mention is made of general approaches to more unusual types of alkene, such as allenes and ketenes.[163]

The standard synthetic route to allenes[164] is via the nucleophilic attack on a propargylic system. The nucleophile may be an organometallic (eqn 140)[165] or a hydride source (eqn 141).[166]

However, other routes to allenes have been reported,[167] including an interesting Stille couping of an allenyl stannane (eqn 142).[168] There has also been interest in the synthesis of cumulenes (butatrienes and higher analogues[169]).

$$R^1\text{–C}\equiv\text{C–CH}(R^2)\text{OTs} \xrightarrow[\substack{-30^\circ\text{C} \\ \text{80-90\% yield}}]{\substack{\text{RMgBr + CuBr} \\ \text{THF}}} R^1(R)\text{C=C=C}(R^2)\text{H} \qquad (140)$$

$$\text{HC}\equiv\text{C–C(Ph)}_2\text{OTs} \xrightarrow[\substack{\text{r.t. 2 min} \\ \text{95\% yield}}]{\substack{\text{Pd(PPh}_3)_4\text{ (1 mol\%)} \\ \text{SmI}_2\text{, }i\text{PrOH THF}}} \text{H}_2\text{C=C=C(Ph)}_2 \qquad (141)$$

$$\text{H}_2\text{C=C=C(H)SnBu}_3 \xrightarrow[\substack{\text{DMF 80}^\circ\text{C 1.5 h} \\ \text{20-71\% yield}}]{\substack{\text{cat. Pd}_2\text{dba}_3\text{ PPh}_3 \\ \text{cat CuI, 3 eq LiCl}}} \text{H}_2\text{C=C=C(H)Ar} \qquad (142)$$

A typical synthesis of a ketane involves the dehalogenation of an α-haloacyl halide, with either zinc[170] or triphenylphosphine (eqn 143),[171] but there are other more specialised methods.[172]

$$\text{Ph}_2\text{CBrCOBr} \xrightarrow[\substack{0^\circ\text{C} \\ \text{82\% yield}}]{\substack{\text{Ph}_3\text{P,} \\ \text{benzene}}} \text{Ph}_2\text{C=C=O + Ph}_2\text{PBr}_2 \qquad (143)$$

References

1. For a review on the Claisen rearrangement, see: Wipf, P. In *Comprehensive Organic Synthesis*; Trost, B. M.; Fleming, I., eds; Pergamon Press: Oxford, **1991**; Vol. 5, p. 827. For the [2,3]-Wittig rearrangement, see: Nakai, T.; Mikami, K. *Chem. Rev.* **1986**, *86*, 855; Marshall, J. A. In *Comprehensive Organic Synthesis*; Trost, B. M.; Fleming, I., eds; Pergamon Press: Oxford, **1991**; Vol. 3, Chapter 3.11. For the Oxy-Cope rearrangement, see: Paquette, L. A. *Angew. Chem., Int. Ed. Engl.* **1990**, *29*, 609.
2. For an overall review of the conversion of carbonyl compounds into alkenes, see: Kelly, S. E. In *Comprehensive Organic Synthesis*; Trost, B. M.; Fleming, I., eds; Pergamon Press: Oxford, **1991**; Vol. 1, p. 729.
3. Maryanoff, B. E.; Reitz, A. B. *Chem. Rev.* **1989**, *89*, 863. Gosney, I.; Rowley, A. G. In *Organophosphorus Reagents in Organic Synthesis*; Cadogan, J. I. G., ed.; Academic Press: New York, **1979**, p. 17.
4. Ager, D. J. *Synthesis* **1984**, 384. Ager, D. J. *Org. React.* **1990**, *38*, 1.

5. See reference 2 and Johnson, C. R.; Kirchhoff, R. A. *J. Am. Chem. Soc.* **1979**, *101*, 3602.
6. See reference 2 and Pine, S. H.; Zahler, R.; Evans, D. A.; Grubbs, R. H. *J. Am. Chem. Soc.* **1980**, *102*, 3270.
7. Alcaraz, L.; Harnett, J. J.; Mioskowski, C.; Martel, J. P.; Le Gall, T.; Shin, D.-S.; Falck, J. R. *Tetrahedron Lett.* **1994**, *35*, 5453.
8. Harnett, J. J.; Alcaraz, L.; Mioskowski, C.; Martel, J. P.; Le Gall, T.; Shin, D.-S.; Falck, J. R. *Tetrahedron Lett.* **1994**, *35*, 2009.
9. Roush, W. R.; Michaelides, M. R.; Tai, D. F.; Lesur, B. M.; Chong, W. K. M.; Harris, D. J. *J. Am. Chem. Soc.* **1989**, *111*, 2984.
10. Mitchell, T. N. *Synthesis* **1992**, 803.
11. McKean, D. R.; Parinello, G.; Renaldo, A. F.; Stille, J. K. *J. Org. Chem.* **1987**, *52*, 422.
12. Oh-e, T.; Miyaura, N.; Suzuki, A. *J. Org. Chem.* **1993**, *58*, 2201 and references therein.
13. Yamamoto, Y.; Asao, N. *Chem. Rev.* **1993**, *93*, 2207.
14. Buse, C. T.; Heathcock, C. H. *Tetrahedron Lett.* **1978**, 1685.
15. Roush, W. R.; Grover, P. T. *J. Org. Chem.* **1995**, *60*, 3806 and references therein.
16. For a review, see: Roush, W. R. In *Comprehensive Organic Synthesis*; Trost, B. M.; Fleming, I., eds; Pergamon Press: Oxford, **1991**; Vol. 2, p. 1.
17. König, K; Neumann, W. P.; *Tetrahedron Lett.* **1967**, 495.
18. Nuss, J. M.; Rennels, R. A. *Chem. Lett.* **1993**, 197.
19. Keck, G. E.; Geraci, L. S. *Tetrahedron Lett.* **1993**, *34*, 7827. See also: Keck, G. E.; Savin, K. A.; Cressman, E. N. K.; Abbott, D. E. *J. Org. Chem.* **1994**, *59*, 7889.
20. Frost, C. G.; Howarth, J.; Williams, J. M. J. *Tetrahedron: Asymmetry* **1992**, *3*, 1089.
21. For a detailed survey of nucleophiles which have been employed in this reaction, see: Godleski, S. A. In *Comprehesive Organic Synthesis*; Trost, B. M.; Fleming, I. eds; Pergamon Press: Oxford, **1991**; Vol. 2, p. 585.
22. See also: Vedejs, E. *Chem. Rev.* **1986**, *86*, 941.
23. Bargar, T. M.; McCowan, J. R.; McCarthy, J. R.; Wagner, E. R. *J. Org. Chem.* **1987**, *52*, 678.
24. Duchêne, A.; Abarbri, M.; Parrain, J.-L.; Kitamura, M.; Noyori, R. *Synlett* **1994**, 524.
25. Russell, C. E.; Hegedus, L. S. *J. Am. Chem. Soc.* **1983**, *105*, 943.
26. Roberts, J. L.; Borromeo, P. S.; Poulter, C. D. *Tetrahedron Lett.* **1977**, 1621. La Clair, J. J.; Lansbury, P. T.; Zhi, B-X.; Hoogsteen, K. *J. Org. Chem.* **1995**, *60*, 4822.
27. Reich, H. J.; Shah, S. K.; Chow, F. *J. Am. Chem. Soc.* **1979**, *101*, 6648.
28. Drewes, S. E.; Roos, G. H. P. *Tetrahedron* **1988**, *44*, 4653.
29. Augé, J.; Lubin, N.; Lubineau, A. *Tetrahedron Lett.* **1994**, *35*, 7947.
30. Brown, H. C.; Zweifel, G. *J. Am. Chem. Soc.* **1959**, *81*, 1512.
31. Brown, H. C.; Basavaiah, D. *J. Org. Chem.* **1982**, *47*, 3808. Zweifel, G.; Fisher, R. P.; Snow, J. T.; Whitney, C. C. *J. Am. Chem. Soc.* **1971**, *93*, 6309.
32. Brown, H. C.; Lee, H. D.; Kulkarni, U. *Synthesis* **1982**, 195.
33. The choice of phosphonium ylid influences the selectivity. For (*Z*)-selective, see, for example: Zhang, X.; Schlosser, M. *Tetrahedron Lett.* **1993**, *34*, 1925. For (*E*)-selective, see, for example: Vedejs, E.; Peterson, M. J. *J. Org. Chem.*, **1993**, 58,

1985. For a summary of recent references, see: Vedejs, E.; Cabaj, J.; Peterson, M. J. *J. Org. Chem.* **1993**, *58*, 6509.
34. Sonnet, P. E. *Tetrahedron* **1980**, *36*, 557.
35. Matikainen, J. K.; Kaltia, S.; Hase, T. *Synlett* **1994**, 817.
36. For a recent example of this, see: Clayden, J.; Collington, E. W.; Warren, S. *Tetrahedron Lett.* **1993**, *34*, 1327.
37. Under certain circumstances, it is possible to convert the *erythro*-adduct into the (*E*)-alkene, see: Lawrence, N. J.; Muhammed, F. *J. Chem. Soc., Chem. Commun.* **1993**, 1187.
38. Kocienski, P. J. In *Comprehensive Organic Synthesis*; Trost, B. M.; Fleming, I., eds; Pergamon Press: Oxford, **1991**; Vol. 6, p. 987.
39. Ihara, M.; Suzuki, S.; Taniguchi, T.; Tokunaga, Y.; Fukumoto, K. *Synlett* **1994**, 859.
40. For an $SnCl_2$-promoted reaction (in place of Zn), see: Lin, R.; Yu, Y.; Zhang, Y. *Synth. Commun.* **1993**, *23*, 271.
41. Jabri, N.; Alexakis, A.; Normant, J. F. *Tetrahedron Lett.* **1981**, *22*, 3851.
42. Wipf, P.; Xu, W. *Tetrahedron Lett.* **1994**, *35*, 5197
43. Lipshutz, B. H.; Wood, M. R. *Tetrahedron Lett.* **1994**, *35*, 6433.
44. For example, see: Cintas, P. *Synthesis* **1992**, 248.
45. Knecht, M.; Boland, W. *Synlett*, **1993**, 837.
46. Nava Salgado, V. O.; Peters, E.-M.; Peters, K.; von Schnering, H. G.; Adam, W. *J. Org. Chem.* **1995**, *60*, 3879.
47. Sonnet, P. E. *Tetrahedron*, **1980**, *36*, 557.
48. Sarel, S. *J. Org. Chem.* **1959**, *24*, 2081.
49. McMurry, J. E.; Lectka, T.; Rico, J. G. *J. Org. Chem.* **1989**, *54*, 3748. Robertson, G. M. In *Comprehensive Organic Synthesis*; Trost, B. M.; Fleming, I., eds.; Pergamon Press: Oxford, **1991**; vol. 2, p. 563.
50. Barton, D. H. R.; Willis, B. J.; *J. Chem. Soc., Perkin Trans. 1* **1972**, 305. Buter, J.; Wassenaar, S.; Kellogg, R. M. *J. Org. Chem.* **1972**, *37*, 4045.
51. Hoogesteger, F. J.; Havenith, R. W. A.; Zwikker, J. W.; Jenneskens, L. W.; Kooijman, H.; Spek, A. L. *J. Org. Chem.* **1995**, *60*, 4375.
52. Ishiyama, T.; Matsuda, N.; Miyaura, N.; Suzuki, A. *J. Am. Chem. Soc.* **1993**, *115*, 11018.
53. Negishi, E.; Ay, M.; Gulevich, Y. V.; Noda, Y. *Tetrahedron Lett.* **1993**, *34*, 1437.
54. Gao, Y.; Harada, K.; Hata, T.; Urabe, H.; Sato, F. *J. Org. Chem.* **1995**, *60*, 290.
55. Deloux, L.; Srebnik, M.; Sabat, M. *J. Org. Chem.* **1995**, *60*, 3276.
56. Ciattini, P. G.; Morera, E.; Ortar, G. *Tetrahedron Lett.* **1992**, *33*, 4815.
57. Zheng, Q.; Yang, Y.; Martin, A. R. *Tetrahedron Lett.* **1993**, *34*, 2235.
58. Farina, V.; Krishnan, B.; Marshall, D. R.; Roth, G. P. *J. Org. Chem.* **1993**, *58*, 5434. A related coupling using an acyclic vinyl triflate was reported to proceed with mainly inversion of the alkene geometry.
59. Yang, G.; Myles, D. C. *Tetrahedron Lett.* **1994**, *35*, 2503.
60. Kocienski, P. J.; Pritchard, M.; Wadman, S. N.; Whitby, R. J.; Yeates, C. L. *J. Chem. Soc., Perkin Trans. 1* **1992**, 3419.
61. Evans, D. A.; Carreira, E. M. *Tetrahedron Lett.* **1990**, *31*, 4703.
62. Ciganek, E. *J. Org. Chem.* **1995**, *60*, 4635.
63. Pelter, A.; Conclough, M. E. *Tetrahedron* **1995**, *51*, 811.
64. Deng, M.-z.; Tang, Y.-t.; Xu, W.-h. *Tetrahedron Lett.* **1984**, *25*, 1797.

65 Basavaiah, D.; Bhavani, A. K. D.; Pandiaraju, S.; Sarma, P. K. S. *Synlett* **1995,** 243.
66. Rabe, J.; Hoffmann, H. M. R. *Angew. Chem. Int. Ed. Engl.* **1983,** *22*, 796.
67. Hanessian, S.; Beaudoin, S. *Tetrahedron Lett.* **1992,** *33*, 7655. See also: Denmark, S. E.; Rivera, I. *J. Org. Chem.* **1994,** *59*, 6887 and references therein.
68. Doris, E.; Dechoux, L.; Mioskowski, C. *Tetrahedron Lett.* **1994,** *35*, 7943.
69. For a review, see: Crandall, J. K.; Apparu, M. *Org. React.* **1983,** *29*, 345. See also: Mordini, A.; Pecchi, S.; Caozzi, G.; Capperucci, A.; Degl'Innocenti, A.; Reginato, G.; Ricci, A. *J. Org. Chem.* **1994,** *59*, 4784. Manna, S.; Yadagiri, P.; Falck, J. R. *J. Chem. Soc., Chem. Commun.* **1987,** 1324.
70. Carruthers, W. *Cycloaddition Reactions in Organic Synthesis*; Pergamon Press: Oxford, **1990**.
71. Pauson, P. L. *Tetrahedron* **1985,** *41*, 5855.
72. Schore, N. E. *Chem. Rev.* **1988,** *88*, 1081. See also: Trost, B. M. *Acc. Chem. Res.* **1990,** *23*, 34.
73. Lee, B. Y.; Chung, Y. K. J. *Am. Chem. Soc.* **1994,** *116*, 8793.
74. Fu, G. C.; Grubbs, R. H. *J. Am. Chem. Soc.* **1992,** *114*, 7324.
75. Fu, G. C.; Nguyen, S. T.; Grubbs, R. H. *J. Am. Chem. Soc.* **1993,** *115*, 9856.
76. Fujimura, O.; Fu, G. C.; Grubbs, R. H. *J. Org. Chem.* **1994,** *59*, 4029. Martin, S. F.; Liao, Y.; Chen, H.-J.; Pätzel, M.; Ramser, M. N. *Tetrahedron Lett.* **1994,** *35*, 6005. Borer, B. C.; Deerenberg, S.; Bieräugel, H.; Pandit, U. K. *Tetrahedron Lett.* **1994,** *35*, 3191.
77. Ukai, J.; Ikeda, Y.; Ikeda, N.; Yamamoto, H. *Tetrahedron Lett.* **1983,** *24*, 4029.
78. Roush, W. R.; Peseckis, S. M. *Tetrahedron Lett.* **1982,** *23*, 4879.
79. Paterson, I.; Schlapbach, A. *Synlett* **1995,** 498.
80. Keck, G. E.; Savin, K. A.; Weglarz, M. A. *J. Org. Chem.* **1995,** *60*, 3194 and references therein.
81. Hanessian, S.; Ugolini, A.; Dube, D.; Hodges, P. J.; Andre, C. *J. Am. Chem. Soc.* **1986,** *108*, 2776.
82. Solladié, G.; Ruiz, P.; Colobert, F.; Carreño, M. C.; Garcia-Ruanao, J. L. *Synthesis* **1991,** 1011.
83. Bonfand, E.; Gosselin, P.; Maignan, C. *Tetrahedron: Asymmetry* **1993,** *4*, 1667.
84. See: McIntosh, J. M.; Sieler, R. A. *J. Org. Chem.* **1978,** *43*, 4431. Lee, S.-J.; Chien, C.-J.; Peng, C.-J.; Chao, I.; Chou, T.-S. *J. Org. Chem.* **1994,** *59*, 4367 and references therein.
85. Corey, E. J.; Herron, D. K. *Tetrahedron Lett.* **1971,** 1641.
86. Dang, H. P.; Linstrumelle, G. *Tetrahedron Lett.* **1978,** 191.
87. Miyaura, N.; Yamada, K.; Suginome, H.; Suzuki, A. *J. Am. Chem. Soc.* **1985,** *107*, 972.
88. Baba, S.; Negishi, E. *J. Am. Chem. Soc.* **1976,** *98*, 6729.
89. Urdaneta, N.; Ruíz, J.; Zapata., A. J. *J. Organomet. Chem.* **1994,** *464*, C33.
90. Roush, W. R.; Warmus, J. S.; Works, A. B. *Tetrahedron Lett.* **1993,** *34*, 4427.
91. Kaga, H.; Ahmed, Z.; Gotoh, K.; Orito, K. *Synlett* **1994,** 607.
92. Paley, R. S.; de Dios, A.; de la Pradilla, R. F. *Tetrahedron Lett.* **1993,** *34*, 2429.
93. Crisp, G. T.; Glink, P. T. *Tetrahedron* **1994,** *50*, 2623.
94. Masters, J. J.; Jung, D. K.; Bornmann, W. G.; Danishefsky, S. J.; de Gala, S. *Tetrahedron Lett.* **1993,** *34*, 7253.
95. For example, see: Hutzinger, M. W.; Oehlschlager, A. C. *J. Org. Chem.* **1995,** *60*,

4595. Trost, B. M.; Müller, T. J. J.; Martinez, J. *J. Am. Chem. Soc.* **1995,** *117*, 1888.

96. Taber, D. F.; You, K. *J. Org. Chem.* **1995,** *60*, 139.
97. Nicolaou, K. C.; Dai, W.-M. *Angew. Chem., Int. Ed. Engl.* **1991,** *30*, 1387. Maier, M. E. *Synlett* **1995,** 13.
98. Yamaguchi, M.; Omata, K.; Hirama, M. *Tetrahedron Lett.* **1994,** *35*, 5689.
99. For example, see: Ikeda, S.; Sato, Y. *J. Am. Chem. Soc.* **1994,** *116*, 5975. Wang, Z.; Wang, K. K. *J. Org. Chem.* **1994,** *59*, 4738.
100. Alami, M.; Crousse, B.; Linstrumelle, G. *Tetrahedron Lett.* **1994,** *35*, 3543. Maier, M. E.; Abel, U. *Synlett* **1995,** 38. Ulibarri, G.; Nadler, W.; Skrydstrup, T.; Audrain, H.; Chiaroni, A.; Riche, C.; Grierson, D. S. *J. Org. Chem.* **1995,** *60*, 2753.
101. Shair, M. D.; Yoon, T.; Danishefsky, S. J. *J. Org. Chem.* **1994,** *59*, 3755.
102. Sonogashira, K.; Tohda, Y.; Hagihara, N. *Tetrahedron Lett.* **1975,** *16*, 4467. Takahashi, S.; Kuroyama, Y; Sonogashira, K.; Hagihara, N. *Synthesis* **1980,** 627.
103. Lipshutz, B. H.; Alami, M.; Susfalk, R. B. *Synlett* **1993,** 693.
104. Nicolaou, K. C.; Chakraborty, T. K.; Piscopio, A. D.; Minowa, K.; Bertinato, P. *J. Am. Chem. Soc.* **1993,** *115,* 4419.
105. Roush, W. R.; Peseckis, S. M. *Tetrahedron Lett.* **1982,** *23,* 4879.
106. Taniguchi, M.; Kobayashi, S.; Nakagawa, M.; Hino, T. *Tetrahedron Lett.* **1986,** *27,* 4763.
107. Paquette, L. A.; Hormuth, S.; Lovely, C. J. *J. Org. Chem.* **1995,** *60*, 4813. Ma, S.; Lu, X.; Li, Z. *J. Org. Chem.* **1992,** *57*, 709.
108. Cousseau, J.; Gouin, L. *J. Chem. Soc., Perkin Trans. 1* **1977,** 1797.
109. Johnson, C. R.; Adams, J. P.; Braun, M. P.; Senanayake, C. B.; Wovkulich, P. M.; Uskokovic, M. R. *Tetrahedron Lett.* **1992,** *33*, 917.
110. Zweifel, G.; Lewis, W.; Os, H. P.; *J. Am. Chem. Soc.* **1979,** *101*, 5101.
111. Brown, H. C.; Bowman, D. H.; Misumi, S.; Unni, M. K. *J. Am. Chem. Soc.* **1967,** *89*, 4531.
112. Swanson, D. R.; Nguyen, T.; Noda, Y.; Negishi, E. *J. Org. Chem.* **1991,** *56*, 2590. Hart, D. W.; Blackburn, T. F.; Schwartz, J. *J. Am. Chem. Soc.* **1975,** *97*, 679.
113. Van Horn, D. E.; Negishi, E. *J. Am. Chem. Soc.* **1978,** *100*, 2252.
114. Chen, J.; Wang, T.; Zhao, K. *Tetrahedron Lett.* **1994,** *35*, 2827.
115. Graven, A.; Jorgenson, K. A.; Dahl, S.; Stanczak, A. *J. Org. Chem.* **1994,** *59*, 3543.
116. Rossi, R.; Bellina, F.; Carpita, A.; Gori, R. *Synlett* **1995,** 344.
117. Tsai, H.-J.; Thenappan, A.; Burton, D. J. *J. Org. Chem.* **1994,** *59*, 7085.
118. Bartlett, P. A.; Otake, A. *J. Org. Chem.* **1995,** *60*, 3107. Boros, L. G.; De Corte, B.; Gimi, R. H.; Welch, J. T.; Wu, Y.; Handschumacher, R. E. *Tetrahedron Lett.* **1994,** *35*, 6033.
119. Matthews, D. P.; Gross, R. S.; McCarthy, J. R. *Tetrahedron Lett.* **1994,** *35*, 1027. Matthews, D. P.; Waid, P. P.; Sabol, J. S.; McCarthy, J. R. *Tetrahedron Lett.* **1994,** *35*, 5177.
120. For a discussion of the formation of enolates, their *O*-substituted derivatives, and the reactions of such compounds see, for example: Carruthers, W. *Some Modern Methods of Organic Synthesis*, 3rd edn, Cambridge University Press: Cambridge, **1986**, and similar general textbooks.
121. Gassman, P. G.; Burns, S. J. *J. Org. Chem.* **1988,** *53*, 5574.

122. Earnshaw, C.; Wallis, C. J.; Warren, S. *J. Chem. Soc., Perkin Trans. 1* **1979**, 3099.
123. Hodgson, D. M.; Witherington, J.; Moloney, B. A.; Richards, I. C.; Brayer, J.-L. *Synlett* **1995**, 32.
124. Casson, S.; Kocienski, P. *Synthesis* **1993**, 1133. Casson, S.; Kocienski, P. *J. Chem. Soc., Perkin Trans. 1* **1994**, 1187.
125. Harwood, L. M.; Moody, C. J. *Experimental Organic Chemistry*; Blackwell: Oxford, **1989**, experiment 53. See references therein for further discussion of the synthesis and chemistry of enamines.
126. Kelly, S. E. In *Comprehensive Organic Chemistry*; Trost, B. M.; Fleming, I., eds; Pergamon Press: Oxford, **1991**; Vol. 1, p. 729.
127. For a recent discussion of the effect of temperature on the (*E*)/(*Z*) ratio of products, see: Pham, T.; Lubell, W. D. *J. Org. Chem.* **1994**, *59*, 3676 and references therein.
128. Barrett, A. G. M.; Graboski, G. G. *Chem. Rev.* **1986**, *86*, 751.
129. Rosini, G.; Ballini, R.; Sorrenti, P. *Synthesis* **1983**, 1014. Denmark, S. E.; Schnute, M. E. *J. Org. Chem.* **1995**, *60*, 1013.
130. Fleming, I.; Dunogues, J.; Smithers, R. H. *Org. React.* **1989**, *37*, 57.
Colvin, E. W. *Silicon Reagents in Organic Synthesis*; Academic Press: London, **1988**.
131. Hiyama, T.; Kusumoto, T. In *Comprehensive Organic Chemistry*; Trost, B. M.; Fleming, I., eds; Pergamon Press: Oxford, **1991**; Vol. 8, p. 769.
132. Takeuchi, R.; Nitta, S.; Watanabe, D. *J. Org. Chem.* **1995**, *60*, 3045.
Takeuchi, R.; Nitta, S.; Watanabe, D. *J. Chem. Soc., Chem. Commun.* **1994**, 1777.
133. Soderquist, J. A.; Colberg, J. C. *Tetrahedron Lett.* **1994**, *35*, 27.
134. Muccioli, A. B.; Simpkins, N. S.; Mortlock, A. *J. Org. Chem.* **1994**, *59*, 5141. Graham, A. E.; Loughlin, W. A.; Taylor, R. J. K. *Tetrahedron Lett.* **1994**, *35*, 7281.
135. Clough, J. M. In *Comprehensive Organic Chemistry*; Trost, B. M.; Fleming, I., eds; Pergamon Press: Oxford, **1991**, Vol. 3, p. 861.
136. Doyle, M. P.; Shanklin, M. S. *Organometallics* **1994**, *13*, 1081.
137. Hodgson, D. M.; Comina, P. J. *Tetrahedron Lett.* **1994**, *35*, 9469.
138. Bodalski, R.; Michalski, T.; Pietrusiewicz, K. M. *ACS Symp. Ser.* **1981**, *171*, 243.
139. Vinyltriphenylphosphonium bromide is called Schweizer's reagent, and is commercially available. See: Zbiral, E. *Synthesis* **1974**, 775.
140. Pietrusiewicz, K. M.; Zablocka, M.; Monkiewicz, J. *J. Org. Chem.* **1984**, *49*, 1522.
141. Minami, T.; Motoyoshiya, J. *Synthesis* **1992**, 333.
142. Jones, G. H.; Hamamura, E. K.; Moffatt, J. G. *Tetrahedron Lett.* **1968**, 5731.
143. Mikolajczyk, M.; Balczewski, P. *Synthesis* **1989**, 101.
144. Hirao, T.; Masunga, T.; Ohshiro, Y.; Agawa, T. *Tetrahedron Lett.* **1980**, *21*, 3595.
145. Simpkins, N. S. *Tetrahedron* **1990**, *46*, 6951.
146. Fiandanese, V.; Marchese, G.; Naso, F. *Tetrahedron Lett.* **1978**, 5131.
147. Craig, D.; Ley, S. V.; Simpkins, N. S.; Whitham, G. H.; Prior, M. J. *J. Chem. Soc., Perkin Trans. 1* **1985**, 1949.
148. Posner, G. H.; Brunelle, D. J. *J. Org. Chem.* **1972**, *37*, 3547.
149. Tanikaga, R.; Tamura, T.; Nozaki, Y.; Kaji, A. *J. Chem. Soc., Chem. Commun.* **1984**, 87.
150. Westwell, A. D.; Rayner, C. M. *Tetrahedron: Asymmetry* **1994**, *5*, 355. Additionally, vinyllithium compounds can be added to the Anderson sulfinate to give

enantiomerically pure vinyl sulfoxides, see: Hulce, M.; Mallamo, J. P.; Frye, L. L.; Kogan, T. P.; Posner, G. H. *Org. Synth.* **1985**, *64*, 196.
151. Murahashi, S.; Yamamura, M.; Yanagisawa, K.; Mita, N.; Kondo, K. *J. Org. Chem.* **1979**, *44*, 2408.
152. Rane, A. M.; Miranda, E. I.; Soderquist, J. A. *Tetrahedron Lett.* **1994**, *35*, 3225.
153. Kuniyasu, H.; Ogawa, A.; Sato, K.-I.; Ryu, I.; Kambe, N.; Sonoda, N. *J. Am. Chem. Soc.* **1992**, *114*, 5902. Bäckvall, J.-E.; Ericsson, A. *J. Org. Chem.* **1994**, *59*, 5850.
154. Kuniyasu, H.; Ogawa, A.; Sato, K.-I.; Ryu, I.; Sonoda, N. *Tetrahedron Lett.* **1992**, *33*, 5525.
155. For a discussion of the literature, see: Greeves, N.; Torode, J. S. *Synlett* **1994**, 537.
156. Soderquist, J. A.; Hassner, A. *J. Am. Chem. Soc.* **1980**, *102*, 1577.
157. Parain, J. L.; Duchene, A.; Quintard, J. P. *Tetrahedron Lett.* **1990**, *31*, 1857.
158. Le Ménez, P.; Firmo, N.; Fargeas, V.; Ardisson, J.; Pancrazi, A. *Synlett* **1994**, 995. For discussion of the rearrangement mechanism, see: Kocienski, P.; Wadman, S. *J. Am. Chem. Soc.* **1989**, *111*, 2363. Takle, A.; Kocienski, P. *Tetrahedron* **1990**, *46*, 4503.
159. Brown, H. C. *Organic Synthesis via Organoboranes*; Wiley-Interscience: New York, **1975**, Pelter, A.; Smith, K.; Brown, H. C. *Borane Reagents*; Academic Press: New York, **1988**.
160. Brown, H. C.; Basavaiah, D. *J. Org. Chem.* **1982**, *47*, 754.
161. Deloux, L.; Srebnik, M. *J. Org. Chem.* **1994**, *59*, 6871.
162. Ishiyama, T.; Nishijima, K.-I.; Miyaura, N.; Suzuki, A. *J. Am. Chem. Soc.* **1993**, *115*, 7219.
163. Patai, S. *The Chemistry of Ketenes, Allenes and Related Compounds*; Wiley: New York, **1980**.
164. Landor, S. R., ed.; *The Chemistry of the Allenes*; Academic Press: New York, **1982**.
165. Vermeer, P.; Meijer, J.; Brandsma, L. *Recl. Trav. Chim. Pays-Bas* **1975**, *94*, 112. See also: Moriya, T.; Miyaura, N.; Suzuki, A. *Synlett* **1994**, 149.
166. Tabuchi, T.; Inanaga, J.; Yamaguchi, M. *Tetrahedron Lett.*, **1986**, *27*, 5237. See also: Crandall, J. K.; Keyton, D. J.; Kohne, J. *J. Org. Chem.* **1968**, *33*, 3655.
167. For a nucleophilic allenyl system, see: Marshall, J. A.; Wallace, E. M. *J. Org. Chem.* **1995**, *60*, 796 and references therein.
168. Badone, D.; Cardamone, R.; Guzzi, U. *Tetrahedron Lett.* **1994**, *35*, 5477. Aidhen, I. S.; Braslau, R. *Synth. Commun.* **1994**, *24*, 789.
169. For a recent example, see: Wang, K. K.; Liu, B.; Lu, Y.-D. *J. Org. Chem.* **1995**, *60*, 1885. See also: Brandsma, L.; Verkruijsse, H. D. *Synthesis of Acetylenes, Allenes and Cumulenes*; Elsevier: New York, **1981**.
170. Hanford, W. E.; Sauer, J. C. *Org. React.* **1946**, *3*, 120.
171. Darling, S. D.; Kidwell, R. L. *J. Org. Chem.* **1968**, *33*, 3974.
172. Tidwell, T. T. *Ketenes*; Wiley: New York, **1995**.

A1

Summary of protocols

This appendix provides a list of the protocols within the book that lead to the synthesis of an alkene. With the aid of the list, the reader should be able to recognise an alkene structure similar to their own desired target structure. Further details of alkene synthesis by considering alkene category are given in Chapter 9, with cross references to other relevant information, including protocols, where appropriate.

Chapter 2, Protocol 4

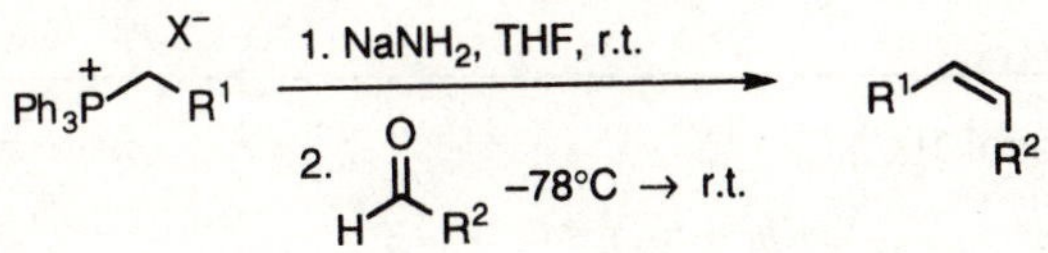

$Ph_3\overset{+}{P}CH_2R^1\ X^-$ → 1. $NaNH_2$, THF, r.t. 2. $HC(O)R^2$, −78°C → r.t. → $R^1CH{=}CHR^2$

Chapter 2, Protocol 5

$Ph_3\overset{+}{P}CH_2CH_3\ Br^-$ → 1. PhLi 2. n-$C_5H_{11}CHO$ 3. PhLi 4. HCl, Et_2O

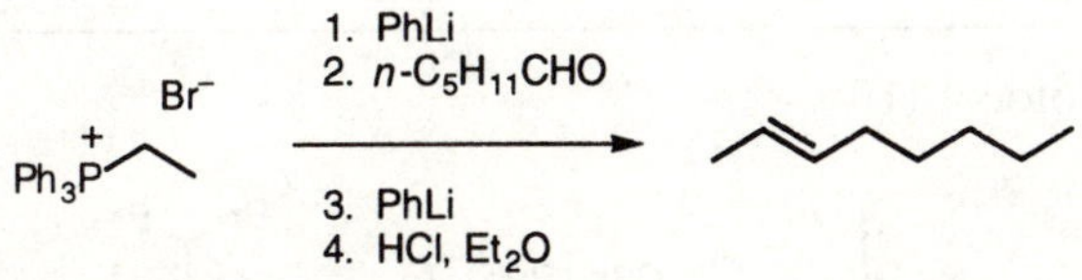

Chapter 2, Protocol 6

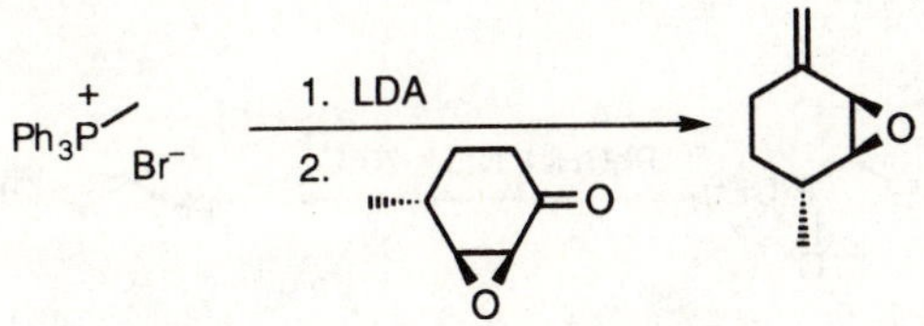

$Ph_3\overset{+}{P}CH_3\ Br^-$ → 1. LDA 2.

Chapter 2, Protocol 7

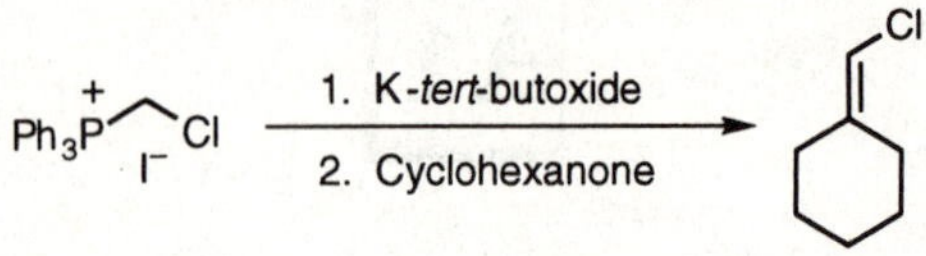

$Ph_3\overset{+}{P}CH_2Cl\ I^-$ — 1. K-*tert*-butoxide; 2. Cyclohexanone → (chloromethylene)cyclohexane

Chapter 2, Protocol 8
(via enol ether)

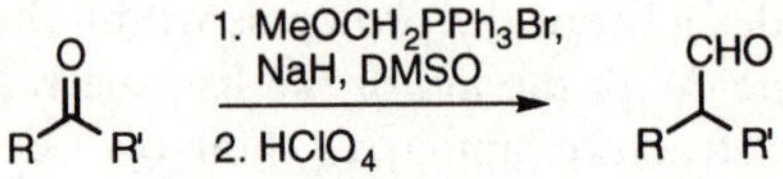

R(C=O)R' — 1. $MeOCH_2PPh_3Br$, NaH, DMSO; 2. $HClO_4$ → R(CHO)R'

Chapter 2, Protocol 9

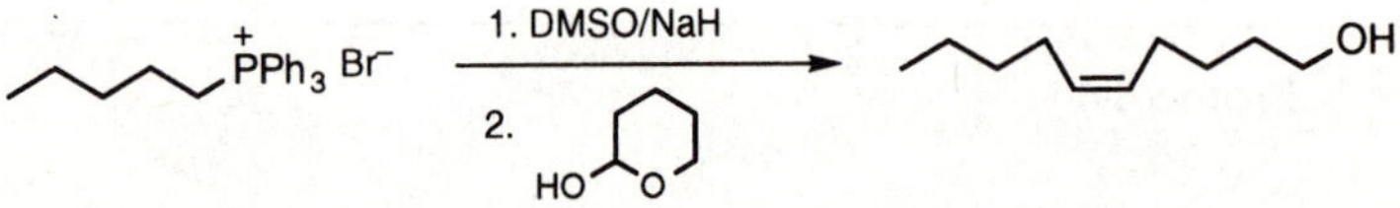

$\overset{+}{P}Ph_3\ Br^-$ — 1. DMSO/NaH; 2. HO (2-hydroxytetrahydropyran) → OH

Chapter 2, Protocol 10

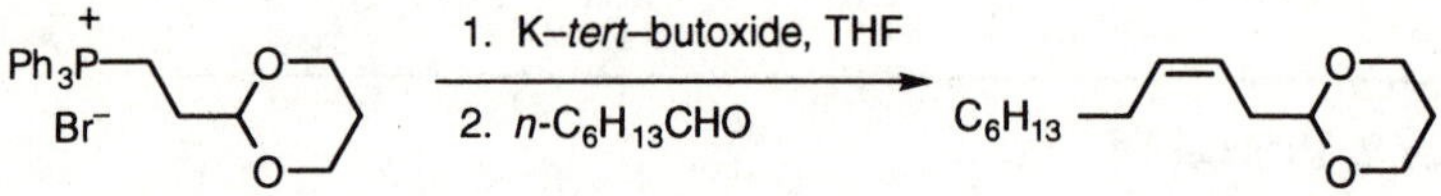

$Ph_3\overset{+}{P}$ Br^- — 1. K–*tert*–butoxide, THF; 2. *n*-$C_6H_{13}CHO$ → C_6H_{13}

Chapter 2, Protocol 11

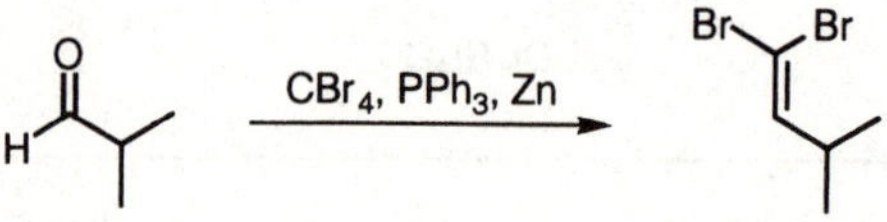

H (isobutyraldehyde) — CBr_4, PPh_3, Zn → Br Br

Chapter 2, Protocol 13

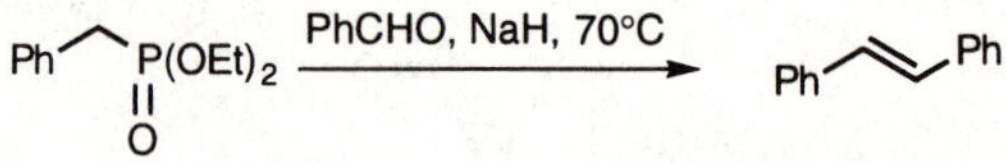

PhCH_2P(O)$(OEt)_2$ — PhCHO, NaH, 70°C → Ph–CH=CH–Ph

Chapter 2, Protocol 14

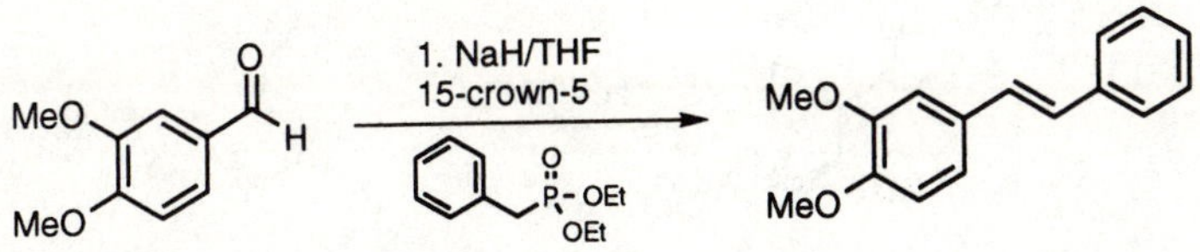

MeO, MeO, O, H

1. NaH/THF
15-crown-5

O, P- OEt, OEt

MeO, MeO

Chapter 2, Protocol 15

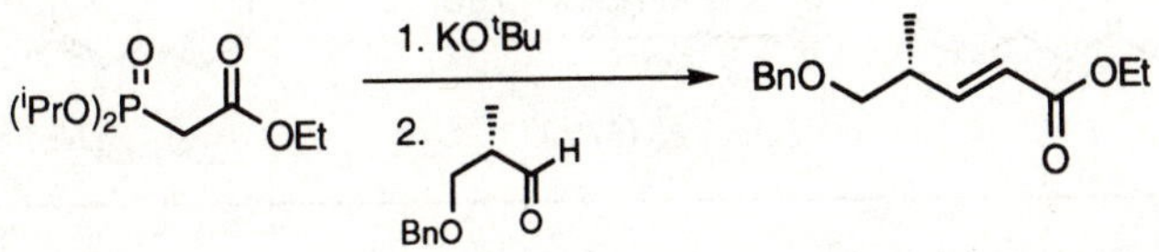

(^{i}PrO)$_2$P, O, O, OEt

1. KO^{t}Bu
2. BnO, H, O

BnO, OEt, O

Chapter 2, Protocol 17

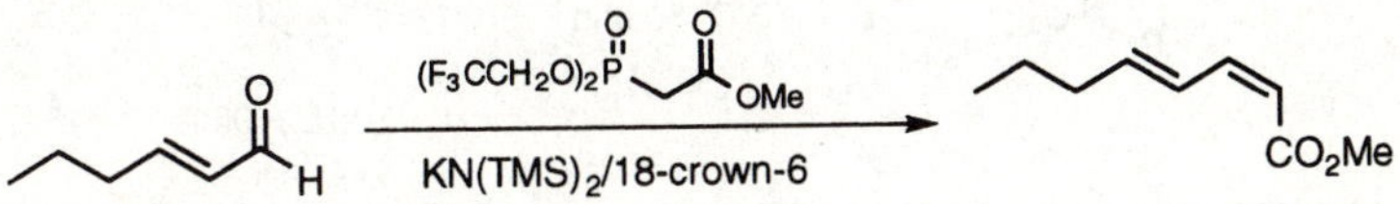

O, H

($F_3CCH_2O)_2P$, O, O, OMe

KN(TMS)$_2$/18-crown-6

CO_2Me

Chapter 2, Protocol 18

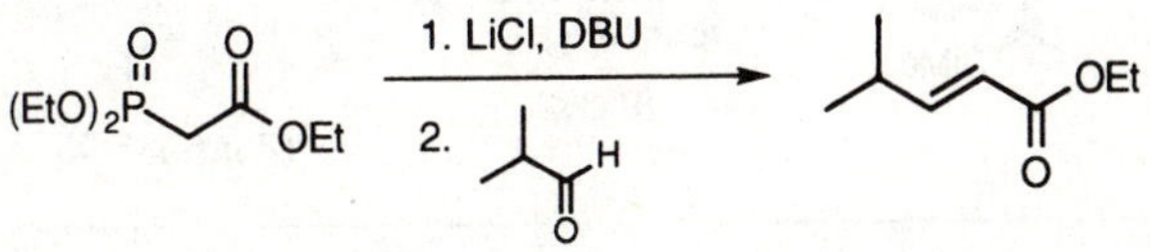

(EtO)$_2$P, O, O, OEt

1. LiCl, DBU
2. H, O

OEt, O

Chapter 2, Protocol 22

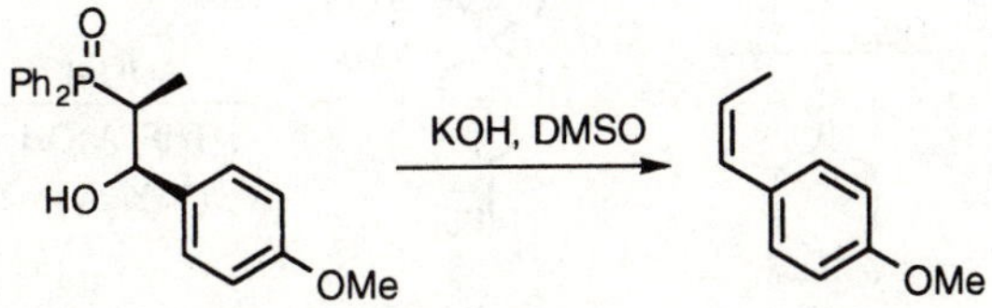

Ph$_2$P, O, HO, OMe

KOH, DMSO

OMe

Chapter 2, Protocol 23

$Ph_2P(O)$–CH(R^1)–CH(OH)(R^2) → NaH, DMF → R^1CH=CHR^2

Chapter 3, Protocol 1

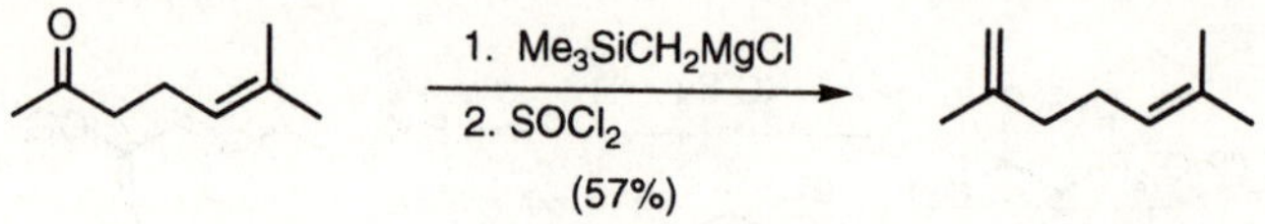

1. Me_3SiCH_2MgCl
2. $SOCl_2$
(57%)

Chapter 3, Protocol 3

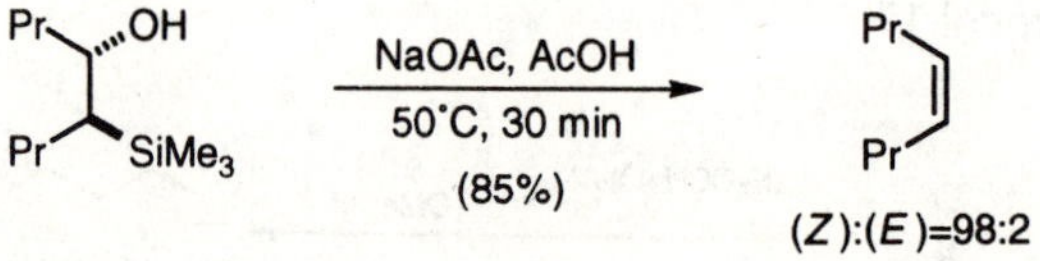

Pr, OH / Pr, $SiMe_3$ → NaOAc, AcOH, 50˚C, 30 min (85%) → Pr–CH=CH–Pr

(*Z*):(*E*)=98:2

Chapter 3, Protocol 4

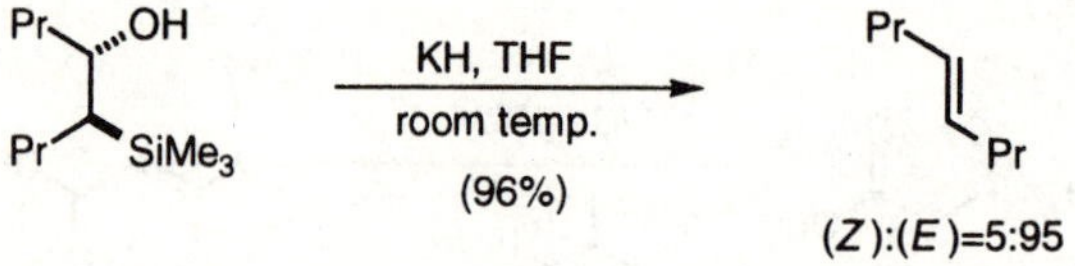

Pr, OH / Pr, $SiMe_3$ → KH, THF, room temp. (96%) → Pr–CH=CH–Pr

(*Z*):(*E*)=5:95

Chapter 3, Protocol 5

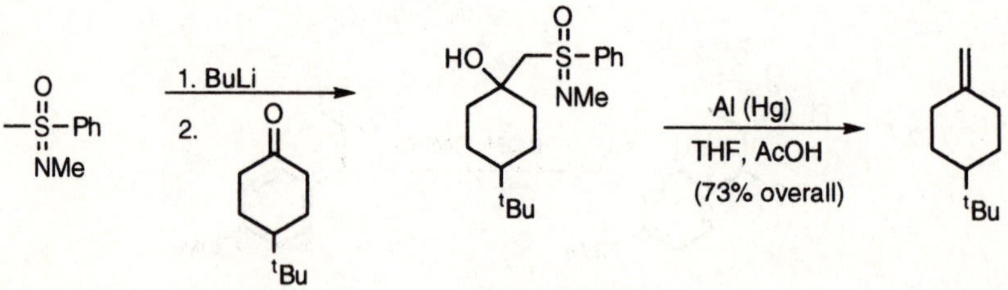

Me–S(O)(=NMe)–Ph → 1. BuLi; 2. 4-^{t}Bu-cyclohexanone → HO, S(O)(=NMe)–Ph, ^{t}Bu → Al (Hg), THF, AcOH (73% overall) → methylene, ^{t}Bu

Chapter 3, Protocol 6

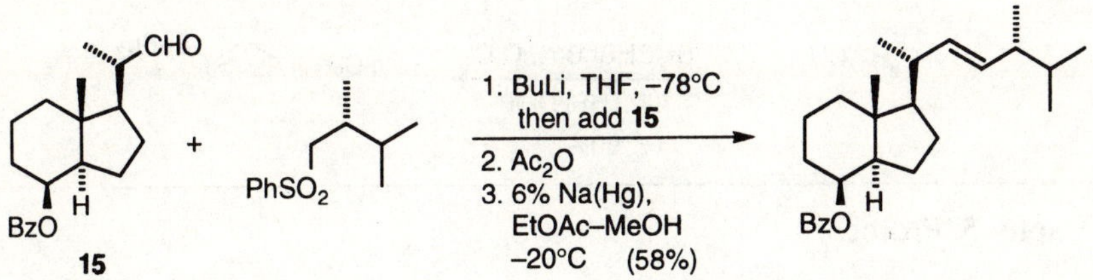

15 + $PhSO_2$ … 1. BuLi, THF, −78°C then add **15** 2. Ac_2O 3. 6% Na(Hg), EtOAc–MeOH −20°C (58%)

CHO, BzO, H, **15**; BzO, H

Chapter 3, Protocol 8

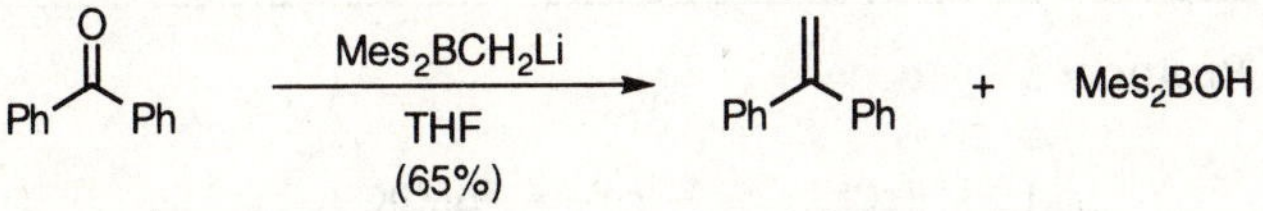

Ph(C=O)Ph → Mes_2BCH_2Li, THF (65%) → Ph(C=CH₂)Ph + Mes_2BOH

Chapter 4, Protocol 1

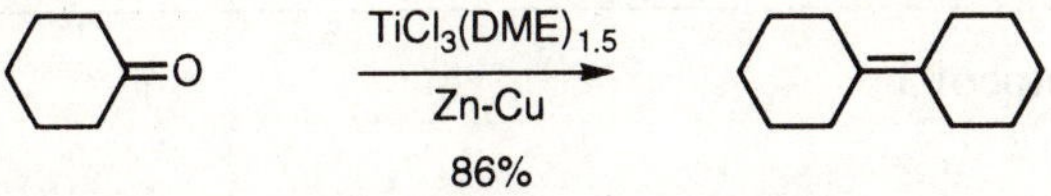

Cyclohexanone → $TiCl_3(DME)_{1.5}$, Zn-Cu, 86%

Chapter 4, Protocol 2

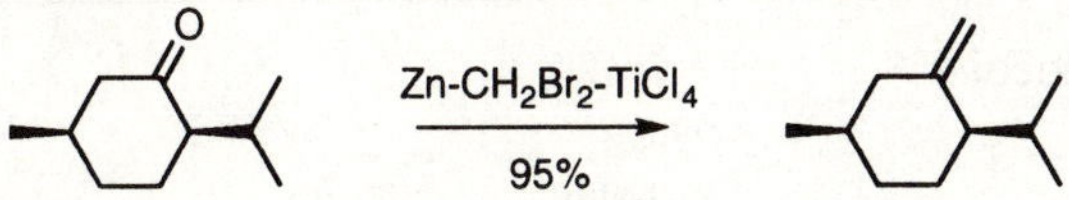

→ Zn-CH_2Br_2-$TiCl_4$, 95%

Chapter 4, Protocol 3

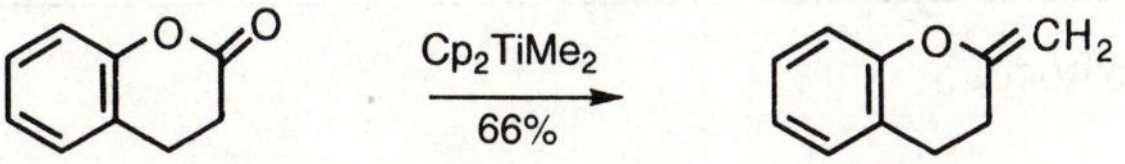

→ Cp_2TiMe_2, 66% → CH_2

Chapter 4, Protocol 4

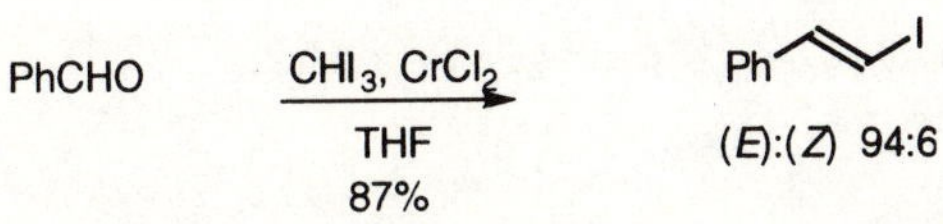

PhCHO → CHI_3, $CrCl_2$, THF, 87% → Ph–CH=CH–I, (*E*):(*Z*) 94:6

Chapter 4, Protocol 5

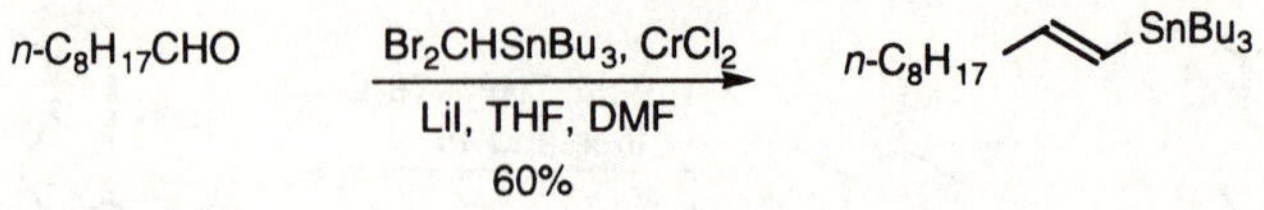

n-$C_8H_{17}CHO$ → ($Br_2CHSnBu_3$, $CrCl_2$; LiI, THF, DMF; 60%) → *n*-C_8H_{17} SnBu$_3$

Chapter 5, Protocol 1

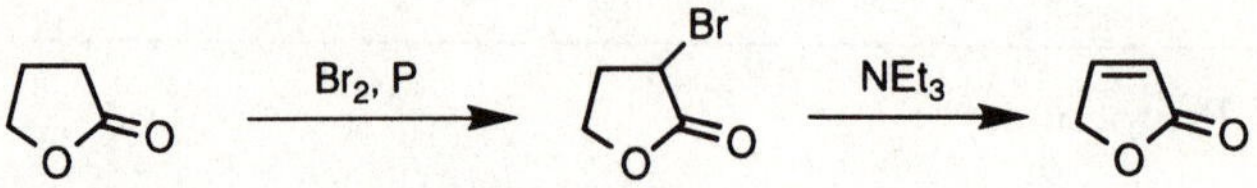

O O → (Br_2, P) → Br O O → (NEt_3) → O O

Chapter 5, Protocol 2

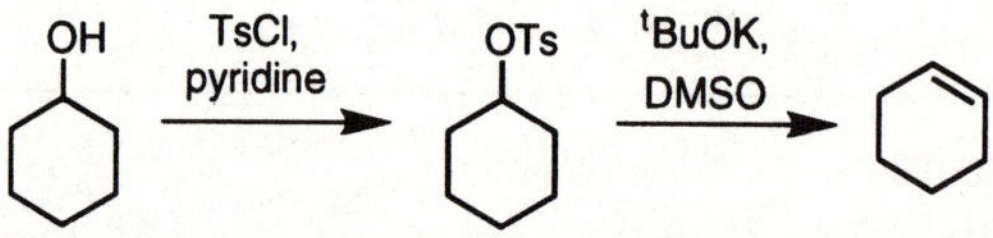

OH → (TsCl, pyridine) → OTs → (^{t}BuOK, DMSO) →

Chapter 5, Protocol 3

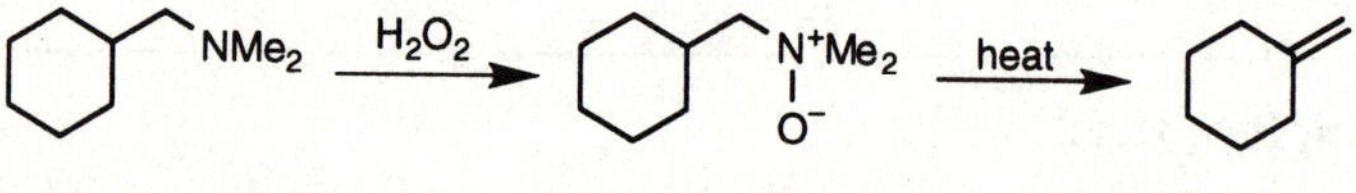

NMe_2 → (H_2O_2) → N^+Me_2 O^- → (heat) →

Chapter 5, Protocol 4

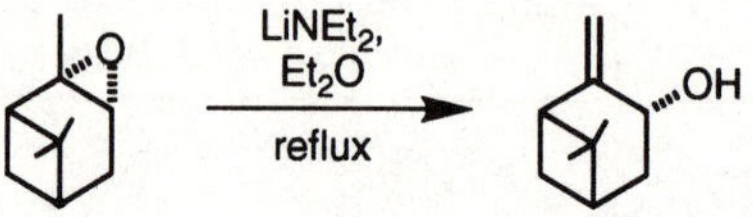

O → ($LiNEt_2$, Et_2O, reflux) → OH

Chapter 5, Protocol 5

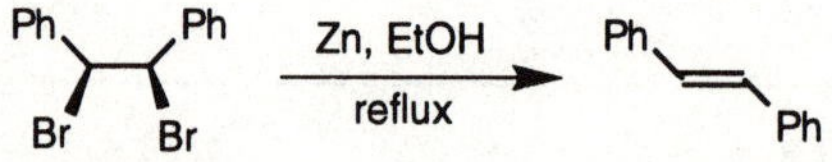

Ph Ph Br Br → (Zn, EtOH, reflux) → Ph Ph

Chapter 5, Protocol 6

$$2PhCHO + CH_3COCH_3 \xrightarrow{NaOH,\ H_2O / EtOH} PhCH{=}CH{-}CO{-}CH{=}CHPh + H_2O$$

Chapter 5, Protocol 7

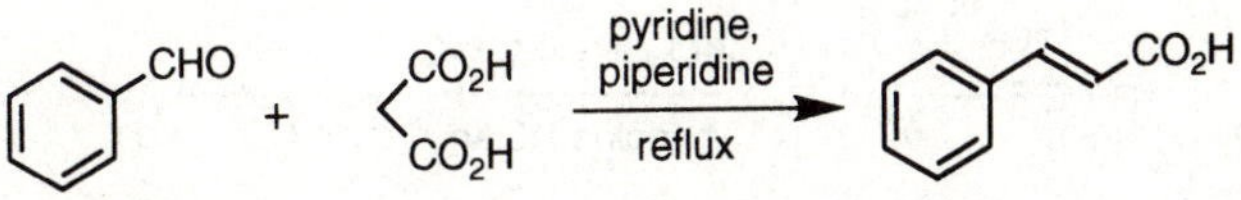

pyridine, piperidine, reflux

CHO, CO_2H, CO_2H, CO_2H

Chapter 5, Protocol 8

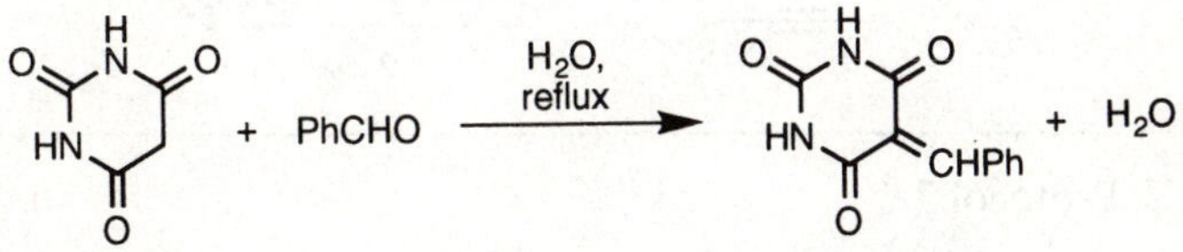

H, O, N, O, HN, O, + PhCHO, H_2O, reflux, H, O, N, O, HN, CHPh, O, + H_2O

Chapter 6, Protocol 2

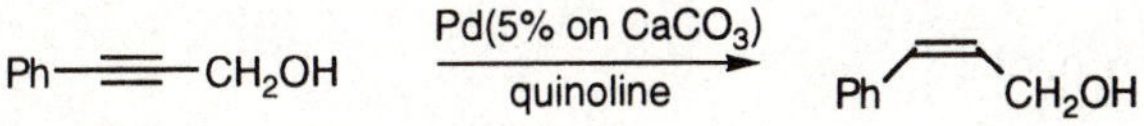

$Ph{-}C{\equiv}C{-}CH_2OH$ — Pd(5% on $CaCO_3$), quinoline → Ph, CH_2OH

Chapter 6, Protocol 3

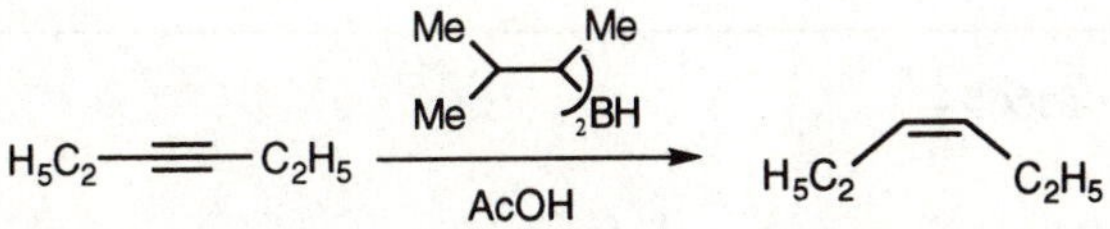

$H_5C_2{-}C{\equiv}C{-}C_2H_5$ — Me, Me, Me, $_2$BH; AcOH → H_5C_2, C_2H_5

Chapter 6, Protocol 4

$C_5H_{11}{-}C{\equiv}C{-}CH_3$ — Na/NH_3 → C_5H_{11}, CH_3

Chapter 6, Protocol 5

Ph—≡—CO_2H $\xrightarrow[DMF/H_2O]{CrSO_4}$ Ph–CH=CH–CO_2H

Chapter 6, Protocol 6

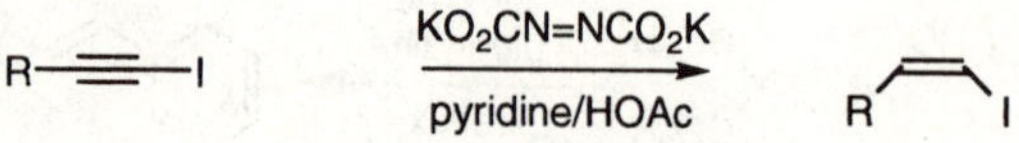

R—≡—I $\xrightarrow[\text{pyridine/HOAc}]{KO_2CN{=}NCO_2K}$ R–CH=CH–I

Chapter 7, Protocol 1

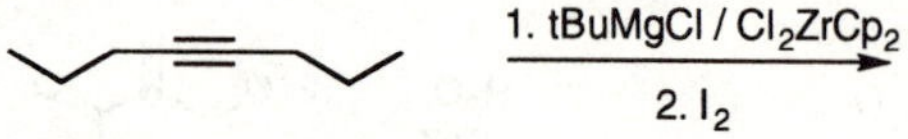

$\xrightarrow[2.\ I_2]{1.\ tBuMgCl\ /\ Cl_2ZrCp_2}$

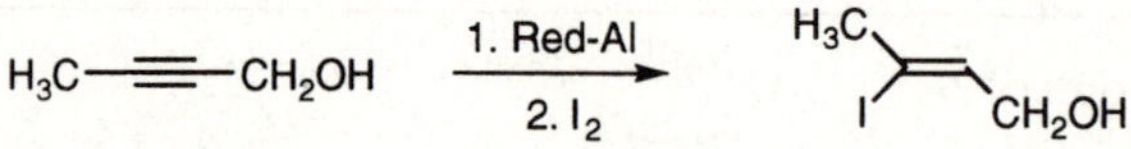

Chapter 7, Protocol 2

H_3C—≡—CH_2OH $\xrightarrow[2.\ I_2]{1.\ \text{Red-Al}}$ H_3C, I, CH_2OH

Chapter 7, Protocol 3

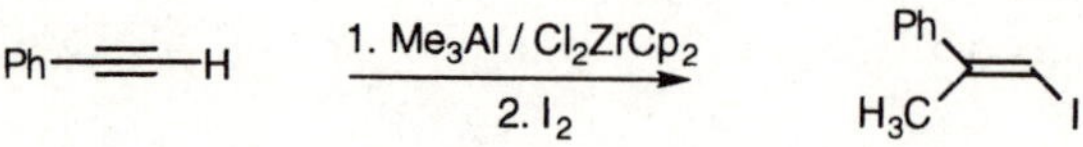

Ph—≡—H $\xrightarrow[2.\ I_2]{1.\ Me_3Al\ /\ Cl_2ZrCp_2}$ Ph, H_3C, I

Chapter 7, Protocol 4

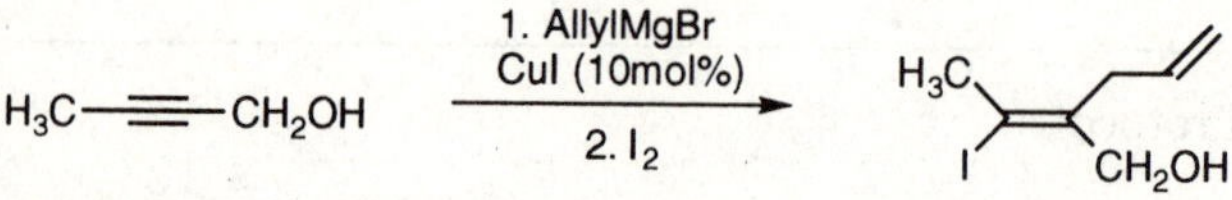

H_3C—≡—CH_2OH $\xrightarrow[2.\ I_2]{1.\ \text{AllylMgBr, CuI (10mol\%)}}$ H_3C, I, CH_2OH

Chapter 8, Protocol 1

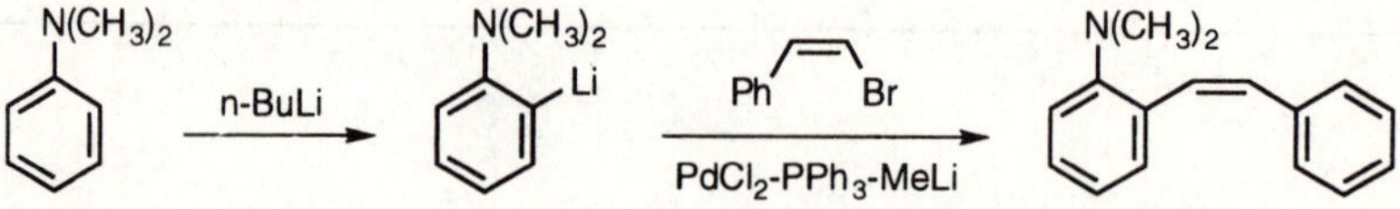

$N(CH_3)_2$ $\xrightarrow{\text{n-BuLi}}$ $N(CH_3)_2$, Li $\xrightarrow[PdCl_2\text{-}PPh_3\text{-MeLi}]{\text{Ph–CH=CH–Br}}$ $N(CH_3)_2$

Chapter 8, Protocol 2

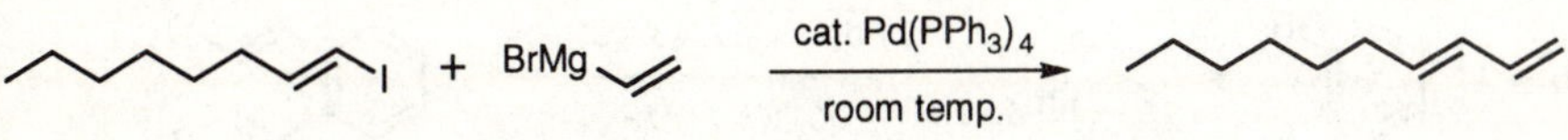

\+ BrMg

cat. $Pd(PPh_3)_4$

room temp.

Chapter 8, Protocol 3

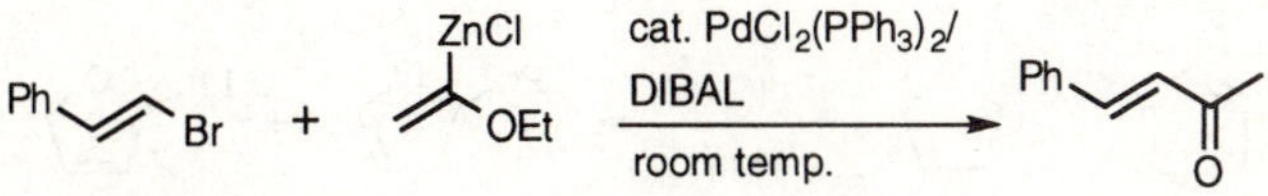

Ph Br + ZnCl OEt

cat. $PdCl_2(PPh_3)_2$/

DIBAL

room temp.

Ph O

Chapter 8, Protocol 4

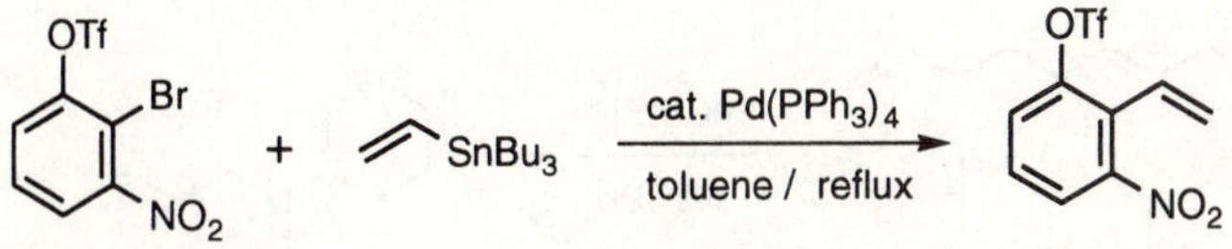

OTf Br NO_2 + $SnBu_3$

cat. $Pd(PPh_3)_4$

toluene / reflux

OTf NO_2

Chapter 8, Protocol 5

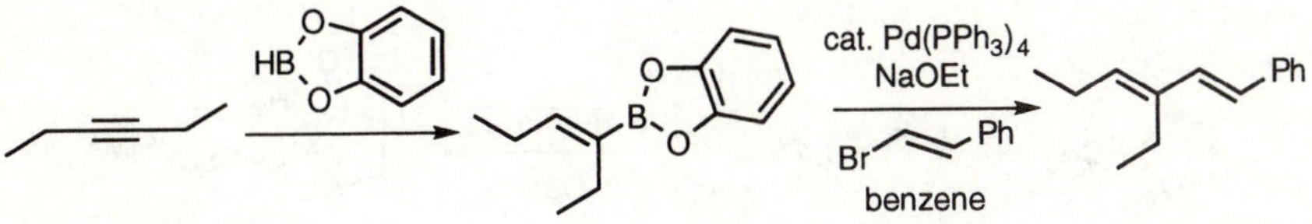

HB O O

B O O

cat. $Pd(PPh_3)_4$

NaOEt

Br Ph

benzene

Ph

Chapter 8, Protocol 6

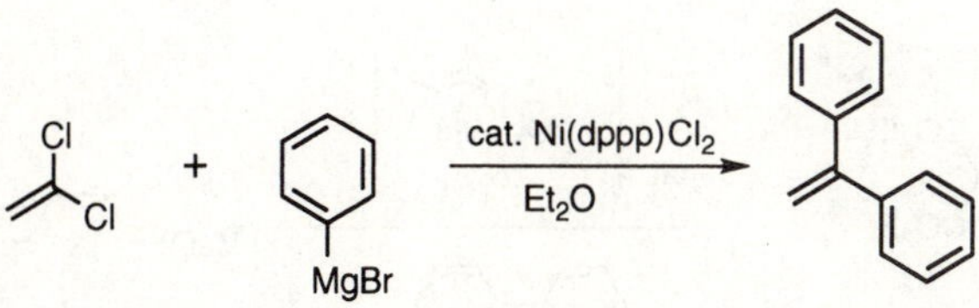

Cl Cl + MgBr

cat. $Ni(dppp)Cl_2$

Et_2O

Chapter 8, Protocol 7

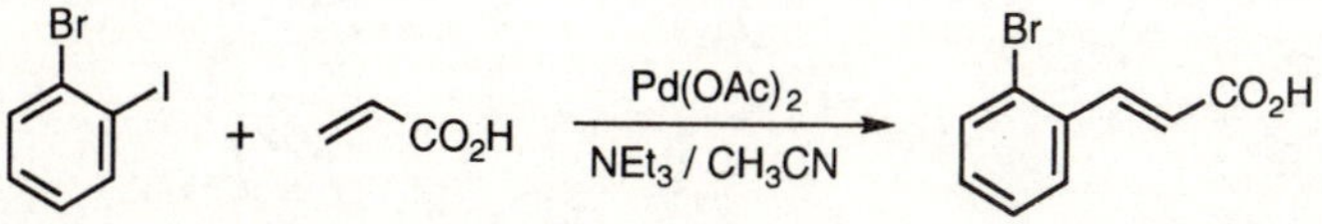

Br I + CO_2H

$Pd(OAc)_2$

NEt_3 / CH_3CN

Br CO_2H

Chapter 8, Protocol 8

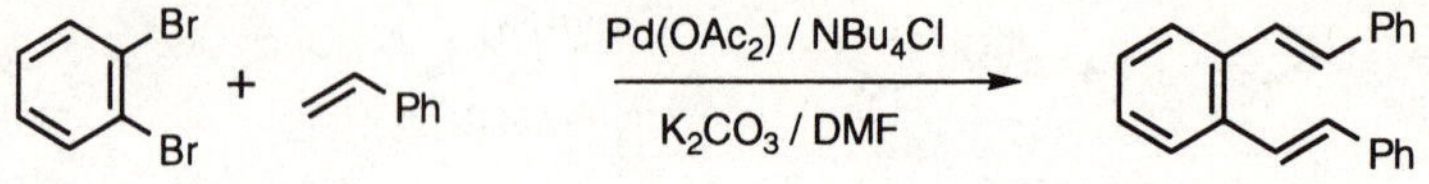

Br, Br, +, Ph, $Pd(OAc_2)$ / NBu_4Cl, K_2CO_3 / DMF, Ph, Ph

Chapter 8, Protocol 9

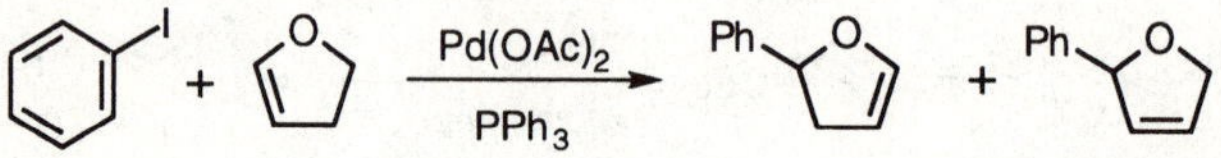

I, +, O, $Pd(OAc)_2$, PPh_3, Ph, O, +, Ph, O

Chapter 8, Protocol 10

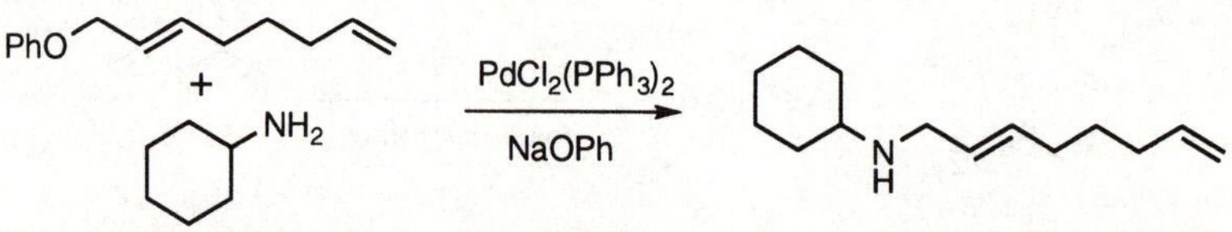

PhO, +, NH_2, $PdCl_2(PPh_3)_2$, NaOPh, N, H

Chapter 8, Protocol 11

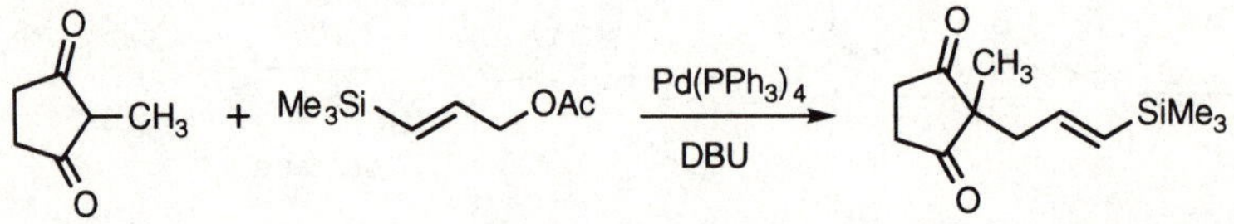

O, CH_3, O, +, Me_3Si, OAc, $Pd(PPh_3)_4$, DBU, O, CH_3, $SiMe_3$, O

Chapter 8, Protocol 12

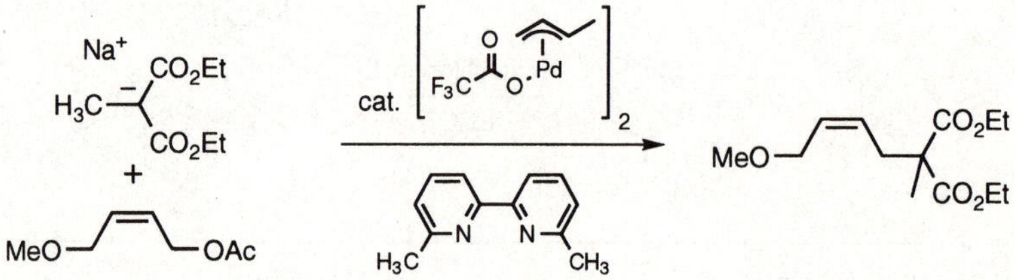

Na^+, CO_2Et, H_3C, CO_2Et, +, MeO, OAc, cat., O, F_3C, O, Pd, 2, N, N, H_3C, CH_3, MeO, CO_2Et, CO_2Et

A2

List of suppliers

Acros Chimica
see Janssen Chimica

Aldrich Chemical Co. Ltd.
France: 80 Rue de Lozais, BP 701, 38070, St Quentin, Fallavier Cedex, LYON. Tel. 74822800
Germany: SAF, Messerschmitt Strasse 17, D-7910 Neu-Ulm. Tel. 0731-9733640
Japan: Aldrich Chemical Co., Inc. (Japan), Kyodo-building Shinkanda, 10 Kandamikura-chou, Chiyoda-ku, Tokyo 101. Tel. 03-54344712
UK: The Old Brickyard, New Road, Gillingham, Dorset SP8 4JL. Tel. 0800-717181
USA: PO Box 355, Milwaukee, WI 53201. Tel. 0414-2733850

Alfa
France: Johnson Matthey SA, BP 50240, Rue de la Perdix, Z1 Paris Nord LL, 95956 Roissy, Charles De Gaulle Cedex. Tel. 1-48632299
Germany: Johnson-Metthey GmbH, Zeppelinstrasse 7, D-7500 Karlsruhe-1. Tel. 0721-840070
UK: Catalogue Sales, Materials Technology Division, Orchard Road, Royston, Herts. SG8 5HE. Tel. 01763-253715
USA: Alfa/Johnson Matthey, PO Box 8247, Ward Hill, MA 01835-0747. Tel. 0508-5216300

BDH
UK: (Head Office and International Sales): Merck Ltd, Merck House, Poole, Dorset BH15 1TD. Tel. 01202-665599

Boulder Scientific Co
USA: 598 3rd Street, PO Box 548, Mead, CO 80542. Tel. 03035354494

Fluka Chemika-BioChemika
France: Fluka S.a.r.l., F-38297 St Quentin, Fallavier Cedex, Lyon. Tel. 74822800
Germany: Fluka Feinchemikalien GmbH, D-7910 Neu-Ulm. Tel. 0731–729670
Japan: Fluka Fine Chemical, Chiyoda-Ku, Tokyo. Tel. 03-32554787
UK: Fluka Chemicals Ltd, Gillingham, Dorset SP8 4JL. Tel. 01747-823097

USA: Fluka Chemical Corp., Ronkonkoma, NY 11779-7238. Tel. 0516-4670980

FMC Corporation, Lithium Division

Japan: Asia Lithium Corporation (ALCO), Shin-Osaka Daiichi-Seimei Building 11F, 5-24, Miyahara 3-Chome, Yodogawa-Ku, Osaka. Tel. 06-3992331

UK and Mainland Europe: Commercial Road, Bromborough, Merseyside L62 3NL. Tel. 0151-3348085

USA: 449 North Cox Road, Gastonia, NC 28054. Tel. 0704-8685300

Heraeus

Germany: Alter Weinberg, D-7500, Karlsruhe 41-Ho. Tel. 0721-4716769

ICN Biomedicals/K and K Rare and Fine Chemicals

France: ICN Biomedicals France, Parc Club Orsay, 4 rue Jean Rostand, 91893 Orsay Cedex. Tel. 1-60193460

Germany: ICN Biomedicals, GmbH, Mühlgrabenstrasse 10, Postfach 1249, D-5309, Meckenheim. Tel. 02225-88050

Japan: ICN Biomedicals Japan Co., Ltd, 8th Floor, Iidabashi Central Building, 4-7-10 Iidabashi, Chiyoda-ku, Tokyo 102. Tel. 03-32370938

UK: ICN Biomedicals, Ltd, Eagles House, Peregrine Business Park, Gomm Road, High Wycombe, Bucks. HP13 7DL. Tel. 04940443826

USA: ICN Biomedicals, Inc., 3300 Hyland Avenue, Costa Mesa, CA 92626. Tel. 0800-8540530

Janssen Chimica (Acros Chimica)

Mainland Europe, Central Offices Belgium: Janssen Pharmaceuticalaan 3, 2440 Geel. Tel. 014-604200

UK: Hyde ParkHouse, Cartwright Street, Newton, Hyde, Cheshire SK14 4EH. Tel. 0613-244161

USA: Spectrum Chemical Mgf. Corp., 14422 South San Pedro Street, Gardena, CA 90248. Tel. 0800-7728786

Johnson Matthey Chemical Products

France: Johnson Matthey SA, BP 50240, Rue de la Perdix, Z1 Paris Nord LL, 95956 Roissy, Charles De Gaulle Cedex. Tel. 48632299

Germany: Johnson-Matthey GmbH, Zeppelinstrasse 7, D-7500 Karlsruhe-1. Tel. 0721-840070

UK: Catalogue Sales, Materials Technology Division, Orchard Road, Royston, Herts. SG8 5HE. Tel. 0763-253715

USA: Alfa/Johnson Matthey, PO Box 8247, Ward Hill MA 01835-0747. Tel. 0508-5216300

Kanto Chemical Co., Inc
Japan: 2-8, Nihonbashi-honcho-3-chome, Chuo-ku, Tokyo 103. Tel. 03-2791751

Lancaster Synthesis
France: Lancaster Synthesis Ltd, 15 Rue de l'Atome, Zone Industrielle, 67800 Bischheim, Strasbourg. Tel. 05035147.
Germany: Lancaster Synthesis GmbH, Postfach 15 18, D-63155 Mülheim am Main. Tel. 0130-6562
Japan: Hydrus Chemical Inc., Tomitaka Building, 8-1, Uchikanda 2-chome, Chiyoda-ku, Tokyo 101. Tel. 03-32585031
UK: Lancaster Synthesis, Eastgate, White Lund, Morecambe, Lancashire LA3 3DY. Tel. 0800-262336
USA: Lancaster Synthesis Inc, PO Box 1000, Windham, NH 03087-9977. Tel. 0800-2382324

Merck
Germany: Promochem, PO Box 101340, Mercatorstrasse 51, D-46469 Wesel. Tel. 0281-530081
Japan: Merck Japan Ltd, ARCO Tower, SF, 8-1, Shimomeguro-1-chome, Merguro-ku, Tokyo 153. Tel. 03-54344712
UK: Merck Ltd, Merck House, Poole, Dorset BH15 1TTD. Tel. 0202-669700
USA: Gallard Schlesinger Companies, 584 Mineola Avenue, Carle Place, NY. Tel. 0516-3335600

Mitsuwa Scientific Corp.
Japan: 11-1, Tenma-1-Chome, Kita-ku, Osaka 530. Tel. 06-3519631

Nacalai Tesque, Inc.
Japan: Karasuma-nishi-iru, Nijo-dori, Nakagyo-Ku, Kyoto 604. Tel. 075-2315301

Organometallics, Inc.
USA: PO Box 287, East Hampstead, NH. Tel. 0603-3296021

Parish Chemical Company
USA: 145 North Geneva Road, Orem. UT 84057. Tel. 0801-2262018

Prolabo
UK etc. : See Rhône-Poulenc
France: 12, Rue Pelee, BP 369, 75526, Paris Cedex 11. Tel. 1-49231500

Rhône-Poulenc
France: Rhône-Poulenc SA, 25 Quai Paul Donmer, F-92408 Courbezoie Cedex. Tel. 1-47681234

Germany: Rhône-Poulenc GmbH, Staedelstrasse 10, Postfach 700862, Frankfurt Am Main 70. Tel. 069-60930
Japan: Rhône-Poulenc Japan Ltd, 15 Kowa Building Annexe, Central PO Box 1649, Tokyo 107. Tel. 03-35854691
UK: Rhône-Poulenc Chemicals Ltd, Laboratory Products, Liverpool Road, Barton Moss, Eccles, Manchester M30 7RT. Tel. 0161-7895878
USA: Rhône-Poulenc Basic Chemicals Co., 1 Corporate Dr., Shelton, CT 06484. Tel. 0203-9253300 or Rhône-Poulenc Inc., Fine Organics, CN 7500, Cranbury, NJ 08512-7500. Tel. 0609-860400

Riedel de Haen
Germany: Wunstorfer Strasse 40, Postfach, D-3016 Seelze 1. Tel. 05137-7070

Strem
France: Strem Chemicals, Inc., 15 Rue de l'Atome, Zone Industrielle, 67800 Bischheim. Tel. 88625260
Germany: Strem Chemicals GmbH, Querstrasse 2, D-7640 Karlsruhe. Tel. 0721-75879
Japan: Hydrus Chemical Co., Tomikata Building, 8-1, Uchkanda, 2-Chome, Chiyoda-ku, Tokyo 101. Tel. 03-2585031
UK: Fluorochem Limited, Wesley Street, Old Glossop, Derbyshire SK13 9RY. Tel. 01457-868921
USA: Strem Chemicals, Inc., Dexter Industrial Park, 7 Mulliken Way, Newburyport, MA 01950-4098. Tel. 0508-4623191

Tokyo Kasei Kogyo Co., Ltd
Japan: 3-1-13, Nihonbashi-Honcho, Chuo-ku, Tokyo 103. Tel. 03-38082821
UK: Fluorochem Limited, Wesley Street, Old Glossop, Derbyshire SK13 9RY. Tel. 01457-868921
USA: PCI America, 9211 North Harbourgate Street, Portland, OR 97203. Tel. 503-2831681

Ventron
See **Alfa**

Wako
Japan: 3-10 Dosho-Machi, Higashi-Ku, Osaka 541. Tel. 06-2033741

Index

Page entries in *italics* refer to protocols on those pages.

Index

Index

Index